Materials Horizons: From Nature to Nanomaterials

Materials are an indispensable part of human civilization since the inception of life on earth. With the passage of time, innumerable new materials have been explored as well as developed and the search for new innovative materials continues briskly. Keeping in mind the immense perspectives of various classes of materials, this series aims at providing a comprehensive collection of works across the breadth of materials research at cutting-edge interface of materials science with physics, chemistry, biology and engineering.

This series covers a galaxy of materials ranging from natural materials to nanomaterials. Some of the topics include but not limited to: biological materials, biomimetic materials, ceramics, composites, coatings, functional materials, glasses, inorganic materials, inorganic-organic hybrids, metals, membranes, magnetic materials, manufacturing of materials, nanomaterials, organic materials and pigments to name a few. The series provides most timely and comprehensive information on advanced synthesis, processing, characterization, manufacturing and applications in a broad range of interdisciplinary fields in science, engineering and technology.

This series accepts both authored and edited works, including textbooks, monographs, reference works, and professional books. The books in this series will provide a deep insight into the state-of-art of Materials Horizons and serve students, academic, government and industrial scientists involved in all aspects of materials research.

Review Process

The proposal for each volume is reviewed by the following:

1. Responsible (in-house) editor
2. One external subject expert
3. One of the editorial board members.

The chapters in each volume are individually reviewed single blind by expert reviewers and the volume editor.

Hamed EL-Mously · Mohamad Midani ·
Eman A. Darwish

Date Palm Byproducts: A Springboard for Circular Bio Economy

Springer

Hamed EL-Mously
Faculty of Engineering
Ain Shams University
Cairo, Egypt

Eman A. Darwish
Faculty of Engineering
Ain Shams University
Cairo, Egypt

Mohamad Midani
Faculty of Engineering and Materials Science
German University in Cairo
Cairo, Egypt

ISSN 2524-5384 ISSN 2524-5392 (electronic)
Materials Horizons: From Nature to Nanomaterials
ISBN 978-981-99-0474-7 ISBN 978-981-99-0475-4 (eBook)
https://doi.org/10.1007/978-981-99-0475-4

This Springer imprint is published by the registered company Springer Nature Singapore Pte Ltd.
The registered company address is: 152 Beach Road, #21-01/04 Gateway East, Singapore 189721, Singapore

Tell the one who left me one day
You will definitely return to me some day
You will realize that Iam an exhaustible fortune
And Iam the Earth inheritance after the oil depletion
I was life's resource and promise to your grandparents
Will its memory survive for your grandchildren?
I was created to raise my head to the sky
But your ignorance forced me to bend
You left me for thirst and negligence
And let me choose between being cut or burnt.
Say to to the one who robbed my greenery
Your offspring would not have survived without my fruit
Iam the palm tree of Adam
Iam the renaissance of next generation
—Feteiah EL-Sharaa Algeria

Preface: Why this Book is Important

This book is concerned with the date palm byproducts. These byproducts have two sources: (i) annual generation in the date palm plantations (e.g. leaves or fronds, petioles, empty fruit bunches, spathes, leaf sheaths and trunks of unproductive palms), or (ii) sites of date palm processing industries (date seeds and waste dates). The amounts of their weights range from hundreds, to thousands or even millions of tons depending on the size of their sources whether palm plantations or industrial sites. These byproducts represent an inexhaustible sustainable treasure of renewable useful materials. Their sources are geographically dispersed being located in 38 countries which are distributed in the five main continents.

Unfortunately, these byproducts are currently viewed in most countries as waste materials and accordingly they are mistreated. They are either burnt in open fields or dumped in landfills. This mistreatment of such a natural treasure not only causes heavy environmental pollution (soil contamination and air pollution) but also inflicts huge economic losses.

Here comes this book to present an alternative fresh look at date palm byproducts as a springboard for bio economy. This book is an invitation to rediscover date palm byproducts as a system of biomaterials. The book chapters will show and exemplify how these sustainable biomaterials can be the base for a wide spectrum of products and uses.

Examples of these numerous products and uses are demonstrated and categorized in the book as five principal uses: (i) pharmaceutical, cosmetics and natural wax, (ii) textiles and composites, cellulose and cellulose derivatives, (iii) timber and wood substitutes, architecture insulation and building materials, (iv) organic fertilizer, compost and soil amendment and water treatment purposes, (v) natural fodder and silage, green fuels and bioenergy.

I hope that the serious and dedicated research endeavors in this book inspire futuristic entrepreneurs around the world and motivate heroes of circular bio economy. I hope this book helps them take pioneer steps to launch their innovative pilot projects relying on the sustainably available—in huge quantities—date palm byproducts.

Cairo, Egypt Hamed EL-Mously

Contents

About the Authors

The Legendary Professor Dr. Hamed EL-Mously is an emeritus professor at the Faculty of Engineering—Ain Shams University, Egypt, and co-founder and chairman of the International Association for Palm Byproducts (ByPalma). Dr. El-Mously is recognized as the founding father of date palm byproducts research and development and one of the warriors of sustainable development of local communities. He received numerous prestigious awards for his work, such as Khalifa International Award for Date Palm and Agriculture Innovation 2013. El-Mously has been working in projects aiming at developing local communities in all villages of Egypt, by applying developing projects using their local resources since 1995. Projects include production of blockboard, parquet and arabesque from palm midribs, production of non-traditional animal feed from agricultural residues, production of organic fertilizers from pruning-products of date palms, doum palms, and mango trees, and production of fig jam. El-Mously contributed to the foundation of several research societies and a member in several strategic consulting and scientific committees, Small Industries and Local Technology Development Center, including the Egyptian Society for Endogenous Development of Local Communities, the Renewable Materials Research Center (RenewMat), the International Association for Palm Byproducts (ByPalma Assn.), the Network of Fiber-Plastic Composites and Tree-Free Wood Innovation, and the Network Foundation for Renewable materials Research, Technology and Applications. He is also a member in several strategic consulting and scientific committees at Ain Shams University. He obtained his Ph.D. from the Metal Cutting Machine Tools Institute, Moscow.

Dr. Mohamad Midani is an assistant professor at the Materials Engineering Department, German University Cairo, and an adjunct assistant professor at Wilson College of Textiles, North Carolina State University. He is also the CEO of VALORIZEN R&I Center and secretary general of the International Association for Palm Byproducts. Dr. Midani has long experience in the textile industry, with a focus on technical textiles and composites, with an ongoing research collaboration and consulting to the industry. His research mostly focuses on natural fibers and their composites, and he is the co-inventor of the long textile fibers from date palm byproducts (PalmFil).

He is also the co-founder of the World Conference on Palm Byproducts and their Applications (ByPalma). He published more than 30 articles in international journals and conferences in addition to three books. Dr. Midani is a reviewer to several international journals from different publishers. He teaches undergraduate and graduate courses on fiber reinforced polymer composites as well as new product development and innovation management. In 2022, Dr. Midani received the distinguished young alumni award from Wilson College of Textiles NC State University in recognition for his impact in the textile field. Dr. Midani has his B.Sc. in Mechanical Engineering from Ain Shams University (2006), MT in Textiles & Apparel Technology and Management (2012), and Ph.D. in Fiber and Polymer Science from the College of Textiles NC State University (2016). Dr. Midani is a certified New Product Development Professional (NPDP) and a member of the Product Development and Management Association (PDMA) and co-founder and member of the board of PDMA Egypt chapter.

Eman A. Darwish is an assistant professor in the Department of Architecture, Faculty of Engineering, Ain Shams University. She is specialized in the design and development of sustainable building materials and elements made from natural materials and agricultural wastes depending on manual manpower and minimal manufacturing in order to maintain high resource and energy efficiencies. Current scopes of research include the implementation of the prefabrication method in the production of versatile building elements made from traditional building materials and agricultural wastes within a multidisciplinary framework for sustainable construction, thermal and acoustic applications. She graduated in 2014, B.Sc. in architectural engineering: Excellent with distinction and top on class. She was awarded M.Sc. in architectural engineering from the faculty of engineering Ain Shams University in 2017. She was awarded Ph.D. in architectural engineering from the faculty of engineering Ain Shams University in 2020. She was awarded best paper in Sustainable Built Environment Conference Series SBE-Cairo 2016 and in the 13th International Engineering, Architecture Conference AEIC-Cairo 2017. She received the best Ph.D. thesis in 2020. She was honored to receive the second-place award in ASU innovates: the researcher track in 2021. She is working currently on rediscovering novel possible utilization fields of local agricultural wastes such as date palm pruning residues.

List of Figures

Part I
Significance of the Date Palm and Its Byproducts

Chapter 1
Cultural and Ecological Significance of the Date Palm

Abstract The date palm has been assimilated as a cultural/spiritual theme in both the Jewish and Christian traditions that have been evolved in the Middle Eastern region. In addition, the date palm has been mentioned—more than any other tree—for 23 times in Quran. The date palm's huge benefits and wide spectrum of uses of its components have been highlighted in Prophet Mohammed's teachings by comparing the date palm with the Muslim. Besides, the date palm components have been intensively used in Medinah (the first Moslem city): as building materials (date palm trunks and midribs in roofing and walling), as well as in household (stuffing of leather pillows with leaf sheaths fibers and weaving carpets from date palm leaflets) and reliance on dates as an essential component of diet. From the ecological point of view, the date palm represents a keystone species and the tree of life creating in arid and semi-arid regions a microclimate that offsets the effects of drought, preserves high soil moisture with low salinity and thus creates a biodiversity hotspot that can shelter a wide range of plant and animal species in hot desert conditions.

Keywords Date palm in Jewish and Christion traditions · Date palm in Koran and Prophet Mohammed's teachings · The date palm as a keystone species in arid and semi-arid regions in the world

1.1 Cultural Significance of the Date Palm

1.1.1 The Date Palm Tree in Jewish and Christian Traditions

Introduction

The date palm tree is an evergreen tree that grows in tropical dry climate. It has a slender stem and many branches. When the date fruit begins to grow, it has a green color which gradually changes into yellow and red until it becomes quite dark. The date palm tree has been a significant component within the existing flora and its associated socioeconomic fabric of the Middle Eastern region. From the earliest

H. EL-Mously et al., *Date Palm Byproducts: A Springboard for Circular Bio Economy*, Materials Horizons: From Nature to Nanomaterials,
https://doi.org/10.1007/978-981-99-0475-4_1

times, the date palm tree has been associated with the Semitic peoples living in the Middle East. In Arabian Desert, the very existence of a human being depends largely upon its presence. In Arabia and Assyria,[1] this tree that appears as a natural habitat has been considered "sacred" in the earliest ages and has historically penetrated the cultural traditions of local communities who have inhabited the sacred lands of the Middle East[2] where Judaism, Christianity and Islam have been incarnated and coexisted for centuries in Historic Palestine, Assyria, Arabian Desert, Ethiopia and Egypt. In this short section, we will focus on how the date palm tree has been assimilated as a cultural/spiritual theme in both the Jewish and Christian traditions that have been evolved in this Middle Eastern region.

1.1.1.1 The Date Palm Tree in Jewish Traditions[3]

The term for date palm tree is commonly mentioned in ancient languages, i.e. the Aramaic, Ethiopic and Hebrew. Tamar is a female name of Hebrew origin. It means "date" or "date palm tree." And due to its beautiful, tall and graceful stem, Hebrews used the term Tamar to describe the beautiful woman. As mentioned in the Old Testament (Torah), Samuel II, Chap. 13, Verse 1 [4]:

> And it came to pass after this, that Absalom the son of King David had a beautiful sister whose name was Tamar ....

During the Sukkot[4] festival (known also as Feast of Tabernacles[5]) that commemorates the years that the Jews spent in Sinai Desert on their way to the "Promised Land" and celebrates[6] the way in which God protected them under difficult desert conditions. The word "sukkot" means huts or "booths" where they

[1] Also called the Assyrian Empire. It was a Mesopotamian kingdom and empire of the Ancient Near East that existed as a state from the twenty-fifth century BC until its collapse between 612 and 609 BC. It has included two major cities—Assur and Nineveh—wedged between state of Babylonia downstream the Tigris river, and states of Mitanni and Hatti upstream the river [refer to "Assyria," Wikipedia].

[2] In Egypt, the tall palm stem forms a constant feature in early Egyptian architecture, and in "Historic Palestine," the palm leaf appears as an ornament upon pottery as far back as 1800 BC [refer to "Palm tree in the Bible Encyclopedia," www.bible-history.com] [1].

[3] Refer to Abdelhamid [2, 3].

[4] Refer to BBC-Judaism: Sukkot [5].

[5] A major Jewish festival beginning on the eve of the 15th of Tishri according to the Hebrew Calendar, and commemorating the shelter of the Israelites during their 40 years in the wilderness [refer to "Feast of Tabernacles"-Dictionary Definition: Vocabulary.com] [6].

[6] On the first day of the Feast of Tabernacles, the Hebrews were commanded to take branches of palms, with other trees, and rejoice before God [refer to Old Testament, Leveticus, Chap. 23, Verse 40] [4].

have used the trunk and branches of the date palm tree to construct a temporary shelter that could protect them in the harsh desert environment:

> Ye shall dwell in "booths" seven days; all that are Israelites born shall dwell in booths. That your generation may know that I made the children of Israel dwell in booths, when I brought them out of the land of Egypt: I am the LORD your God (Old Testament, Leviticus, Chap. 23, Verses 42–43) [4]

Date palm tree has been considered a symbol of the Kingdom of Israel and was mentioned in connexion with the architectural ornaments of Solomon's Temple:

> And the house which king Solomon built for the LORD, the length there of was three score cubits, and the breadth there of twenty cubits, and the height there of thirty cubits……………….And he carved all the walls of the house round about with carved figures of cherubim[7] and "palm trees" and open flowers, within and without" (Old Testament, Kings I, Chap. 6, Verses 2 & 29) [4]

Meanwhile, we find the same usage of date palm tree as an ornamental motif within the architecture of the holy city/temple that was spiritually revealed in a "revelation" to Ezekiel the Prophet during the time of captivity of the Jewish people:

"In the visions of God brought he me into the land of Israel, and set me upon a very high mountain by which was as the frame of a city on the south" (Old Testament, Ezekiel, Chap. 40, Verse 2) [4]. Then the prophet was shown the details of the Holy Temple and its architectural ornaments:

> Afterwards he brought me to the temple and measured the posts, six cubits broad on the one side, and six cubits broad on the other side which is the breadth of the tabernacle………The door posts, and the narrow windows, and the galleries round about on their three stories, over against the door, cieled with wood round about, and from the ground up to the windows, and the windows were covered……..And it was made with cherubim and palm trees, so that a palm tree was between a cherub and a cherub; and every cherub had two faces: so that the face of a man was toward the palm tree on the one side, and the face of a young lion toward the palm tree on the other side. It was made through all the house round about" (Old Testament, Ezekiel, Chap. 41, Verses 1, 16, 18, 19) [4]

The symbolic power of the date palm tree within the Jewish cultural/spiritual traditions is revealed through the biblical narrative in Book of Judges, Old Testament (Torah) on Prophetess Deborah (the fourth Judge of premonarchic Israel[8]) who has dispensed religious/political judgment to the people of Israel from under a date palm tree:

> And Deborah, a prophetess, the wife of Lapidoth, she judged Israel at the time. And she dwelt under the "Palm Tree of Deborah" between Ramah and Bethel in mount of Ephraim;

[7] Plural of "cherub," which refers to a celestial winged being with human, animal, or birdlike characteristics who functions as a throne bearer of the Deity. Derived from ancient Middle Eastern mythology and iconography, these celestial beings serve important litugical and intercessory functions in the hierarchy of angels. The term most likely derives from the Akkadian verb karabu meaning "to pray" or "to bless" [refer to https://www.britannica.com/topic/cherub] [7].

[8] Refer to "Deborah," Wikipedia.

and the children of Israel came up to her for Judgement (Old Testament, Judges, Chap. 4, Verses 4–5) [4]

▪ When Prophet Moses was guided by God to show him the Promised Land that he has given to the people of Israel, Jericho the city of palm trees was there waiting:

And Moses went up the plains of Moab unto the mountain of Nebo, to the top of Pisgah that is over against Jericho. And the LORD showed him all the land of Gilead, unto Dan, and all Naphtali, and the land of Ephraim, and all the land of Judah, unto the utmost sea. And the south, and the plain of the valley of Jericho, the "city of palm trees", unto Zoar (Old Testamen, Deutronomy, Chap. 34, Verses 1–3) [4]

Zionist believers claim that Jericho's date palm trees have roots that stretch toward a source of freshwater that has turned a desert into a garden. According to their belief system, the perennial spring—that supported the city for centuries and provided a splendid irrigation system distributing water to the plain as well as to all travelers in antiquity—has been originally purified—according to the Old Testament (Torah) [4]—by Prophet Elisha.[9]

▪ Date palm tree's branch is used in Old Testament to signify the "head," the highest of the people, as contrasted with the "rush" or the "tail," the humblest of the people:

Therefore the LORD will cut off from Israel head and tail, branch and rush, in one day (Old Testament, Isaiah, Chap. 9, Verse 14) [4]

1.1.1.2 The Date Palm Tree in Christian Traditions

The Palm Tree Christian[10]

When contemplating the photographs of the strongest hurricanes in history leaving so much destruction, it is surprising to see that there is always something standing, and it is the palm tree.

Palm trees are thin and slender, that is why many come to consider it a "fragile tree." However, these trees with the appearance of weakness are able to remain standing after storms. There are two essential things that Christians can learn from this wonderful tree:

1. A palm tree has been designed to overcome storms. One of the qualities of palm trees is that they are very flexible. Instead of offering resistance to hurricane-force winds, they allow themselves to be pushed and shaken. This is a

[9] Refer to Old Testament, Kings II, Chap. 2, Verses 19–22. See also; Wayne Stiles, "Sites and Insights: Jericho, city of palms," Jerusalem Post, October 18, 2012 [8].

[10] Refer to [Are you a "palm tree Christian"?], Bibliatodo Reflection, October 10, 2020, [www.bibliatodo.com] [9].

mechanism that they have developed over time (refer to New Testament, I Corinthians, Chap. 10, Verse 13) [10]:

> There hath no temptation taken you but such as is common to man. But God is faithful who will not suffer you to be tempted above that ye are able; but with the temptation also make a way to escape that ye may be able to bear it.

2. Even if a "Palm Tree Christian" fall, he must learn how to stand up again. Another important quality that Christians can learn of the palm tree is that it has managed to put down deep roots. This quality allows it to bend until its branches touch the ground, but after the storm the Palm Tree Christian stands up again. So, when you are going through a difficult time, remember—if you are a Christian believer—the qualities of a palm tree:

> The righteous shall flourish like the "Palm Tree"; he shall grow like a cedar in Lebanon. Those that be planted in the house of the LORD shall flourish in the courts of our God" (Old Testament, Psalms 92, Verses 12–13) [4]

Triumphal Entry of Jesus into Jerusalem[11] (known as Palm Sunday)

> On the next day much people that were come to the feast, when they heard that Jesus was coming to Jerusalem, took branches of "palm trees" and went forth to meet him. And cried, Hosanna: Blessed is the King of Israel that cometh in the name of the LORD. And Jesus, when he had found a young ass, sat thereon as it is written. Fear not daughter of Sion: behold, thy King cometh, sitting on an ass's colt (New Testament, John, Chap. 12, Verses 12–15) [11]

Palm Sunday[12] is a Christian moveable feast that falls on the Sunday before Easter. The feast commemorates the triumphal entry of Jesus into Jerusalem on a donkey that symbolizes his arrival in peace rather than a "war-waging king" arriving on a horse. For adherents of Nicene Christianity,[13] Palm Sunday marks the first day of the Holy Week.[14] In most liturgical churches, Palm Sunday is celebrated

---

[11] Refer to "Triumphal entry into Jerusalem," Wikipedia.

[12] Refer to "Palm Sunday," Wikipedia.

[13] It is named for the city of Nicaea (present day Iznik of Turkey) where the Nicene Creed was originally adopted by the First Ecumenical Council in the year 381 AD. The Nicene Creed was adopted to resolve the Arian controversy, as Arius (a clergyman of Alexandia) objected to Bishop Alexander's apparent carelessness in blurring the distinction in nature between the Father and the Son in the Christian faith. Bishop Alexander and his supporters created the Nicene Creed to clarify the key tenets of Christian faith in response to the widespread adoption of Arius' doctrine. It explicitly affirms the co-essential divinity of the Son.

[14] The "Holy Week" is the week when Jesus Christ was historically put to trial by the Jewish leaders, condemned to death and crucified on Good Friday, and rose from the dead on Easter Sunday.

by the blessing and distribution of palm branches representing the palm branches which the crowd scattered in front of Christ as he rode into Jerusalem.

Palm Trees in the "New Jerusalem"

> And I saw a new heaven and a new earth: for the first heaven and the first earth were passed away and there was no more sea. And I John saw the holy city "new Jerusalem" coming down from God out of heaven, prepared as a bride adorned for her husband (New Testament, Revelation of Saint John the Divine, Chap. 21, Verses 1–2) [11]

It was extraordinary that while Saint John the Divine was describing the "new Jerusalem" where God would dwell by himself in the midst of his faithful people, he discovered that these godly people that were salvaged from the great tribulation and were arrayed in white robes were rejoicing before the throne and before the Lamb[15] while holding branches of palm trees in their hands:

> After this I beheld, and lo, a great multitude which no man could number, of all nations, and kindreds, and people, and tongues, stood before the throne, and before the Lamb, clothed with white robes, and "palms" in their hands. And cried with a loud voice, saying Salvation to our God which sitteth upon the throne, and unto the Lamb (New Testament, Revelation of Saint John the Divine, Chapt 7, Verses 9–10) [11]

1.1.1.3 Date Palm Tree in Qur'an and Sunnah

Date Palm Tree in Qur'an

Although the Holy Quran is a religious book dealing with many aspects of life, almost one sixth of its verses are of a scientific nature, reviewing diverse scientific topics such as biology, embryology, geology and archeology.

The plant kingdom was hugely celebrated in the Holy Qur'an. Almost 22 identifiable plants belonging to 17 plant families are cited in the Holy Quran some of which are denoted holy. Keywords like plants, seeds, grains, gardens, trees, fruits and herbs are mentioned very often to denote a plant, plant part, type and/or places where plants normally grow. Typical biological issues such as biodiversity, seed germination, photosynthesis and diverse uses of plants are interpreted from various surahs of the Quran. A comprehensive list of surahs and ayahs, where a distinct plant or a keyword indicating plants, is included [12].

This diversity is described in the Qur'an as follows: "And in the earth are neighbouring tracts, and gardens of vines, and green crops (fields), and date palms, growing into two or three from a single stem root, or otherwise (one stem root for every palm), watered with the same water; yet some of them We make more excellent than others to eat. Verily! In these things, there are signs for the people who understand." (Surat Ar-Ra'd 13:3–4).

---

[15] That represent Jesus Christ.

Table 1.1 References of palm tree in the Quran [14]

1	Surah-2. Al-Baqara, Part 2, Verse 266:	Would any of you wish to have a garden with *date-palms* and vines [7]
2	Surah-6. Al-An'âm, Part 7, Verse 99:	And out of the *date-palm* and its spathe come forth clusters of *dates* hanging low and near [7]
3	Surah-6. Al-An'âm, Part 7, Verse 141:	And it is He Who produces gardens trellised untrellised and *date palms[l]*
4	Surah-13. Ar-Ra'd, Part 13, Verse 4:	And in the earth there are Neighboring tract, and gardens of vines and green crops (fields), and *date palms* [7]
5	Surah-16. An-Nahl, Part 14, Verse 11:	With it He produces for you com, olives, *date-palms,* grapes and every kind of fruits: [15]
6	Surah-16. An-Nahl, Part 14, Verse 67:	And from the fruits of *date-palms* and grapes, you drive strong drink and goodly provision [7]
7	Surah-17. Al-Isrâ, Part 15, Verse 91:	Or you have a garden of *date-palms* and grapes and cause rivers to gush forth in their midst abundantly [7]
8	Surah-18. Al-Khaf, Part 15, Verse 32:	And put forward to them the example of two Men: unto one of them We had given two gardens of grapes and We had surrounded both with *date-palms;* and had put between them green crops (cultivated fields) [7]
9	Surah-19. Maryam, Part 16, Verse 23:	And the pain of childbirth drove her to the trunk of a *date-palm[l]*
10	[a]Surah-19. Maryam, Part 16, Verse 25:	And Shake the trunk of the *date palm* towards you, it will let fall fresh *ripe-dates* upon you [7]
11	Surah-20. Tâ-Hâ, Part 16, Verse 71:	So I will surely cut off your hands and feet On opposite sides and 1 will crucify you on the trunks of *date-palm* [7]
12	Surah-23.A1-Mu'minun, Part 18, Verse 91:	Then We brought forth for you therewith gardens of *Date palms* and grapes [7]
13	Surah-26. Ash-Shu'ara, Part 19, Verse 148:	And green crops (fields) ans *date palms* with soft spadix [7]
14	Surah-36. YâSîn, Part 22, Verse 34:	And we have made therein gardens of *date-palms* and grapes [16]
15	Surah-36 YâSîn, Part 22, Verse 39:	And the moon, We have measured for it mansions (to traverse) till it turns like the old dried curved *date* stalk[7]
16	Surah-50. Qâf, Part 26, Verse 10:	And tall (and stately) *palm-trees,* with shoots of fruit-stalks, piled one over another" [15]
17	Surah-54. Al-Qamar, Part 27, Verse 20:	Plucking out men as if they were uprooted Stem *date-palm* [7]
18	Surah-55. Ar-Rahmân, Part 27, Verse 11:	Therein are fruits, *date-palms* producing sheathed fruit-stalks (enclosing dates) [7]
19	Surah-55. Ar-Rahmân, Part 27, Verse 68:	In them (both) will be fruits and *date-palms* and pomegranates [7]
20	Surah-59. Al-Hashar, Part 28, Verse 5:	What you (O Muslims) cut down the *palm trees* (of the enemy), or you left them standing on their stems, it was by leave of Allah and in order that He might disgrace the Fâsiqun [7]
21	Surah-69. Al-Hâqqah, Part 29, Verse 7:	Which Allah imposed on them for seven nights and eight days in succession, so that you could see men lying overthrown (destroyed), as if they were hollow trunks of *date-palm*! [7]
22	Surah 80.Abasa, Part 30, Verses 27–29:	And We cause therein the grain to grow and grapes and clover plants (i.e. green fodder for the cattle) and olives and *date-palm* [7]

[a]Date 2 times mentioned in verse no. 25 of Sura Maryam

The date palm tree, mentioned more than any other fruit-bearing plant in the Qur'an, is regarded as holy and a symbol often associated with Islam and Muslims, see Table 1.1 [13].

The significance and versatility of the date palm are borne out by the fact that this fruit and its blessed palm find mentioned in the Qur'an for 23 times [17].

In the second chapter of the Quran, it was quoted "Would one of you like to have a garden of palm trees and grapevines underneath which rivers flow in which he has from every fruit?..." (Al-Baqarah 2:266)

> And it is He who sends down rain from the sky, and We produce thereby the growth of all things. We produce from it greenery from which We produce grains arranged in layers. And from the palm trees—of its emerging fruit are clusters hanging low. And [We produce] gardens of grapevines and olives and pomegranates, similar yet varied. Look at [each of] its fruit when it yields and [at] its ripening. Indeed, in that are signs for a people who believe. (Al-An'am 6:99)

It was also mentioned "It is He Who has brought into being gardens- the trellised and untrellised - and the palm trees, and crops, all varying in taste, and the olive and pomegranates, all resembling one another and yet so different. Eat of their fruits when they come to fruition and pay His due on the day of harvesting. And do not exceed the proper limits, for He does not love those who exceed the proper limits. (Al-An'am 6:141).

> And from the fruits of the palm trees and grapevines you take intoxicant and good provision. Indeed, in that is a sign for a people who reason (An-Nahl 16:76)

Many parts of date palm have been mentioned by the Qur'an including: its spadix, trunk and the nucleus. **Palm spadix** is described in many verses as soft, producing clusters of dates hanging low and near and being arranged and layered one above another, "and the date-palms that stand tall with piled spathes, one [cluster of dates] sitting on top of the other" (Qaf 50:10). **Palm tree trunk** was many times referred to in the Quran "...so that you could see men lying overthrown (destroyed), as if they were hollow trunks of date-palms!" (Al-Haqqah 69:7). Three different elements were defined in the **palm stone**:

1. Al-fatil: a scalish thread in the long slit of a date stone. "...And they will not be dealt with injustice even equal to the extent of a fatila." (An-Nisa' 4:49)
2. An-naqir: refers to the fovea in the nucleus from which the date palm grows. "Or have they a share in the dominion? Then in that case they would not give mankind even a naqira." (An-Nisa' 4:53)
3. Al-qitmir: refers to the thin membrane over the date stone. "And those, whom you invoke or call upon instead of Him, own not even a qitmir." (Fatir 35:13) [18].

The use of palm trees as a windbreak was brought up in the Quran as well. "And present to them an example of two men: We granted to one of them two gardens of grapevines, and We bordered them with palm trees and placed between them [fields of] crops" (Al-Kahf 18:32).

The palm tree was where Maryam (peace be upon her) gave birth to Prophet Esa (peace be upon him) according to the Quran. "And the pains of childbirth drove her to the trunk of a palm tree. She said, "Oh, I wish I had died before this and was in oblivion, forgotten" (Maryam 19:23). Additionally, the date was the food that Maryam was told to eat from to regain her strength after giving birth. "And shake toward you the trunk of the palm tree; it will drop upon you ripe, fresh dates. So eat and drink and be contented!" (Maryam 19:25–26).

> Have you not considered how Allah presents an example, [making] a good word like a good tree, whose root is firmly fixed and its branches (high) in the sky? Producing its fruit all the time by the permission of its Lord. And Allah presents examples for the people that perhaps they will be reminded (Ibrahim 14:24–25)

Here, the good word was described to be similar to the good tree, and the meaning in conformity with the view adopted by many of scholars of Tafsir is that the good word in this verse means the declaration that there is no deity worthy of worship but Allah (i.e. Shahadah) and the good tree is the date palm tree [19].

Palm Tree in the Life of Prophet Mohammed

The palm tree had a very significant status in the life of the Prophet Mohammed. Its huge benefits and wide spectrum of usefulness were highlighted by comparing it to the believer in the hadeeth narrated in Sahih Al-Bukhari hadith 6122 by Ibn`Umar: "While we were with Allah's Messenger, peace be upon him, he said: «Tell me of a tree which resembles a Muslim man. Its leaves do not fall and it does not, and does not, and does not, and it gives its fruits every now and then». Then he said: «It is the date-palm tree» [20–22] He and in another hadith He said "The believer is like a palm tree, whatever you take from it will benefit you."

The palm tree was a prominent source of building materials used extensively by Prophet Muhammad as he established the very first foundations of what is called now Islamic Architecture.

When Prophet Muhammad arrived in Madinah from Makkah, the first task relating to the built environment that he embarked on fulfilling was building the city's central mosque (also called the Prophet's Mosque). When completed, the form of the mosque was extremely simple. It consisted of an enclosure with walls made of mud bricks and an arcade on the qiblah side made of palm trunks used as columns to support a roof of palm leaves and mud. The mosque was shaded by erecting palm trunks and wooden cross beams covered with date palm leaves [3].

The houses of Prophet Muhammad bare witness of the significance of date palm trees as a source of materials with diverse potentials. A partial description of the Prophet's houses is given by Ibn Sa'd in his *al-Tabaqat al-Kubra,* "There were four houses of mud brick, with apartments partitioned off by palm leaves plastered with mud, and five houses made of palm leaves plastered with mud and not divided into rooms. Roofs were made of palm leaves [23].

When ‘Umar ibn al-Khattab visited the Prophet, though he was moved to tears by the simplicity of the Prophet’s living, yet he reported that he found the Prophet in one of his houses in his attic to which one must climb by means of a ladder made of date palm midribs [24].

Inside his house, he had a mat made of date palm tree leaflets with nothing between him and the mat. Underneath his head, there was a leather pillow stuffed with date palm sheaths fibers and leaflets as narrated by Ibn-Abbas in hadith 4913 in Sahih Al-Bukhari [21].

Prophet Muhammad considered date palm trees as valued asset that he said the best asset a man can have is a road of lined up pollinated date palm trees as narrated in Hadeeth 387 in Al Tarikh Al Kabir by Al-Bukhari.

Prophet Muhammad had a deep spiritual connection to the date palm trees. The Prophet used to stand by a date palm trunk on Friday. Then an Ansari woman or man said. “O Allah’s Messenger (ﷺ)! Shall we make a pulpit for you?” He replied, “If you wish.” So, they made a pulpit for him and when it was Friday, he proceeded towards the pulpit (for delivering the sermon). The date palm trunk cried like a child! The Prophet (ﷺ) descended from (the pulpit) and embraced it while it continued moaning like a child being quietened. The Prophet (ﷺ) said, “It was crying for (missing) what it used to hear of religious knowledge given near to it.” [20].

Planting date palm trees was tremendously encouraged by Prophet Muhammad and the reward of which will continue on even after the person’s death in hadith 3602 [25] …. to an extent that it was narrated that Prophet Muhammad said “If the Final Hour comes while you have a shoot of a date palm in your hands and it is possible to plant it before the Hour comes, you should plant it.”

Dates were a main component of Prophet Mohammad’s Diet. He stressed on the importance of dates as a major food item, saying to his wife: “O’ Aishah, a house in which there are no dates, it’s people will go hungry” as mentioned is Hadith 5337 in Sahih Muslim [26].

Anas ibn Malik reported: The Prophet, peace and blessings be upon him, would break his fast before praying with fresh dates, if there were no fresh dates, then with dry dates, as mentioned in hadith 696 in Sunan Al-Tirmithi [19]. On the basis of this Hadith and many other similar ones, Muslims insist on breaking their fasts with dates till this time.

Prophet Muhammad also taught his disciples that the date was not only an antidote to poison but also an effective defense against black magic. “Whoever eats seven Ajwah dates in the morning will not be harmed by any poison or witchcraft that day.” as mentioned in Sahih Muslim [26].

1.2 Ecological Significance of the Date Palm

1.2.1. The date palm can tolerate temperature extremes and prolonged periods of drought [27, 28]. Maximum temperatures around 50 °C do not harm the date palm [29]. The date palm is well adapted to arid conditions where annual precipitation rarely exceeds 250 mm with a strong summer heat of about 50 °C and cold winter—10 °C [30].

1.2.2. The date palm can tolerate water salinity better than any other fruit tree. Date palms can produce full crop if irrigated with saline water up to 2000 ppm [15]. They can tolerate levels of salinity up to 6000–7000 [15] and 12-BdSm^{-1} [14].

1.2.3. The date palm enjoys a low transpiration rate due to the waxy layer covering the date palm leaves (midribs and leaflets) [31].

1.2.4. The date palm plays a significant role in struggle with desertification, sand encroachment, sand dunes movement and desert winds [29, 32–34]. The date palms act as well as a windbreak protecting the plants beneath the date palms from the winds [32].

1.2.5. The date palms protect the plants beneath it from the direct sun rays especially in summer leading to the "burning" of leaves and green off shoots and falling of juvenile buds and fruits [11].

1.2.6. The date palm provides shade beneath it (15 m^2/palm [31]). This decreases the evaporation beneath it. The potential evapotranspiration (PET) reaching 2500 mm/year decreases only to 1200 mm/year under the date palms [34] contributing to the creation of a favorable microclimate for agriculture.

1.2.7. The date palm root system (up to 10 m in length) [35] contributes to the maintenance of soil from erosion [31]. The date palm root system and rhizosphere soil reveal a complex diversity that enclose a reservoir of plant growth promoting bacteria involved in the regulation of plant homeostasis [36]. Date palm roots shape endophytic bacteria communities that are capable to promote plant growth under drought conditions and thus contributing an essential ecological service to the entire oasis ecosystem [16].

1.2.8. The date palm is a precious keystone species and the tree of life [18] creating in arid and semi-arid regions a microclimate that can allow agriculture and offsets the effects of drought [16]. The microclimate created by the date palms preserves high soil moisture with low salinity. This in turn creates a biodiversity hotspot that shelters a wide range of plant and animal species located in the thick of hot desert conditions [32]. The date palms occupying small area and being high (up to 30 m [10]) allow the planting of fruit trees (citrus, mango and guava [15, 37–39], olive [40], pomegranate, fig [40], pomegranates and fig [8], olive, mango, citrus, fig, pomegranate guava, mango, olive, citrus and grape [39, 41]), coffee, pomegranates, mango, citrus, maize alfalfa, vegetables and cereals [42], in addition to wheat, onion, beans, forage [38, 40], legime in winter, forge crops in summer [15], Henne, vegetables, fodder crops cereals, legumes, forages[43] and breeding of livestock [8, 34].

1.2.9. The date palm could be viewed as a sustainable source of a whole system of renewable materials empowering the poor rural communities in arid and semi-arid regions in the world [43]. Its trunk furnishes timber; leaves provide roofing materials, leaf midribs material for crates and furniture, the leaflets are manufactured into baskets and mats; the leaf bases are used for fuel; the fruit stalks provide rope and fuel; the leaf sheaths fibers are processed into cordage and packing material and the seeds are ground and used as stock feed. Syrup, alcohol and vinegar can be processed from wasted dates. The sap is also used as a beverage, either fresh or fermented. When palm trees are felled, the tender terminal bud can be eaten as a salad. Dates are conserved rather easily; to the benefit of small farmers' ability to gain maximum returns from their crop.

1.2.10. The date palm cultivation is a labor-intensive industry making a valuable contribution to improving and sustaining livelihood in desert rural communities [34]. Thus, the date palms provide rural communities with food resources and livelihood opportunities for employment in drylands. Besides, it provides an added value product (dates) to generate income [28, 32]. The date fruit products like date juice, liquid sugar, protein yeast and vinegar, marmalade, chocolate, wine, alcohol, organic acids and so on made the fruit an important input for food and related industries that create good opportunity for agropastoralists to improve their income and thus their livelihoods [28]. Thus, the date palm played an important role in settlement of rural communities in arid and semi-arid regions and decrease of migration to urban areas. The date palm has played a pivotal role in the history of the Arab countries in helping nomads to settle, organize communities and begin farming [42]. It can be estimated that the date palm production has bestowed jobs for 50 million, 35 of them in south Mediterranean countries.

References

1. Al-Albani MN (1988) Al Gami Al Saghir, Beirut: Al Maktab Al Islami
2. Khan MM (1997) The translation of the meanings of Sahih Al-Bukhari, vol 4
3. Abdelhamid MM (2020) The significance of the date palm as a decorative motif in the synagogues of Cairo, Egypt. Int J History Cult Stud 6(1):1–13
4. Mohamed N (n.d.) Plants of the Qur'an: the date palm. https://islamonline.net/en/plants-of-the-quran-the-date-palm/
5. Stiles W (2012, Oct 18) Sites and insights: Jericho, city of palms. Jerusalem Post
6. Cherif H, Marasco R, Rolli E, Ferjani R, Fusi M, Soussi A, Mapelli F, Blilou I, Borin S, Boudabous A, Cherif A, Daffonchio D, Ouzari H (2015) Oasis desert farming selects environment-specific date palm root endophytic communities and cultivable bacteria that promote resistance to drought: oasis palm endophytes promote drought resistance. Environ Microbiol Reports 7(4):668–678. https://doi.org/10.1111/1758-2229.12304
7. Othman AMA (2017) Development of palm cultivation and date production in Sudan. Khalifa International Award for Date Palm, the blessed tree (in Arabic)
8. Proposal for an International Year of Date Palm (2020) Committee on Agriculture. FAO
9. Are you a "palm tree Christian"? (2020) www.bibliatodo.com
10. Al-Khattab N (2007) English translation of Sahih Muslim, vol 5. Darussalam, Riyadh
11. Palm tree in the Bible Encyclopedia (n.d.) www.bible-history.com

12. Khafagi I, Zakaria A, Dewedar A, El-Zahdany K (2006) A voyage in the world of plants as mentioned in the Holy Quran. Int J Botany 2(3)
13. Khan MM (1997c) The translation of the meanings of Sahih Al-Bukhari. In: The translation of the meanings of Sahih Al-Bukhari, vol 8, pp 85–86. Riyadh, Darussalam
14. Marwat SK (2016) Ethnobotanical and religious importance of the Quranic Plant-Nakhl (*Phoenix dactylifera* L.) review (16th ed). American-Eurasian J Agri Environ Sci
15. El-Sherbasy SF (2021) The use of date palms and dates in the horticultural intensification used (a modern agricultural model), p 50. Khalifa International Award for Date Palm and Agricultural Innovation (in Arabic)
16. BBC-Judaism: Sukkot (n.d.)
17. Mahjoub MM (2017) Challenges of cultivation and production of dates in Sudan. Khalifa International Award for Date Palm, the blessed tree (in Arabic)
18. Johnson DV, Al-Khayri JM, Jain SM (2015) Introduction: date production status and prospects in Asia and Europe. In: Al-Khayri JM, Jain SM, Johnson DV (eds) Date palm genetic resources and utilization. Springer, Netherlands, pp 1–16. https://doi.org/10.1007/978-94-017-9707-8_1
19. Feast of Tabernacles—Dictionary Definition: Vocabulary.com. (n.d.)
20. Ibrahaim KM (n.d.) The role of date palm in improvement of the environmental. Collage of Science, Al-Nahrain University, Iraq
21. Lemlem A, Alemayehu M, Endris M (2018) Date palm production practices and constraints in the value chain in Afar Regional State, Ethiopia. Adv Agri 2018:1–10. https://doi.org/10.1155/2018/6469104
22. El-Juhany LI (2010) Degradation of date palm trees and date production in Arab countries: causes and potential rehabilitation. Aust J Basic Appl Sci 4(8):1991–8178
23. Mihi A, Tarai N, Chenchouni H (2019) Can palm date plantations and oasification be used as a proxy to fight sustainably against desertification and sand encroachment in hot drylands? Ecol Ind 105:365–375. https://doi.org/10.1016/j.ecolind.2017.11.027
24. Al Samrani MR (2009) The palm tree in the civilization of the Mesopotamia Valley in Iraq. Khalifa International Award for Date Palm, the blessed tree(in Arabic)
25. Nazri MKNZ (2016) The descriptions of date palms and an ethnomedicinal importance of dates mentioned in the Quran, vol 7. Mediterranean Journal of Social Sciences
26. Workshop on "Irrigation of date palm and associated crops" (2008). FAO and Faculty of Agriculture, Damascus University
27. Arab L, Kreuzwieser J, Kruse J, Zimmer I, Ache P, Alfarraj S, Al-Rasheid KAS, Schnitzler J-P, Hedrich R, Rennenberg H (2016) Acclimation to heat and drought—lessons to learn from the date palm (*Phoenix dactylifera*). Environ Exp Bot 125:20–30. https://doi.org/10.1016/j.envexpbot.2016.01.003
28. El Hadrami I, El Hadrami A (2009) Breeding date palm. In: Breeding plantation tree crops: tropical species. Springer, pp 191–216
29. International Fund for Agricultural Development (1997) An analytical study of the agricultural systems in the date palm areas and an assessment of the economic and social repercussions of the technical obstacles facing the palm sector in the Republic of Sudan [analytical study]. The Arab Center for the Studies of Arid Zones and Dry Lands, Palm Research and Development Network (in Arabic)
30. Kreidly H (2008) The true collection SUNAN AL-TIRMIRHI, vol 1. Dar Al-Kotob Al-Ilmiyah, Beirut
31. The Holy Bible, New testament, King James Version (n.d.)
32. Khan MM (1997b) The translation of the meanings of Sahih Al-Bukhari, vols 6, 8. Riyadh: Darussalam
33. Shri MJ (2015) Introduction: date production status and prospects in Asia and Europe
34. The Holy Bible, Old testament, King James Version (n.d.)
35. Ali HG (2010) Development of date palm cultivation and its role in sustainability of agriculture in Oman. Acta Horticulturae 882:29–35. https://doi.org/10.17660/ActaHortic.2010.882.1

36. de Grenade R (2013) Date palm as a keystone species in Baja California peninsula, Mexico oases. J Arid Environ 94:59–67. https://doi.org/10.1016/j.jaridenv.2013.02.008
37. Ahmed A, Faki HHM, Ahmed KH (2018, October) Economics and constraints of palm date production and marketing in Sudan. In: 2nd International conference for date palm (ICDP 2016). At: College of Agriculture and Veterinary Medicine, Qassim University, Qassim, Kingdom of Saudi Arabia, Qassim, Kingdom of Saudi Arabia
38. Omer S (2010) Some lessons from Prophet Muhammad (SAW) in architecture: the Prophet's Mosquein MadÊnah, vol 18. Intellectual discourse
39. Omer S (2011) Housing lessons from the life of prophet Muhammad (PBUH): the form of the house. https://medinanet.org/2011/09/housing-lessons-from-the-life-of-prophet-muhammad-pbuh-the-form-of-the-house/
40. Omer S (2018) The houses of the prophet. https://www.islamicity.org/7989/the-houses-of-the-prophet/
41. Britannica (n.d.) https://www.britannica.com/topic/cherub
42. El Hadrami A, Al-Khayri JM (2012) Socioeconomic and traditional importance of date palm. Emir J Food Agric:371–385
43. Yaish MW, Kumar PP (2015) Salt tolerance research in date palm tree (*Phoenix dactylifera* L.), past, present, and future perspectives. Front. Plant Sci 6. https://doi.org/10.3389/fpls.2015.00348

Chapter 2
The Date Palm Byproducts: Description, History of Utilization and Associated Technological Heritage

Abstract The date palm abundantly exists in North Africa, the Arab Peninsula and Iran. The first emergence of date palm dates back to 4000 BC in Mesopotamia. This long history, along with the high renewability rate, has led to the accumulation of a rich technical heritage associated with the utilization of all the secondary date palm products, including whole leaves (midribs, leaflets), petioles, spadix stems, leaf sheaths fibers, date kernels and trunks. Midribs have been used in roofing, fencing, furniture making, and manufacturing crates and coops. Leaflets have been used in making mats, baskets and bags. Leaf sheaths fibers have been used in making ropes, nets, bags, brooms and fly whiskers. Spadix stems have been used in making brooms and household sieves. In addition, fibers obtained from spadix stems have been used for tying agricultural crops. The palm trunk has been used as windows lintels, beams and columns in construction. Moreover, trunks have been used as a wood substitute in furniture making. This chapter reveals the technical heritage associated with several traditional uses of the secondary date palm products to satisfy the human needs in the Arab region. In addition, geometrical description of these secondary products and the procedures of their preparation are included.

Keyword Date palm · Traditional handicrafts · Midribs · Leaf sheaths fibers · Spadix stem

2.1 Introduction

The local materials are the material milieu by virtue of which cultures were able to express themselves. Proceeding from the historical perspective, the different cultures of the world were born and developed in company with different materials. Who could deny the relation between the ancient Egyptian culture and papyrus, lotus, limestone and granite, nor between the Asian cultures and bamboo, rattan and rice? It is extremely important to capture the relation between culture and local materials as an important asset for development. The linking between development and local materials means that you are building on the existing culture of interaction with these materials; i.e. you are not beginning development from a zero datum, but with what people—members of each local community—have at hands (the local materials),

H. EL-Mously et al., *Date Palm Byproducts: A Springboard for Circular Bio Economy*, Materials Horizons: From Nature to Nanomaterials,
https://doi.org/10.1007/978-981-99-0475-4_2

as well as in minds (psychological familiarity with these materials, and technical heritage, associated with their production, manufacture and use in the different walks of life). In this concern, the date palm (*Phoenix Dactylifera* L.) represents an eloquent example. It is an authentic element of the region's flora, which accompanied our historical march for thousands of years.

As a resource, the date palm could be seen as a system of renewable materials including the primary and secondary products. The date palm is associated with a very rich technical heritage being a product of thousands of years of accumulated expertise of interaction between the diverse local communities and the date palm material system for the satisfaction of various basic material needs.

This technical heritage includes appropriate technologies that may be of contemporary value for peasants in rural areas such as preservation of date in clay jars as food for the whole year and the use of palm midribs and trunks in roofing. In addition, the technical heritage represents a software of its own right, revealing a mode of adaptation with the environment and cultural expressiveness, and thus inspiring to think, imagine and innovate in harmony with the environment and culture.

2.2 Date Palm: A Basic Element of the Flora of the Arab Region

It may be difficult to record the first emergence in history of the date palm (*Phoenix dactylifera* L.), but it was well-known in 4000 years BC, where it was used to build the moon temple near to Ore, south of Iraq [1]. The second proof of the deep-rootedness of the date palm comes from the Nile valley, where the date palm was taken as the symbol of the year and the palm midrib as a symbol of the month in the hieroglyphic Egyptian language. But the cultivation of the date palm in Egypt was 2000–3000 years later than Iraq.

The date palm was one of the pivots of economic and, hence, social and cultural life in this region from ancient times. In ancient Egypt, the heads of pillars in temples were made resembling the growing top of the date palm. The date palm appeared frequently on walls of temples in different contexts revealing its significance in life in Egypt. The palm leaves were fundamental in ancient Nubian and Upper Egypt houses. The roofs were constructed by split palm trunks and leaves, and the interior walls were covered by palm leaves ornaments [2]. Until now, date palm constitutes a basic element in several surviving traditions in Nubia and South Egypt, where a palm tree is planted every time a child is born [3]. So when he becomes an adult, the date palm will have grown into many palms that will be the basis of his new life after marriage. As a result, date palm has played a major role in the formation of the culture and heritage in Egypt until the present day [4].

Economically, date palms are a major part of the life-supporting plantations in every village in Upper Egypt [3, 5]. Moreover, the annual byproducts of date palm are being utilized in many traditional crafts by the cultivators and craftsmen in Egypt

[6], thus playing a huge role in sustaining the rural societies against the immigration to urban cities, as date palm-related crafts and cultivation support over one million families in Egypt [5].

Thus, the significance of date palm does not only depend on the multiple uses of the fruit in food, spirits, pharmaceutical, cosmetic and medicinal products, but also on the large number of the secondary products that have been widely used in construction and handicrafts [5]. Palm midribs and trunks have been used for roofing in a fashion that still survives in the western oases and the poor rural areas in Egypt [6, 7]. Hence, the technical heritage associated with the products of date palm pruning is still thriving as their cheapness and abundance qualified them to be the favorable raw materials for several traditional industries with a know-how that goes back to ancient Egypt [6].

This technical heritage thrives only because of the high adaptability of date palm in the Arab region environment. The date palm can survive in a wide range of temperature from − 15 to 60 °C [8]. Direct sunlight helps palm leaves become stronger, taller and thicker and helps them grow faster [9, 10]. Moreover, a date palm provides shade and protection for crops and tolerates high levels of heat and salinity as date palm cultivation is found along the seashore in Egypt [3]. In addition, date palms need less water and maintenance and are less prone to diseases than other fruit trees [3]. Figure 2.1 illustrates traditional forms of utilization of date palm byproducts.

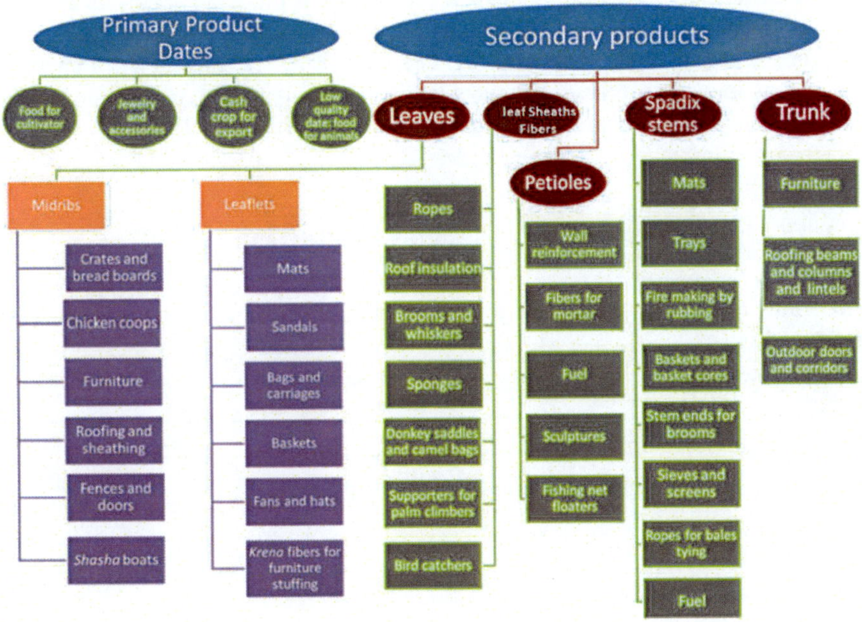

Fig. 2.1 Traditional forms of utilization of date palm byproducts

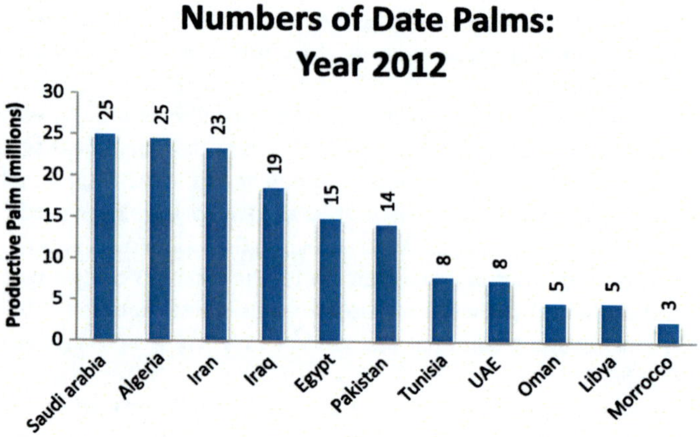

Fig. 2.2 Distribution of date palms in the world [12]

2.3 Distribution of Date Palms in the World

The historical roots of date palms cultivation still have a huge impact on the present situation of date palm distribution in the world. Historically, date palm cultivation originated in Iraq [11]. Now, the Sahara, North Africa, the Arab Peninsula and Iran acquire the most dense date palm plantations in the world as shown in FAO world map of the annual date production [8]. The latest FAO statistics of the number of date palms showed that Saudi Arabia, Algeria, Iran, Iraq and Egypt hold the highest ranks in date palm numbers in the world as shown in Fig. 2.2 [12].

2.4 Estimation of the World Dates Production

According to Lemlem et al. [13], the world dates production (million tons) can be estimated as:-

1961	1985	2001	2017
1.8	2.8	5.04	8.06

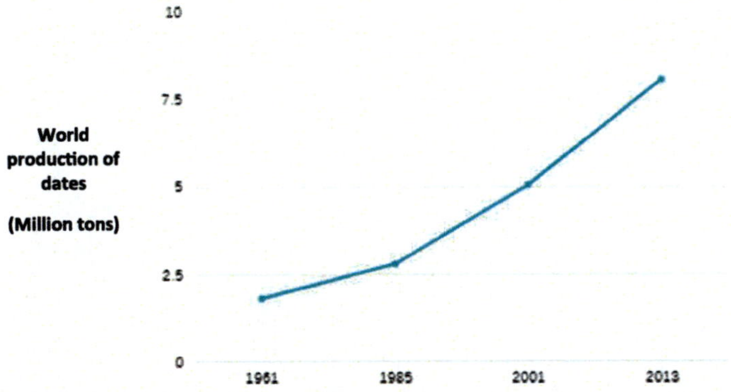

Fig. 2.3 Relationship between the world production of dates and years of production [13]

Figure 2.3 gives a presentation of the relationship between the world production of dates and years of production.

[1]The best-fit quadratic function is

$$Y = 2.93562 - 0.1185899X + 0.00314954X^2,$$

where Y is the dates production and X is the difference between the year and 1950.

Therefore, for the year 2050 the estimated world dates production is 22.57 million tons.

2.5 Date Palm Pruning

Date palms (Fig. 2.4) can live up to 100 years and over, reaching the height of maximum 24 m. Date palms are traditionally pruned annually.

2.5.1 Benefits of Pruning

On average, 13 leaves, 13 petioles and 7 bunches are cut per date palm in the annual pruning process [14]. Annual pruning is necessary for the following reasons [4]:

- Achieving the most suitable symmetry to guarantee the upright standing of the palm.
- Removing abnormal and dead tissues that may take the nutrition from the fruits.
- Stimulating fruit production and flowering necessary for pollination.
- Decreasing the threat of catching fire when the leaves become dry.

[1] This equation has been kindly found by Prof.Dr. Mamdouh Abdel-Hamid.

Fig. 2.4 A date palm

- Getting rid of dry and yellow leaves especially if they were infected.
- Removing the thorns and excess leaves that would obstacle the processes of pollination or harvesting.
- Allowing the sunlight to reach the fruits for high quality of the photosynthesis process.
- Collecting the products of pruning that represent abundant raw materials for several traditional or modern forms of utilization.

2.5.2 Timing and Procedure of Pruning

The annual time of pruning varies from a place to another, but is mainly one of those 3 timings: in autumn after the harvest, in the beginning of spring in the pollination time, and in the ripening time of the leaves in the summer.

Special and trained workers usually perform the annual process as shown in Fig. 2.5. It begins with the removal of the 3-year-old dry leaves using a sharp knife. The cutting should be 10–12 cm above the petiole and the cutting direction should be down-up so the slope of the petiole would expel rainwater.

2.5.3 Products of Pruning of the Date Palm

The products of the annual pruning process of the date palm include:

2.5.3.1 Date Palm Leaves

12–15 new leaves are formed annually by a date palm [8]. The life of each leaf ranges from 3 to 7 years [10]. The length of the palm leaves ranges from 3 to 6 m [10]. Naturally, each fruit cluster of weight of 8–10 kg is supported on one leaf [10]. Annual pruning procedures remove the dry leaves in order to provide better access for the crown for harvesting, in addition to save more nutrition for the fruits [6]. A whole date palm leaf is shown in Fig. 2.6.

At the location of the palm leaf near to the trunk, there are sharp spines with lengths that can reach up to 20 cm [10]. These spines are usually used as sewing needles in traditional weaving [8, 15]. The spines taken from three leaves are shown in Fig. 2.7.

2.5.3.2 Date Palm Midribs

The dominance of date palm midribs over the total quantities of the products of pruning granted them a well-developed surviving technical heritage in traditional handicrafts and architecture in Egypt and the Arab region [6, 16]. Midribs are the main ribs of the whole leaves. They extend from their root at the trunk to the last leaflet [8]. The palm midrib cross-section begins at the base with a triangular shape, and the cross-section becomes narrower and less triangularly shaped with higher density toward the upper end of the palm midrib [8, 17]. A date palm midrib is shown in Fig. 2.8. Curved bases of the midribs are often trimmed and used as fuel resource [8]. The curved base and its cross-sections are shown in Figs. 2.9, 2.10, 2.11.

Fig. 2.5 Pruning of a date palm by a service worker of date palm head (Nakkal in Arabic), supported by a climbing belt made from leaf sheath fibers

Fig. 2.6 A date palm leaf (dimension in centimeters)

Fig. 2.7 Spines taken from 3 leaves (dimension in centimeters)

Fig. 2.8 A date palm midrib (dimension in centimeters)

2.5.3.3 Date Palm Leaflets

Each leaf in a date palm contains 120–240 leaflets [8]. Leaflets are used in woven baskets, ropes, mats, fans and sandals. Date palm leaflets are shown in Fig. 2.12. Date palm leaflets collected from three leaves are shown in Fig. 2.13.

2.5.3.4 Date Palm Spadix Stems

Spadix stems are the trimmed stalks of an empty date bunch [3]. The stems grow carrying the relatively heavy weight of the dates [8]. As a result, the stems adapt by acquiring a notable high tensile strength and high fiber ratio [8]. In addition, the fibers

Fig. 2.9 Base of midrib (dimension in centimeters)

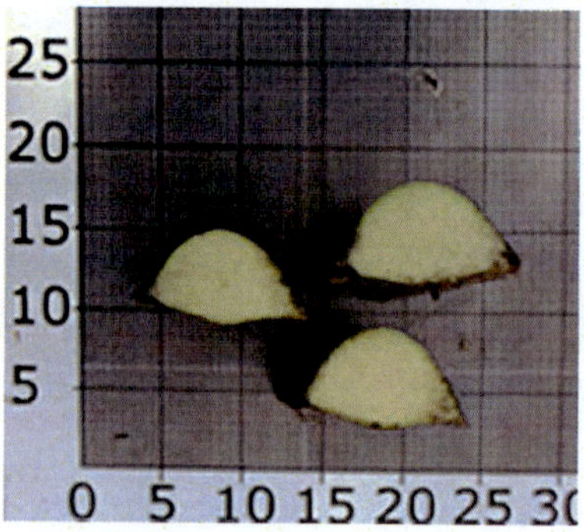

Fig. 2.10 Cross-section at the beginning of the base of the midrib (dimension in centimeters)

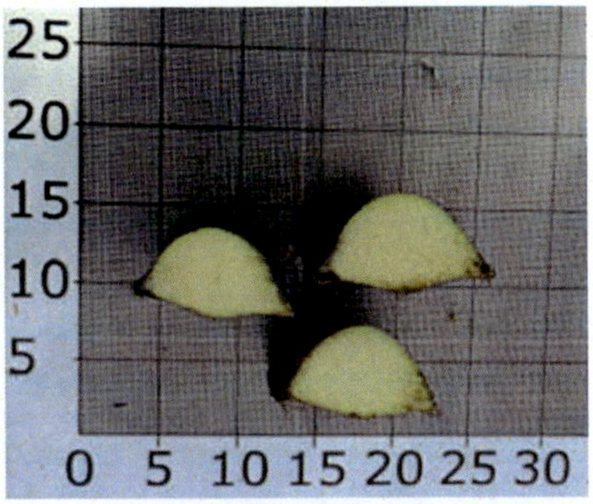

Fig. 2.11 Cross-section at the end of the base of the midrib (dimension in centimeters)

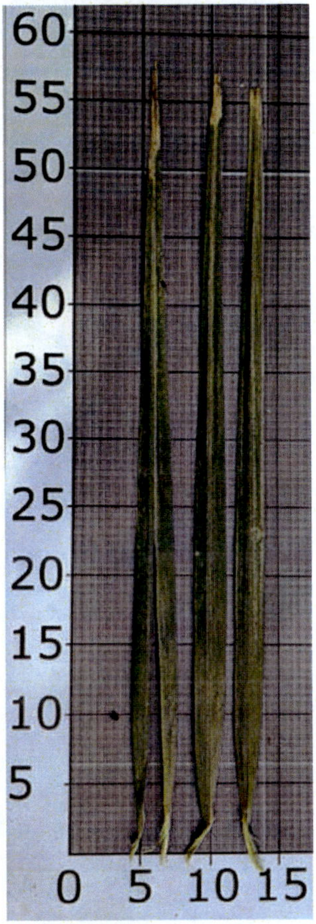

Fig. 2.12 Date palm leaflets (dimension in centimeters)

Fig. 2.13 Date palm leaflets collected from three leaves (dimension in centimeters)

in a spadix stem are long and preferred as the main material for several traditional uses [3]. A date palm spadix stem and a cross-section of the spadix stem are shown in Figs. 2.14, 2.15, respectively.

2.5.3.5 Date Palm Leaf Sheaths Fiber

The leaf sheaths fibers are in the form of tissues that cover the new date palm leaves as they come out and grow [8]. After the growth, these tissues remain attached to the trunk of the palm. These tissues turn into a brownish coarsely woven fabric, after drying and can be torn away during the annual pruning [8]. They are used for protecting the newly planted offshoots, shadings, brushes and fishnets. Date palm leaf sheaths fiber tissues are shown in Fig. 2.16.

2.5.3.6 Date Palm Petioles

A petiole is the base of the leaf that is left after pruning on the trunk [8]. This leaf base is usually trimmed and removed after drying in the next year. The petioles lack high density that is needed for durable applications [10]. A date palm petiole and cross-sections are shown in Figs. 2.17, 2.18, 2.19.

2.5.3.7 Date Kernel

Date kernels are the pits of the dates. They constitute about 10% of the weight of the fruit [18]. Their sizes and colors vary according to the type of cultivar [8]. Fresh seeds are used for breeding and propagation, animal feed for their high dietary fiber content such as phenolic acids and flavonoid [19, 20]. The palms grown by seeds are of unknown species. They represent approximately 27% of the whole number of palms in Egypt [5]. A pair of date kernels is shown in Fig. 2.20.

2.5.3.8 Date Palm Trunk

The availability of palm trunks depends on the end of the useful life cycle of the tree. The trunk is the vertical and cylindrical stem of the palm. It consists of tough vascular bundles glued together with cellular tissues [10]. Hence, the trunk is covered with the bases of the old dry petioles; however, the surface of old trunks is mostly softened by the weather [10]. Date palm trunks are shown in Fig. 2.21.

Fig. 2.14 A date palm spadix stem (dimension in centimeters)

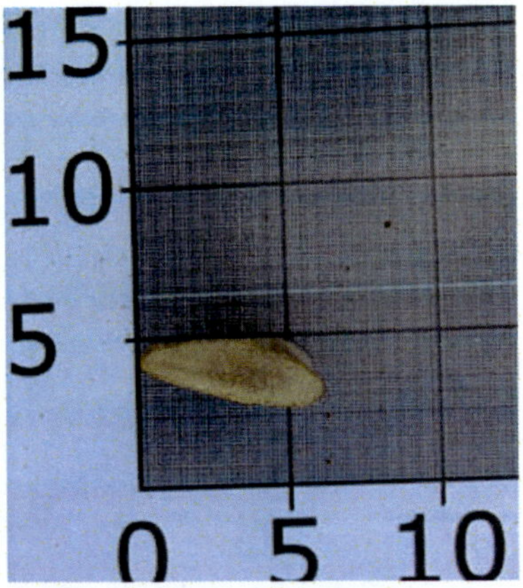

Fig. 2.15 Cross-section of a date palm spadix stem (dimension in centimeters)

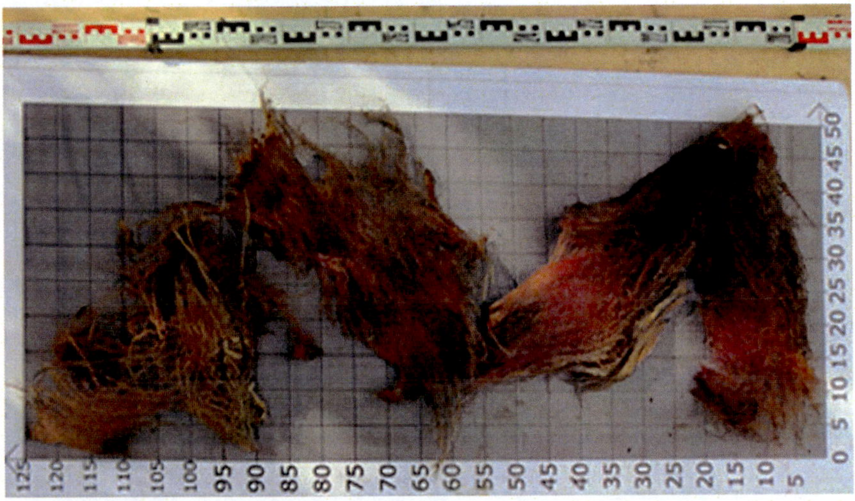

Fig. 2.16 Date palm leaf sheaths fibers tissues (dimension in centimeters)

2.5.4 *Estimation of the Quantities of the Annual Pruning of a Date Palm*

Date palm products of annual pruning include the midribs, the leaflets, the spadix stems, the leaf sheaths fiber tissues and the petioles [17]. In addition, date palm trunk is considered a valuable byproduct that is mostly used as a substitute of timber [21]. In the Arab Gulf region, the annual pruning process produces the average of 6–8 leaves. The weights of the whole leaf, petiole and spadix stem are 0.43 kg, 0.50 kg

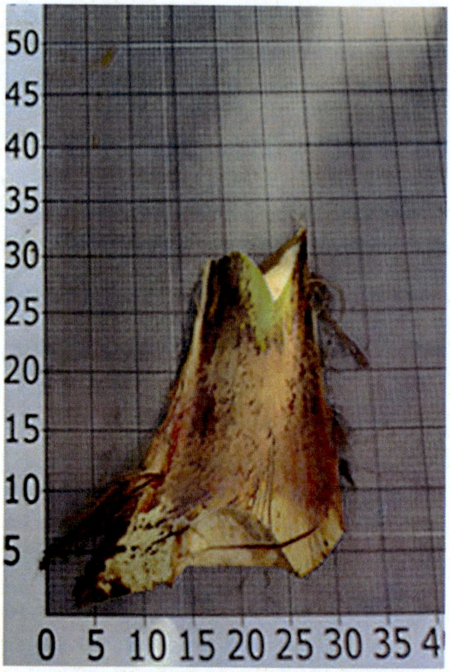

Fig. 2.17 A date palm petiole (dimension in centimeters)

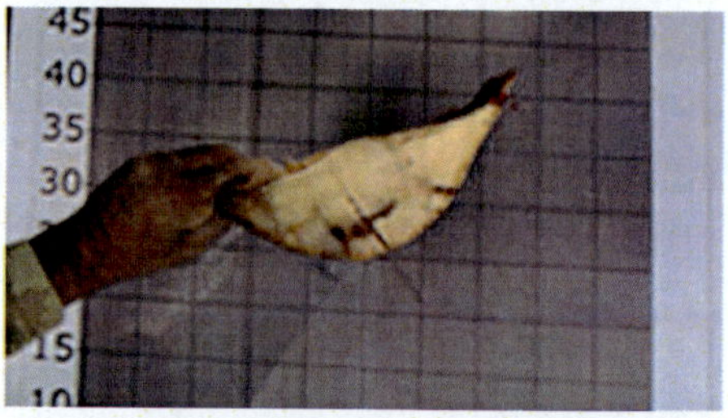

Fig. 2.18 A cross-section at the beginning of the petiole (dimension in centimeters)

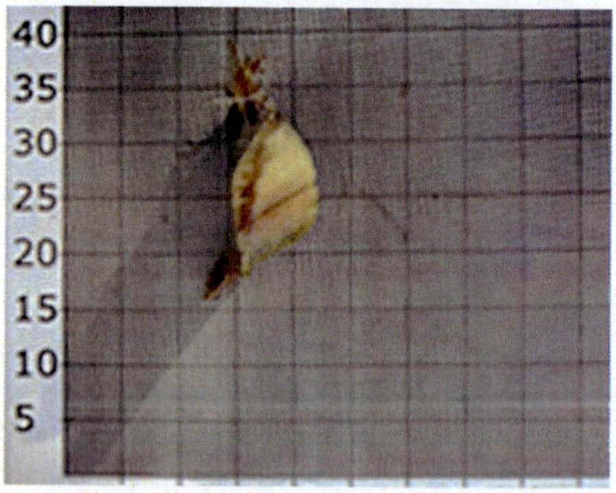

Fig. 2.19 A cross-section at the end of the petiole (dimension in centimeters)

Fig. 2.20 A pair of date kernels

and 0.50 kg, respectively. In Egypt, the quantities of the products of the annual pruning and percentages are shown in Table 2.1.

According to the data shown in Table 2.1, the quantities of the products of annual pruning of date palms in Egypt pruning are estimated to be approximately 810 thousand tons (air-dried weight), which represents a huge material base for a wide spectrum of industries.

2.6 Traditional Forms of Date Palm Leaves Utilization

2.6.1 Traditional Wickerwork Wall Construction

Previous studies predicted that the roofs of the small houses of the workers in ancient Egypt were made of mats of whole date palm leaves and rows covered with a paste of mud that was so thick that it could be rain proof [2, 6]. This wickerwork technique is still used till now in some rural houses in Egypt.

The plan of the early Egyptian house of a worker was about 3 * 5 m, with walls of a wickerwork of palm leaves that were coated on the inside and outside with mud paste. The maintenance of the gaps that would form as the mud gets old was filled over and over with more layers of mud until the thickness of the walls would reach 20–30 cm as shown in the excavations [2]. In a similar manner, the roof was covered with whole date palm leaves and straw, with a cover of beaten earth mud [6]. The sole function of the roof was to shelter from heat, sun and dusty winds, regardless of the rain which was considered too scarce to build a more costly type of roof [2, 22]. This type of roofing and walling can still be found in some houses in poor rural villages in Egypt.

Fig. 2.21 A date palm trunk

2.6.2 *Simple Outdoor Sheathing*

The roots of using date palm leaves, in layers over a secondary net for roof and wall sheathing, extend back in traditional rural huts that can be seen today in Egypt and UAE [6].

In that type of wall sheathing, structural nets, made from reeds or date palm midribs are fixed between the structural poles of the hut. Then, the whole date palm leaves are connected together using threads to make mats called *Sedda* or *hassir* (Fig. 2.22). Finally, each mat is fixed by threads to the nets to create a dense sheathing which

Table 2.1 Products of the annual pruning of date palm (Siwi species) (10% moisture content-air-dried mass) in Egypt [3]

Quantity available annually	Palm midribs	Palm leaflets	Spadix stems	Leaf sheaths fiber	Petiole	Total (kg/palm)
Per palm, air-dried kg (mature female)	15	14.6	9	1.56	14	54.2
Percentage (%)	27.6	26.9	16.6	2.8	25.8	100

offers a highly efficient heat conservation method for the indoor environment of the hut [6, 22].

For roof sheathing, the method relatively resembles thatching technique. The *Sedda* mats are fixed by thread in accumulative layers over the sloped roofing structural grid or a stiff net made from date palm spadix stems or reeds (Fig. 2.23).

Fig. 2.22 Whole date palm leaves Sedda

Fig. 2.23 Sheathing by whole date palm leaves over spadix stem nets supported by wooden poles

2.6.3 *Sheds and Partitions*

Fences, simple sheds and privacy partitions have been simply built by planting the date palm leaves vertically in the soil and tying them together with two horizontal rows of leaves bundles by ropes as shown in Fig. 2.24. In addition, leaves have been used in roofs by laying them across the ceiling beams that are usually made from palm trunks. The thickness of layer of the leaves may reach up to 20–30 cm, and then mud is poured above this layer in the present day [6] (Fig. 2.25). This method is clearly inspired by the ancient wickerwork walls discussed earlier.

2.7 Traditional Forms of Palm Midribs Utilization

2.7.1 *Preparation of Midribs*

Firstly, the leaflets are stripped from the midribs. The leaflets are to be used later in stuffing furniture. In Egypt, the leaflets are removed from the midribs manually. However in the Arabian Gulf region, the green leaves are laid on the ground, where the goats and sheep would feed on the leaflets, leaving the midribs stripped completely from the leaflets. Then, the midribs are laid vertically for 2–3 weeks to dry with the cut facing down to expel all humidity by gravity. For other purposes where long and

Fig. 2.24 Traditional palm leaves fence

straight midribs are needed, the midribs are laid horizontally for 4–6 weeks. Finally, the dried midribs are gathered in bundles: bundles of 5 for handicrafts and bundles of 25 for construction [22] (Fig. 2.26).

2.7.2 Traditional Crates and Bird Coops

Generally, the practical quality of the crate is more important for the crate maker than the aesthetic value. The piercing technique used in making crates dates back to the late Roman period in Egypt [23]. A professional crate maker uses his big toe to hold a palm midrib as he punctures and drives a tin tube to create holes into the midrib piece. A light breadboard uses only 2 midribs maximum, while heavy-duty chicken coops and fruit crates use 4 to 5 midribs [6, 15]. The majority of the used midribs in crate making in Egypt depend on the Nile Valley in Upper Egypt for the hardness and durability of the palm midribs there.

The used midribs for breadboards, crates and coops are usually only 3 weeks old to have the proper ductility needed for work. Then, they are sprayed lightly with water to gain moderate flexibility and the midribs are shaved lightly by a knife to clean any thorns or bumps [15]. Then the midribs are cut to the desired sizes by a wide sharp knife over a wooden chopping block. Then the crate maker marks the

Fig. 2.25 Date palm leaves over wooden poles for roofing in a storage house in Menya, Egypt

Fig. 2.26 Drying of date palm midribs. 1: vertical drying. 2: horizontal drying

points where the holes are to be punctured. These holes are driven in the midrib pieces using a sharp thin iron hollow pipe and a mallet [6, 8]. The assembly of the crate is usually from the bottom up, where the punctured horizontal elements are laid on the ground and vertical members are fixed upright in those holes [6, 8, 15]. Additional punctured horizontal elements are driven down the vertical elements repeatedly as beams of the box to the top [15]. Finally, all the edges are hammered to level them [8]. The horizontal elements are almost green, whereas the vertical members are almost dry. With the drying of the horizontal members, their joints with the vertical members become very tight providing rigidity to the crate. Details of making a standard chicken coop and a bread board are shown in Figs. 2.27, 2.28, respectively.

Fig. 2.27 Details of a standard date palm midribs chicken coop. 1: the vertical members are fixed through the holes of the bottom horizontal members at the base. 2: secondary prepunctured horizontal members are driven down to tie the vertical members. 3: the vertical members are leveled by the top horizontal members

Fig. 2.28 Details of a date palm breadboard. 1: the longitudinal members are cut to standard shapes and punctured at a specific spacing. 2: the transverse members are hammered through the punctured holes of the longitudinal members by friction. 3: additional members are added in the middle to prevent excessive deformation under the loads in the middle

This method is adopted in making crates, coops, cages and sometimes sliding doors [6, 8]. Sometimes, the crates are lined with palm leaflets in the cases of their use for delicate products. This art is developed in more artistic and sophisticated products such as ornaments and furniture.

2.7.3 Traditional Handmade Furniture

Historically, date palm midribs were used as girders, fixed across the timber frame to build beds in ancient Egypt, above which a woven mat made from spadix stem fibers was fixed to make the mattress as shown in Fig. 2.29.

Later, furniture makers, often called “artists,” give high aesthetic value to their products as the piercing technique becomes more sophisticated [15, 23]. Unlike crates, furniture design has no specific standard design and may vary from a place to another. Generally, the midribs are cut according to the desired elements and sizes of the design. The legs of a chair are usually cut from the wide section of the midrib, and the frame and the latticework are made of stiff dry midribs, while the armrests and seats are made from green midribs to facilitate bending [6, 15]. Then the elements of the armrest, seat and back are cut, respectively.

The same tools of crates making are used here also to make the lattices and the arabesque forms of the armrests and back. Then, all the plates of armrest, seat and back are fixed over the frame that is made of repeated vertical posts and horizontal beams upon which the seat is to be fixed with nails as seen in the figures. 4 cm nails are used to fix the armrests together and 10 cm nail is used to fix the armrests to the seat [6, 15]. High-quality products often use special keys from the midribs in order

Fig. 2.29 An ancient Egyptian bed, New Kingdom. 1: date palm midribs girders. 2: date palm spadix stem woven mat, Egyptian Museum in Cairo

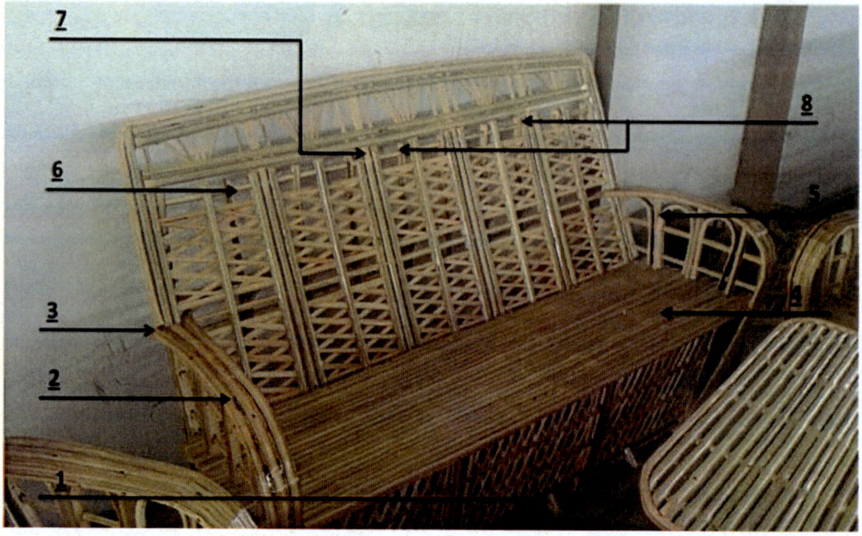

Fig. 2.30 Details of a standard date palm midribs chair. 1: the legs that bear the chair weight are assembled by horizontal beams through prepunctured holes along the legs. 2: the armrest members are bent and joined using nails, then fixed to the legs on both ends. 3: fixation of the armrest to the rear legs using nails. 4: the seat is assembled by nails over cantilever beams protruding from the legs. 5: lattice with bent midribs to fix the seat to the armrests. 6: horizontal beams to join main legs and secondary lattice columns. 7: the back is pre-assembled as a lattice using prepunctured holes. 8: the secondary lattice columns fix the back to the legs by friction through the lattice

to assemble the green members firmly so that they do not disassemble with drying. The whole set of plates is produced in maximum 3 days and is assembled to make a chair or a table in less than a day [15]. Details of making a standard chair are shown in Fig. 2.30.

2.7.4 Rural Wall and Roof Sheathing

Most of the traditional forms of utilization of date palm midribs focused on their use in sheathing, such as in Arish Houses in UAE [24]. Date palm midribs are connected by three rows of ropes to make *Sedda*. This *Sedda* mat has been used to be sheathing between the main structural system elements which used to be made of wooden poles or palm trunks [6]. The natural narrowing geometry of the midribs controls the design of *Seddas*. The assembly of a *Sedda* depends on laying each midrib where its wider end is between the narrower ends of two adjacent midribs.

The wall sheathing mats are usually fixed by ropes to the columns and side bracings of timber along the mats with maximum spacing of 3 m to ensure the planarity and verticality of the mat [6]. In simple roofing, the midribs are laid in perpendicular layers

Fig. 2.31 Roofing using palm midribs over a series of tree branches as beams, Fayoum, Egypt

over a series of local tree branches as beams as shown in Fig. 2.31. Furthermore in roofing of an outdoor corridor, shown in Fig. 2.32, these *Seddas* are supported by timber beams by ropes and covered with a thick mud layer to increase the thermal and moisture resistance of the roof.

2.7.5 Doors and Windows

In a manner that is similar to wall sheathing by date palm midribs, date palm midribs have been used in door-making using the technique of binding by rope [23]. In ancient Egyptian doors and windows (Fig. 2.33), the midribs were bound by ropes to form a single layer that was reinforced with diagonal midrib bracings. Loam remains can still be found over the bracings to enhance the coherence between the bracings and the doors. The teeth of the wooden locks were reported to be made from date kernels.

Traditional date palm midribs doors that can still be found now in rural areas in Egypt are clearly inspired by their precedents as shown in Fig. 2.34. However, the main differences are: wooden posts are used as a frame to which the midribs are fixed by nails, the bracings are fixed onto the midribs layers by nails instead of ropes, and no loam is used to stick down the bracings to the door.

Fig. 2.32 Ceiling of an outdoor corridor made of date palm midribs supported by wooden planks, North Sinai, Egypt

2.7.6 Fencing

Fencing by date palm midribs *Seddas* is one of the simplest and most spontaneous methods of building fences. Fences made by date palm midribs can be classified into simple and lattice fences. Simple fences depend on the planting of the midribs *Seddas* in the soil and fixing the fences using ropes to wooden poles at the corners (Fig. 2.35). This kind of fences works only as a temporary outdoor partition element without resisting any loads [6].

On the other hand, lattice fences are stiffer and sturdier because they consist of date palm midribs lattices that are tied together with ropes. In the UAE, the date palm midribs lattices have a heritage of being used as walls in traditional summer Arish Houses [16, 24]. This lattice is used in relatively heavier weight of fencing between internal structural columns, made of steel or timber, along the walls and in the corners to ensure verticality as shown in Fig. 2.36. The Egyptian version of lattice fences is heavily inspired by the traditional diagonal lattice of the reed huts in Manzala lake region in Northern Egypt [25]. The midribs are arranged in a diagonal lattice and fixed with nails to a timber frame as shown in Fig. 2.37.

This type of fencing can carry the loads of additional wall sheathing and can remain durable and functional for relatively longer periods of time [6, 16].

Fig. 2.33 An ancient Egyptian date palm midribs door, Old Kingdom, Egyptian Museum in Cairo. 1: the binding technique using leaf sheaths fibers ropes. 2: wooden lock with date kernels teeth. 3: using loam as a cohesive between the bracings and the door

2.7.7 Boats

Ancient evidences have been found that indicate the use of date palm midribs in boats that go back to the Early Bronze Age in a fashion that is still common to the present day in the Arab Gulf region, known as the date palm midribs-based *Shasha* raft boats [26]. Evidences of ancient Magan boats, named after the ancient name of the Arab gulf, were found in the coasts of Kuwait in 1980s. The evidences showed solidified chunks of bitumen, dating back to the Ubaid period (5300–4700 BC) in Kuwait, with the impression of thin parallel longitudinal grooves [27], suggesting that the bitumen was used as a waterproof cover over the date palm midribs or reeds used in building the boats [28]. Ur III text (dating back to 2100 BC) listed the use of palm-fiber ropes cured with fish oil to tie the reed bundles used in building Mesopotamian boats which are finally also caulked with bitumen [29].

Fig. 2.34 A traditional date palm midrib door in Fayoum, Egypt

Traditional *Shasha* boats (Fig. 2.38), small raft boats made from date palm midribs, are still used to the present day for fishing and short-distance travelling and racing sports in the UAE [27]. The basic material of the boats is date palm midribs after removing the leaflets. The midribs, 150 midribs required for 1 *Shasha* boat, are soaked in saline water for a week until they are manageable and then they are dried overnight [27]. Secondly, they are tied together with date palm leaf sheaths fibers ropes to make a mat as shown in Fig. 2.39. Then the mats are fixed tightly to a frame made from local acacia wood with cross beams and side beams to create the hull. Thirdly, the base of the boat is lined with date palm petioles to create buoyancy. Fourthly, more date palm midribs are fixed over the petioles with date palm leaf sheaths fibers ropes to create the deck of the *Shasha* [27].

Fig. 2.35 A simple date palm midribs fence

Fig. 2.36 A lattice date palm midribs fence, Ain, UAE

2.7.8 Bats and Discs

A traditional household utensil in the Egyptian village is the baking bat, *Matraha*. Baking bats were used to spread the bread dough and bring it into and from the

Fig. 2.37 Diagonal date palm midribs lattice in an outdoor fence, New Valley, Egypt

Fig. 2.38 A *Shasha* boat sailing in the sea

oven. Greco-Roman baking bats depended on the sewing technique which used parallel strings of date palm leaflets by a needle and thread over a network made of shredded and pierced date palm midribs using the piercing technique [23]. The midribs also were used for the bat handles and reinforcement. Such technique was clearly demonstrated in discs, fans and baking bats as shown in Fig. 2.40. This ancient technique of baking bats is still inherited to the present day. Traditional baking bats now depend on the piercing technique. Rigid midribs are connected by perpendicular

Fig. 2.39 Building a *Shasha* boat from date palm midribs

midribs that are driven through pierced holes, without the need of sewn leaflets as shown in Fig. 2.41.

The same piercing technique is used in simple doors and fences, where the vertical midribs are pierced and bounded together by horizontal midribs passing through the pierced holes as shown in Fig. 2.42.

Fig. 2.40 Baking bats made from sewn date palm leaflets fixed onto a date palm midrib frame, Greco-Roman period, Egyptian Museum in Cairo

Fig. 2.41 A traditional baking bat made from pierced date palm midribs

2.7.9 *Miscellaneous Uses*

Other miscellaneous uses of midribs are for making fishing rods and supporters for growing grape vines [8]. In addition, bent midribs that cannot be used in crates or furniture have been used traditionally as a source of light charcoal, especially the thick petioles at the base of the midrib [8].

Fig. 2.42 A simple door made from pierced date palm midribs, New Valley, Egypt

2.8 Traditional Forms of Palm Leaflets Utilization

Date palm leaflets are the secondary product of the preparation of the midribs as discussed earlier. As the second most abundant pruning residue, date palm leaflets have acquired a widely spread technical heritage that still thrives to be one of the main sources of income of many families in Fayoum, Sinai and Nubia [15]. The well-sustained rich cultural background of Fayoum, Sinai and Nubia has led to the continuity of the date palm leaflets heritage in the field of handicrafts to the present day.

Nubia, in South Egypt and North Sudan, is one of the regions where the art of weaving date palm leaflets still thrives. The weaving technique of date palm leaflets has been used since the ancient Egyptian traditions in making many traditional products such as hats, baskets and mats [8]. This technique still thrives because it only needs braiding the leaflets with hand without any other special tools [15].

2.8.1 *Traditional Preparation of Date Palm Leaflets*

The leaflets are manually removed from the midribs and are laid in the sun for 2 to 3 days to get rid of fungus and insects, while in summer it is enough to lay the leaflets in the shade to prevent discoloring [15]. Then, each leaflet is split by fingernails into several strips and soaked in saline water for a day to become more flexible. This water may be dyed with vegetable dyes to make various colored strips. Another method used now for dying the strips is soaking the strips in dyed boiling water with a small amount of salt in it until the desired hue is achieved [15].

2.8.2 *Bags, Mats and Baskets*

In ancient Egypt, several techniques were employed to use date palm leaflets in making various household accessories as shown in Fig. 2.43.

Fig. 2.43 Ancient Egyptian household utensils made from date palm leaflets, Old Kingdom, Egyptian Museum in Cairo. 1: protective discs made by the sewing technique. 2: a box made by coiling technique. 3: protective disc made by looping technique. 4: a basket lid made by looping technique

After preparation, leaflets are plaited and interwoven together to produce the desired shape according to one or more of the following techniques.

2.8.2.1 Plaiting Technique

Plaiting is the ancient Egyptian technique where several strands are woven into fabrics by interlacing them with a set of perpendicular strands [23]. The ends of the strands were usually folded into the fabric. This method was widely employed in ancient Egypt basically in making bags and sandals as shown in Fig. 2.44.

Another traditional form of using date palm leaflets that are deeply rooted in the Bedouin culture in South Sinai in Egypt is the *Sousel,* shown in Fig. 2.45. The *Sousel* consists of two plaited tubes that are designed to contain two dates, so that every morning, the children would present the *Sousel* with the two dates to their parents as they awaken them in the morning.

This traditional and artistic method in Egypt is more evolved today in making Hassir, depending on weaving the leaflets strips together along their lengths on a planar desk [15]. *Hassir*, a handmade mat, is made from natural fibers such as reeds and palm leaflets using the traditional plaiting technique. These strips are arranged in 2 diagonal perpendicular grids, then the strips are woven together (Fig. 2.46).

Fig. 2.44 Ancient Egyptian sandals from the Middle Kingdom. Egyptian Museum in Cairo

Fig. 2.45 A *Sousel*, made from plaited date palm leaflets

This type of mats depends extensively on high artistic qualities and on high-quality leaflets that are preferably just pruned to gain less brittle fibers with high elasticity in work [8].

In a more modern method, leaflets strips are used after soaking in water to increase their flexibility. Then, every 3 strips are used to make plaited strand that is sewn by machines with thread side to side with the other strands until the wanted area is completed [15]. Finally, the edges are bent and sewn to secure the ends of all the strands (Fig. 2.47). Hence, the used leaflets are not required to be as fresh as in the first type of mats. This type of mat is much lighter than the first type. Therefore, the first type is used for flooring mats and fans as shown in Fig. 2.48, while the second type is used for ornaments, bags and hats as shown in Fig. 2.49.

2.8.2.2 Coiling Technique

The coiling technique depends on creating a coil using a stiff material on which the strands are to be wrapped around [23]. In the ancient Egyptian period, the coiling technique was used mostly in making baskets and plates. Light plates (Fig. 2.50) **were made by** wrapping full-size leaflets around coils made from coiled leaflets bundles.

Fig. 2.46 Rolled date palm leaflet Hassir mats

Traditional trays inherited this technique using cores mode from spadix stem strands on which full-size leaflets are wrapped as shown in Fig. 2.51.

Fig. 2.47 Machined date palm leaflet stable mat

2.8.2.3 Looping Technique

When the strands wrapped around the coiled core are linked and intertwined in loops, the technique used is called Looping [23]. This ancient Egyptian technique was widely employed in making sturdy baskets and sandals as shown in Figs. 2.52, 2.53, 2.54, 2.55.

This technique is inherited to the present day in making sturdy vase-shaped baskets. Sturdy baskets, shown in Figs. 2.56, 2.57, are made by using dense cores made from date palm spadix stem fibers [8]. These cores are coiled in a spiral form according to the desired shape of the basket. Then, shredded leaflets are wrapped

Fig. 2.48 Fans made from woven date palm leaflets with wool embroidery

Fig. 2.49 A hat made from plaited date palm leaflets

around the spiral cores continuously to link them together while also being intertwined. The handles of the baskets are also made from shredded leaflets, wrapped over spadix stem cores, as in the simple coiling technique, in order to provide adequate support while carrying. The bottoms of heavy baskets and heavy-duty plates are made using the same method but with additional sewing in order to fasten the looping of the leaflets over the core made of leaflets bundles as shown in Figs. 2.58, 2.59.

Fig. 2.50 Light date palm leaflet plates, Middle Kingdom, Egyptian Museum in Cairo

Fig. 2.51 Traditional trays, made from date palm leaflets using the coiling technique

2.8.2.4 Sewn-Plaits Technique

Tri-plaited strands of date palm leaflets are made according to the needed length (Fig. 2.60). Then, the strands are sewn together by a large needle using a strong Doum palm *(Hyphaene thebaica)* fiber-based thread or from fibers extracted from the leaflets [8, 23] (Fig. 2.61). The skills of the craftsman significantly affect the quality of the product; the stiffness of a bag increases as long as the plaiting is fine

Fig. 2.52 Date palm leaflets baskets dating back to the New Kingdom, Egyptian Museum in Cairo

Fig. 2.53 A date palm leaflets basket dating back to the Middle Kingdom, Egyptian Museum in Cairo

and the strands are narrow with tight sewing [8]. Therefore, using freshly pruned leaflets is highly preferred. This technique can also be used in making modern bags (Fig. 2.62) and mats (Fig. 2.63).

Finally when the strands are stacked spirally and sewn together, ropes made from date palm leaf sheaths fibers are added as handles. These bags come in different shapes and volumes, where the largest bags, *Quffa*, (diameter of 50 cm at the bottom and the height of 75 cm) can be used for vegetables transportation with up to 35 kg

Fig. 2.54 Ancient Egyptian sandals made by looping technique, New Kingdom, Egyptian Museum in Cairo

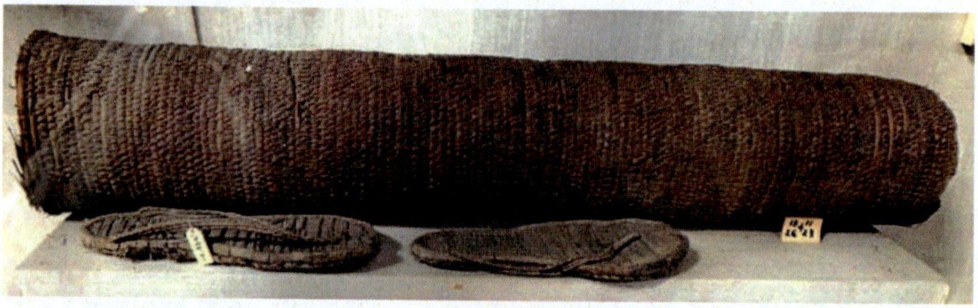

Fig. 2.55 Rolled mat and sandals made from date palm leaflets using the looping technique, Old Kingdom, Egyptian Museum in Cairo

capacity [8]. A *Quffa* requires a plaited strand of 10 cm width and 15 m length [8]. Therefore, the bottoms of these heavy-duty bags require reinforcement by date palm spadix stem fibers.

Fig. 2.56 A reinforced basket made from date palm leaflets and spadix stem cores

2.8.3 *Krena Fibers*

Low-quality leaflets have been traditionally used as stuffing material for bedding, cushions and mattresses, known as *Krena* [8]. The whole leaves here are dried on the ground and then, the leaflets are collected and soaked in water to soften. The soaked leaflets are then fed into a rippling machine in order to make them into fine threads in order to be dried and baled for later uses [8]. These bales can be used for stuffing of furniture or for thick ropes.

2.8.4 *Miscellaneous Uses*

Leaflets have been arranged and tied to make simple hand brooms and fly whisks as shown in Fig. 2.64. Heavy-duty ropes are also made using high-quality leaflets (Fig. 2.65). The leaflets here are shredded into 2–3 mm wide strips [22]. Those strips are soaked and made into a strand. Then the strands are plaited to make the final ropes [8].

Fig. 2.57 Reinforced plates and trays made from date palm leaflets and spadix stem cores

Fig. 2.58 Sewing the shredded leaflets around the core in a basket bottom

Fig. 2.59 A baby-basket with handles made from shredded leaflets around leaflet bundles core using the looping technique

2.9 Traditional Forms of Palm Spadix Stem Utilization

Spadix stems acquire recognizable tensile strength because of their natural function of carrying the weight of date through the season. Therefore, several handcrafts' products that require durability depend on date palm spadix stems as the main raw material.

2.9.1 Preparation of Date Palm Spadix Stem

Spadix stems are firstly soaked to soften the stems and then they are hammered with broad-faced hammers to loosen the fibers. Then, the fibers are stripped away longitudinally by hand from the basal end of the stalk to the other end [8].[2]

[2] A modern method of preparation that is used now in Egypt for faster products is laying the spadix stems on the asphalt roads to be run over by cars and trucks ground in order to disassemble the fibers of the stems for further use.

Fig. 2.60 Plaiting a palm leaflets strand

2.9.2 *Household Accessories*

The ancient Egyptian twining technique depended on extracting strands or fibers from twisting rows of leaflets or ropes around spadix stems to create a stiff disc [23]. Such technique is demonstrated in the sieve shown in Fig. 2.66.

Utilizing the same technique of making spadix stem *Hassir*, smaller woven patches are made using the shredded fibers from the spadix stems on looms. Those patches are sturdy on their own and can be used to make tissues boxes, tablecloths, bags (Fig. 2.67) and lamp shades (Fig. 2.68).

2.9.3 *Sturdy Baskets*

Being stiffer than leaflets, stronger type of baskets (Fig. 2.69) can be made from thick coiled date palm spadix stems fibers that are sewn and twined with wool threads.

Fig. 2.61 Sewing the spiral strands in a Quffa

Fig. 2.62 Spiral strands in a date palm leaflets bag

Fig. 2.63 Spiral strands in a date palm leaflets mat

2.9.4 Heavy-Duty Mats

The method used in making mats is inspired by the ancient Egyptian weaving technique that depended on interlacing strands, which were tensioned on a loom, with perpendicular strands [23]. The fibers of the spadix stems are passed and pressed in between the tensioned threads in the machine. Consequently, the fibers are woven in an orthogonal net until the needed area is completed. The resultant mat (Fig. 2.70) is highly durable and can be used directly over the soil and in the outdoors.

2.9.5 Decorative Trays

The disassembled spadix stem strands can be gathered to create a durable bottom for household trays and saucers (Fig. 2.71). The sides of the trays are created by wrapping shredded leaflets over fixed decorative spadix stems strands. The method of making the handles for the spadix stem trays, shown in Fig. 2.72, is clearly inspired by the ancient Egyptian simple coiling technique discussed earlier in 2.8.2.2.

2.10 Traditional Forms of Palm Petioles Utilization

Date palm Petioles have been used as bordering walls around open wells when the usual brick is not available [30]. In this method, the petioles are sharpened at the

Fig. 2.64 A broom made from shredded leaflets, fourth century AD, Egyptian Museum in Cairo

thinner end and hammered closely until a firm and dense wall is formed. The low density of petioles led to their use as floaters for the fishermen nets and traditional *Shasha* boats [8, 27, 31]. In construction, petioles have been used as vertical sticks that are hammered into the ground to line and stiffen the bond between mortar and mud walls [8, 31]. In handicrafts, date palm petioles offer a suitable soft medium to make distinctive sculptures as shown in Figs. 2.73 and 2.74. However, the most prominent use of petioles in the present day is using them as a fuel [8].

Fig. 2.65 Date palm leaflets ropes, New Kingdom, Egyptian Museum in Cairo

2.11 Traditional Forms of Date Palm Leaf Sheaths Fibers Utilization

Ropes made from date palm leaf sheaths fibers are traditionally known to acquire sufficient strength [8] as they were the favorite type of ropes for sailing in the UAE [24], although no reliable data has been found regarding the actual mechanical properties of this type of ropes. These ropes can be made in different diameters for various uses such as tying, handling and binding [8]. Moreover, ropes can be made into nets that can carry heavy loads for transportation over camels in rural areas [8].

Other miscellaneous traditional uses of date palm leaf sheaths fibers include being a fuel source, making fishnet, basket handles, brushes, bedding and shading live plants and offshoots [8]. In addition, they are used as stuffing for the spaces between the date palm trunk beams and midribs in the roofing of traditional rural houses in Egypt [7, 32]. They increase the overall thermal insulation of the roof which enhances the indoor air quality [32]. Furthermore, they represent a basic element in the mixture

Fig. 2.66 A sieve made from date palm spadix stem fibers, Roman period. Egyptian Museum in Cairo

Fig. 2.67 Tables cloths, tissues boxes and bags made from woven spadix stem fibers

of the traditional wickerwork mortar and plaster in the rural houses in Egypt [25]. In addition, the poor in Upper Egypt uses these fibers in stuffing pillows in order to save cotton to decrease the costs of the pillows.

Fig. 2.68 Lamp shades made from woven spadix stem fibers

2.11.1 Plaited Ropes and Bags

Date palm leaf sheaths fibers were used as the main source of fibers for ropes by plaiting technique since ancient Egypt as shown in Fig. 2.75 [23]. The same technique is still employed to the present day as shown in Fig. 2.76.

Plaited ropes made from leaf sheaths fibers were employed in miscellaneous uses such as hangers (Fig. 2.77), fire wicks (Fig. 2.78), balances (Fig. 2.79), fishing nets (Fig. 2.80) and bags (Fig. 2.81) in ancient Egypt. In the Roman period, the leaf sheaths fibers were simply rolled by hand to make wigs as shown in Fig. 2.82.

Fig. 2.69 A basket made from coiled date palm spadix stems fibers

Fig. 2.70 Rolled date palm spadix stem fibers *mats*

Fig. 2.71 A decorative cup saucer

Fig. 2.72 Decorative trays made from woven spadix stems strips

2.11.2 In Wicks

Wicks are used in oil lamps or candles. Figure 2.78 illustrates wicks made of date palm leaf sheaths fibers with traces of oil, Middle Kingdom, Egyptian Museum in Cairo.

Fig. 2.73 A sculpture from date palm petioles

2.11.3 *In Wigs*

Wigs are used as covering for the head. Figure 2.82 illustrates a wig, made from date palm leaf sheaths fibers from the Roman Period, Egyptian Museum in Cairo.

2.11.4 *In Cattle Accessories*

The well-developed ancient Egyptian expertise in making the ropes from date palm leaf sheaths fibers made them strong enough to be used in bags and agricultural plows. The weaving technique, interlacing strand using a loom, was used in ancient Egypt in saddles, blindfolds and bridles for cattle as shown in Fig. 2.83. The same technique is still used in the present day in making camel bags, donkey saddles, supporting belts for palm climbers and bird traps as shown in Figs. 2.84, 2.85, 2.86, 2.87, respectively.

Fig. 2.74 A sculpture from date palm petiole

2.11.5 In Belts for Date Palm Climbers

Figure 2.86 illustrates a belt for palm climbers made from date palm leaf sheaths fibers.

2.11.6 In Bird Catches

Figure 2.87 illustrates a bird catcher, made from date palm leaf sheaths fibers and date palm spadix stem.

Fig. 2.75 A rope made of date palm leaf sheaths fibers, New Kingdom period, Egyptian Museum in Cairo

Fig. 2.76 Ropes made from date palm leaf sheaths fibers

Fig. 2.77 A wall hanger made from date palm leaf sheaths fibers, Old Kingdom, Egyptian Museum in Cairo

2.12 Traditional Forms of Date Kernels

Preparation procedures of kernels vary according to the type of use. They can also be pressed to yield edible oil, roasted to make coffee, and heated to make charcoal [8, 27]. Prior to using kernels in jewelry, the kernels are extracted, washed, soaked in dyes and dried 48 h [19]. Then, the kernels are holed and stringed together by thread to make jewelry, necklaces, prayer beads (Fig. 2.88) and bags (Fig. 2.89) [27, 29].

2.13 Traditional Forms of Palm Trunks Utilization

Date palm trunks have been known to be strong and durable, which qualified the trunks to be used as timber substitutes. In Fig. 2.90, a worker is confidently ascending to a truck while using a trunk as stairs. Moreover, the worker is carrying a bundle of twenty midribs, with a total weight of approximately 40 kg. Yet, the natural coarse surface of the trunk offers the needed friction for his bare feet. As a result, date palm trunks were found to be used in the fields which require high durability and stiffness such as construction and furniture elements.

Fig. 2.78 A wicks made of date palm leaf sheaths fibers with traces of oil, Middle Kingdom, Egyptian Museum in Cairo

2.13.1 Traditional Construction

Date palm trunks have also been used as columns and beams in traditional housing in many rural areas in Egypt and the Arab region. The most primitive forms of using date palm trunks in construction were as simple door, beams in outdoor corridors and simple roofing as shown in Figs. 2.91, 2.92, 2.93.

In Siwa Oasis in the western desert in Egypt, the technical heritage of using date palm trunks in construction evolved to demonstrate a spontaneous cleverness in terms of changing the assembly of the trunks according to the covered span [32]. In this method, date palm trunks were halved or quartered to work as beams supported by load-bearing walls made from Kershef soil in the Siwa oasis [7], or from mud bricks in the Nile Valley. The types of roofs were as follows:

- Primary roof: for rooms with spans of 2–3 m, roofing depended on using planks from the palm trunks supported by the walls in the transverse direction. Then the spaces between the trunks were filled with 10–20 cm thick mortar and date palm leaf sheaths fibers for intermediate floorings or roofing [32].

Fig. 2.79 Ropes in a balance made from date palm leaf sheaths fibers, Middle Kingdom, Egyptian Museum in Cairo

Fig. 2.80 Date palm leaf sheaths fibers fishing products, Middle Kingdom, Egyptian Museum in Cairo. 1: fishing net. 2: bait bag. 3: safety ropes. 4: darts

Fig. 2.81 A plaited bag made from date palm leaf sheaths fibers, Old Kingdom, Egyptian Museum in Cairo

Fig. 2.82 A wig, made from date palm leaf sheaths fibers from the Roman Period, Egyptian Museum in Cairo

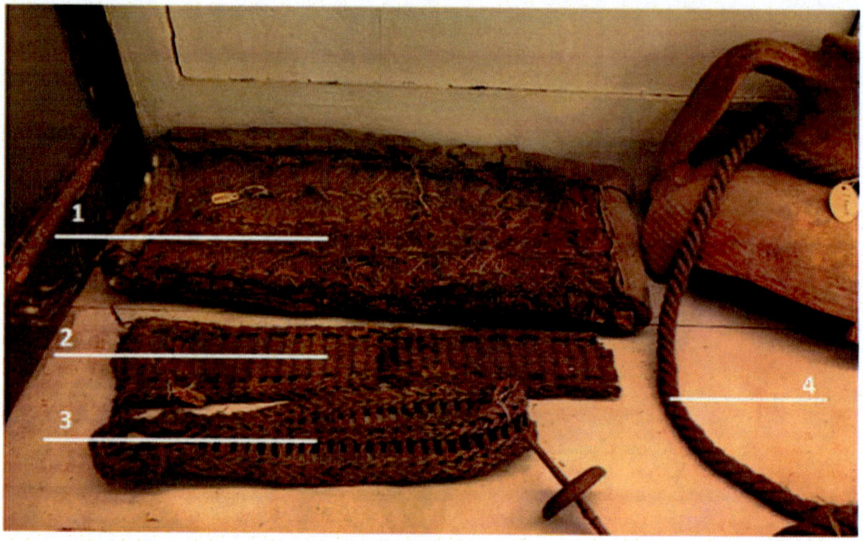

Fig. 2.83 Cattle accessories made from date palm leaf sheaths fibers, Old Kingdom, Egyptian Museum in Cairo. 1: a buffalo saddle. 2: blindfold. 3: bridle. 4: plaited ropes for collecting water utensil

Fig. 2.84 Weaving a camel bag from date palm leaf sheaths fibers using a loom, Fayoum, Egypt

Fig. 2.85 A donkey saddle made from woven date palm leaf sheaths fibers

Fig. 2.86 Belts for palm climbers made from woven date palm leaf sheaths fibers

Fig. 2.87 A bird catcher made from woven date palm leaf sheaths fibers and spadix stem

- Secondary roof: for rooms with spans of 4–5 m, additional beams are added where each beam consists of two halves of a trunk laid adjacently on the curved side. Above these beams, longitudinal planks of trunks are laid together on which the roofing layers are added as described earlier [32].
- Tertiary roof: for halls with spans 5–8 m, main full trunk beams are supported on walls and piers, above which the beams of the secondary roofs and roofing layers are added [32].

In a simpler and more recent fashion, date palm trunks are used as visible columns and beams in light huts. In this method, special workers peel the outer tough surface to achieve an organized surface to handle during construction (Fig. 2.94). Then, the columns are made of whole trunks (Fig. 2.95), while beams generally consist of quartered trunks that may be arranged back-to-back as in the traditional roofing method in the western oases in Egypt (Fig. 2.96), or the trunks are shaped into rectangular cross-sections for beams and columns to achieve a modern and sophisticated design as shown in Fig. 2.97.

Fig. 2.88 A prayer bead made from date palm kernels

2.13.2 Traditional Furniture

Trunks are shaped into blocks that act as the armrests and supporters of tables and chairs [3]. The fixation methods used are very similar to those used in timber furniture. The coarse vascular structure of the trunk gives the furniture a rustic look (Fig. 2.98) that is desirable in various touristic projects.

2.14 Conclusion

It is clear from the aforementioned that the date palm byproducts enjoyed a long history of utilization in many regions in the world, especially in the Arab region, extending for thousands of years. Relying basically on the periodical pruning (palm service) activity they represented a sustainable material base for the satisfaction of basic material needs of the local populace: in shelter, furniture, agricultural equipment, transportation and household utensils.

Fig. 2.89 A bag made from date palm kernels

Fig. 2.90 A worker ascending on a date palm trunk, carrying his own weight and the weight of a 20-midribs bundle, Assiut, Egypt

Fig. 2.91 A traditional door made from date palm trunk planks, EL-Kharga, New Valley, Egypt

The technical heritage, associated with the date palm byproducts reveals a generic perception of the date palm as a whole resource, whereby all the elements of the resource were-according to the available level of technology-efficiently used for the satisfaction of the basic human needs. It also reveals, though implicitly, a huge body of traditional knowledge about the properties and behavior-under different environmental and loading conditions-of these byproducts. The wide spectrum of uses of different palm byproducts reveals a high degree of innovation and, sometimes, high levels of skills seemingly unattainable at our present contexts.

It is also clear from the aforementioned that this technical heritage operated- and still operates-as a force of inspiration to develop new techniques for processing of date palm byproducts and innovative products made from them. But as mentioned in the introduction the most valuable in the technical heritage is that it acts as a

Fig. 2.92 Using date palm trunks as beams in an outdoor corridor, Arish, Egypt

Fig. 2.93 Using date palm trunks as beams with date palm midribs ceiling in a simple roof, Arish, Egypt

Fig. 2.94 Peeling and squaring a date palm trunk piece

Fig. 2.95 Using date palm trunks as columns and rafters in a date palm midrib hut

Fig. 2.96 Quartered date palm trunk beams that support date palm midribs roof

Fig. 2.97 Squared date palm trunk beams and columns in a modern date palm midribs hut

software for discovering in future new environmentally and culturally tuned routs of utilization. Rediscovery of the date palm byproducts as a sustainable material base opens endless opportunities for innovation in the interface between understanding the physical, chemical and mechanical properties of these byproducts and the human needs on the local, national and international levels.

Fig. 2.98 Furniture made from date palm trunk: Tunis

References

1. Johnson DV (2011) Introduction: date palm biotechnology. In: Date palm biotechnology. Springer
2. Azzam OA (1960) The development of urban and rural housing in Egypt. Swiss Federal Institute of Technology, Zurich. https://www.research-collection.ethz.ch/bitstream/handle/20.500.11850/134539/1/eth-33671-01.pdf
3. El-Mously H (2001) The industrial use of the date palm residues: an eloquent example of sustainable development. United Arab Emirates University, El Ain, 21
4. Bekheet S (2013) Date palm biotechnology in Egypt (Review article). Appl Sci Rep 3:144–152
5. Bekheet S, Elsharabasy SF (2015) Date palm status and perspective in Egypt. In: Date palm genetic resources and utilization. Springer Science + Business Media, Dordrecht
6. Darwish EA, Mansour Y, Elmously H, d Abdelrahman A (2019b) The technical heritage of date palm leaves utilization in traditional handicrafts and architecture in Egypt & the Middle East. By-products of palm trees and their applications. Mater Res Proc
7. Ahmed RM (2014) Lessons learnt from the vernacular architecture of Bedouins in Siwa Oasis, Egypt. In: The 31st international symposium on automation and robotics in construction, Sydney. https://doi.org/10.22260/ISARC2014/0123
8. Barreveld WH (1993) Date palm products. FAO. http://www.fao.org/docrep/t0681e/t0681e00.htm#con
9. El-Mously HI (2016, May 3) Innovating green products as a mean to alleviate poverty in Upper Egypt. In: 1st International joint symposium on "product development and innovation," Ain Shams University
10. Zaid A, de Wet PF (2002) Botanical and systematic description of the date palm. In: Date palm cultivation. FAO plant production and protection paper (p 156), Rome, Italy. http://www.fao.org/docrep/006/y4360e/y4360e05.htm

11. Munier P (1973) Le palmier-dattier. Maisonneuve Larose, Paris
12. El-fadda S, Abu Ayana R (2017) Palm secondary products: types and economic value (second). Al Rajhi endowment. http://www.rajhiawqaf.org/ar/mediacenter/DocLib3/%D8%A7%D9%84%D9%85%D9%86%D8%AA%D8%AC%D8%A7%D8%AA%20%D8%A7%D9%84%D8%AB%D8%A7%D9%86%D9%88%D9%8A%D8%A9%20%D9%84%D9%84%D9%86%D8%AE%D9%8A%D9%84.%20%D8%A7%D9%86%D9%88%D8%A7%D8%B9%D9%87%D8%A7%20%D9%88%D8%A3%D9%87%D9%85%D9%8A%D8%AA%D9%87%D8%A7%20%D8%A7%D9%84%D8%A5%D9%82%D8%AA%D8%B5%D8%A7%D8%AF%D9%8A%D8%A9.pdf
13. Lemlem A, Alemayehu M, Endris M (2018) Date palm production practices and constraints in the value chain in afar regional state, Ethiopia. Adv Agri 2018:1–10. https://doi.org/10.1155/2018/6469104
14. Agoudjil B, Benchabane A, Boudenne A et al (2011) Renewable materials to reduce building heat loss: characterization of date palm wood. Energy Build 43:491–497
15. El-Batraoui M (2016) The traditional crafts of Egypt. The American University in Cairo Press, Cairo, New York
16. Eldeeb A (2017) Recycling agricultural waste as a part of interior design and architectural history in Egypt. Acad Res Community Publ:1–7. https://doi.org/10.21625/archive.v1i1.116
17. Elmously HI (2005) The palm fibres for the reinforcement of polymer composites: prospects and challenges. In: The First Ain Shams international conference on environmental engineering, Cairo, Egypt
18. Almana HA, Mahmoud RM (1994) Palm date seeds as an alternative source of dietary fiber in Saudi bread. Ecol Food Nutr 32:261–270. https://doi.org/10.1080/03670244.1994.9991406
19. Mirghani MES, Al-Mamun A, Daoud JI, Mustafa SM (2012) Processing of date palm kernel (Dpk) for production of edible jam. Austr J Basic Appl Sci:22–29
20. Peterson J, Dwyer J (1998) Flavonoids: dietary occurrence and biochemical activity. Nutr Res 18:1995–2018. https://doi.org/10.1016/S0271-5317(98)00169-9
21. Elmously H (2019) Rediscovering date palm by-products: an opportunity for sustainable development, pp 3–61. https://doi.org/10.21741/9781644900178-1
22. Darwish EA, Mansour Y, Elmously H, Abdelrahman A (2019) Development of sustainable building components utilizing date palm midribs for light wide-span multi-purpose structures for rural communities in Egypt. J Build Eng. https://doi.org/10.1016/j.jobe.2019.100770
23. Wendrich WZ (2009) Basketry. In: Ancient Egyptian materials and technology. Cambridge University Press
24. Piesik S (2012) Arish: palm-leaf architecture. Thames & Hudson
25. Helal A (1989) Spontaneous architecture in Egypt: analytical study of reed architecture in Manzala lake region. Faculty of Fine Arts, Helwan University, Cairo
26. Vosmer T, Potts DT, Naboodah HA, Hellyer P (2003). The naval architecture of early Bronze Age reed-built boats of the Arabian Sea. In: Proceedings of the first international conference on the archaeology of the UAE, pp 152–157
27. Johnson D (2016). Unusual date palm products: Prayer beads, walking sticks and fishing boats. Emir J Food Agric:12–28. https://doi.org/10.9755/ejfa.2015-11-1021
28. Carter R (2002) Ubaid-period boat remains from As-Sabiyah: excavations by the British archaeological expedition to Kuwait. In: Proceedings of the seminar for Arabian studies. Archaeo Press, Edinburgh, pp 13–30
29. Ray HP (2013) The archaeology of seafaring in ancient South Asia. Cambridge University Press
30. Dowson VHW, Aten A (1978) Food and agriculture organization of the United Nations. In: Dates: handling, processing and packing. FAO
31. Popenoe P (1973) The date palm. In: Field H (ed) Field research projects. Coconut, Miami
32. El-tawil H (1989) Environment and architecture in Siwa. Faculty of Fine Arts, Alexandria University, Alexandria

Part II
Future Applications of Date Palm Byproducts in Circular Bioeconomy

Chapter 3
Date Palm Byproducts in Enzymes, Food, Beverage, Pharmaceuticals, Cosmetics and Natural Wax

Abstract Peroxidase—an important enzyme of a high value-added—has been successfully extracted from date palm leaflets, usually treated as waste! Seeds from waste dates have been used as a source for functional ingredient in food systems enjoying a high economic feasibility. Date palm leaflets and midribs have been successfully used as a substrate for microbial protein production. Carotenoids—natural colors of a very high commercial value—have been extracted from date wastes. Date palm leaflets have been used as a substrate for growth of pleurotus fungi. Bakers' yeast and citric acid have been extracted from date wastes. Glucose and citric acid have been extracted from date palm fronds, petioles and leaf sheaths. Lactic acid, successfully produced from date palm sap, has revealed the existence of a diversity of microflora including yeasts, coliforms and lactic acid bacteria, which can be used as a starter culture for the production of fermented beverage. Insoluble fibers were extracted from waste dates being a good dietary source rich in mineral contents and can be used as an excellent food ingredient. Date palm kernel extract was found to exhibit antiaging effect suggesting its use in antiaging skin care products. Seeds oil extracted from date palm seeds was successfully used in cream, liquid shampoo and bar shaving soap. Natural wax was successfully extracted from date palm leaflets to be used in cosmetics and health care products.

Keywords Date palm byproducts · Peroxidase · Functional ingredient in food systems · Microbial protein · Carotenoids · Pleurotus fungi · Bakers' yeast · Citric acid · Glucose · Lactic acid · Insoluble fibers · Remedy for skin wrinkles · Cosmetic cream · Liquid shampoo · Bar shaving soap · Natural wax

3.1 Peroxidase from Date Palm Leaflets

Date palm leaflets contain highly active soluble and ionically wall-bound peroxidases. Peroxidase [1] from date palm leaflets was purified to homogeneity and characterized biochemically. The purification factor for purified date palm peroxidase was 17 with 5.8% yield. Date palm peroxidase was assayed at optimum pH in

H. EL-Mously et al., *Date Palm Byproducts: A Springboard for Circular Bio Economy*, Materials Horizons: From Nature to Nanomaterials,
https://doi.org/10.1007/978-981-99-0475-4_3

different temperatures ranging between 25 and 80 °C for 5 min. The results indicate that the date palm peroxidase is a heat resistant enzyme. It can be speculated that peroxidase enzyme preparation from date palm leaflets can be used at higher temperatures ranging from 60 to 70 °C. In conclusion, the properties of date palm leaflets peroxidase, such as high activity and good stability over a wide pH range, excellent thermostability, wide substrate specificity and stability in the presence of high ionic metals concentration, could be a promising tool in many industrial and medical applications.

3.2 Protein as a Functional Ingredient in Food System from Date Palm Seeds

The continuous increase of demand on proteins on the world level from animal sources such as meat, egg and dairy products has pushed the food industry to search for plant alternative sources of proteins to be used as functional ingredients in formulated foods. The date palm is one of the main fruit crops in dry and semi-dry regions extending from Morocco in the far west of Africa to Iraq and Saudi Arabia in the east. The seed of the date representing ~10.3% (w/w) of the total mass of the fruit is treated most dominantly as waste. These seeds contain 5–7% protein by weight [2]. This study has been devoted to the extraction of proteins from date seed, characterization of these proteins using mass spectrometry and testing of their emulsifying properties.

Seeds were removed from 40 kg of whole Deglet Nour dates, washed in water and then air-dried for a week and dried overnight at 40 °C in a drying oven. The seeds were then milled using a hammer mill to a particle size that could pass through a 1–2 mm sieve screen and then stored at − 20 °C until further preparation. After extraction of oil, the defatted date palm seed powder was left to dry overnight to allow the hexane to evaporate and was then kept at − 20 °C to obtain date seed protein concentrate.

The protein from the date seed protein concentrate was extracted. Of the 300 proteins, 90 have been identified with high confidence (~85% of them having metabolic functions and 15% are storage proteins). The date seed protein concentrate emulsifying properties were found comparable with those for soy protein isolate. Therefore, the date seed protein concentrate could be potentially used as a functional ingredient in food systems. Proceeding from the fact that date palm seeds are almost priceless (of no value) in the date processing industries, the use of waste date palm seeds, as a source for a functional ingredient in food systems will enjoy a high economic feasibility.

3.3 Use of Date Palm Leaflets and Midribs as a Substrate for Microbial Protein Production

The date palm is one of the most widespread palm varieties in the world. The annual pruning of the date palm results into huge quantities of byproducts (e.g. midribs, leaflets, petioles, leaf sheaths fibers and spadix stems) most dominantly treated as waste in palm plantations. Therefore, it is essential to find new avenues of economic utilization of these huge quantities of renewable materials. One of these avenues is to hydrolysis cellulosic fibers to make fermentation substrates.

In one of these studies [3], the date palm leaflets were separated from the midribs. The leaflets and midribs were dried at 100–105 °C and cut into small pieces and thoroughly ground to get a homogenous powder. To hydrolyze the cellulose fibers, the samples have been treated with NaOH (0.5% and 1%) for 1 h at 25 °C and H_2SO_4 (5%) for 4 h: at 25 °C and 90 °C. The results show that Trichoderma viride was able to utilize a part of the palm-tree leaflets for its growth and formation of microbial protein. Leaflets treated with 5% H_2SO_4 at 90 °C gave the best result as a substrate. The experiments, conducted on midribs gave similar results, but the final protein content was less, and the midribs were less suitable as a fermentation substrate. It could be thus concluded that the date palm leaflets and midribs could be hydrolyzed with alkali or acid forming substrates with good suitability for the growth of Trichoderma viride and the formation of protein with qualitatively normal amino-acid content.

3.4 Carotenoids from Date Wastes

Carotenoids are natural colors of a very high commercial value as essential components in different industrial applications: in pharmaceutical, chemical, food and feed industries. There is a growing demand on natural carotenoids, since the chemically synthetic carotenoids are restricted due to their toxicity. The date palm fruit is one of the widely cultivated crops, particularly in the Arab region with a total world production of 6.9 million tons [4]. The date palm harvesting is associated with great losses during: picking, sorting and processing. Therefore, it is important to find out avenues of economic utilization of this wasted resource. In an important study [4], three strains of *Lactobacillus plantarum*, isolated from bakery's yeast, olive fermentation and cheese have been used for the production of carotenoids from date wastes. A yield of 16–21 mg/kg dry cell of carotenoid has been realized using a date syrup at 5% concentration. This yield has been elevated to 54.89 mg/kg dry cell by supplementation with other salts and organic nitrogen sources together with the optimization of initial pH (7.0) and temperature (30 °C).

3.5 Use of Date Palm Leaflets as a Substrate for the Growth of Pleurotus Fungi

In one of the researches [5], wheat straw and date palm leaflets were milled to pass a 1 mm sieve and used as substrates for the growth of pleurotus fungi (*Pleurotus florida, Pleurotus ostreatus* A., *P. ostreatus* M and *P. ostreatus* T.). The specimens were sprayed with water in a ratio of 3 to 1 (water/substrate) and left 24 h at room temperature. Then all plates were autoclaved for 20 min at 121 °C in 1.5 bar pressure. After the first and second weeks of incubation at 28 °C, all plates were controlled for growth rate and probability of contamination using direct visual method.

The research results show that *P. ostreatus* A and *P. ostreatus* T. fungi have a rate of growth ($p < 0.05$) on date palm leaflets higher than wheat straw. This clearly indicates that the date palm leaflets could be used to grow fungi faster than wheat straw to produce higher crude protein.

3.6 Bakers' Yeast and Citric Acid from Date Wastes

The date is one of the major fruit crops in North Africa and the Arab countries. A considerable amount of the date crop is not suitable for marketing for consumers and is thus treated as waste. One of the solutions to this problem is the transformation of date waste by biotechnological processes for the production of high value-added products. The worldwide demand on Bakers' yeast and citric acid is steadily increasing. These products are mostly imported by Arab countries, thus representing a heavy burden on their balance of payment. A research has been conducted [6] to study the potentiality of use of date wastes as a substrate for the production of Bakers' yeast and citric acid using strains of *Saccharomyces cerevisiae* ATCC 1102 and *Aspergillus niger* ATCC 16404.

The results of this study show that the date wastes could serve as an appropriate substrate for the growth of *S. cerevisia* and *A. niger*, which produced considerable amounts of Bakers' yeast and citric acid. The optimum yield of Bakers' yeast was obtained with a rate of dilution of 0.22 h^{-1} and the use of ammonium phosphate as a nitrogen source. As far as the citric acid production is concerned, the optimal fermentation period was 144 h. The maximum citric acid production was obtained at 30 °C and at 150.0 gL^{-1} of sugar. The addition of methanol at concentration up to 30% resulted in a marked increase in the citric acid production. In addition, the optimum pH for maximum citric acid production was 3.5. The aeration rate kept at a level of 1.0L/L/min was found to be optimum. Finally, the best results were observed when 2.5 gL^{-1} ammonium nitrate and 2.5 gL^{-1} potassium phosphate were added into the medium. In summary, a maximum citric acid production, i.e. 126.4 gL^{-1} was obtained at these optimal conditions. These results indicate that the date wastes could serve as a great potential for the production of Bakers' yeast and citric acid production, and there is a high potential for the development of endogenous technologies in the Arab world for the production of Bakers' yeast and citric acid.

3.7 Glucose and Lactic Acid from Date Palm Fronds, Petioles and Leaf Sheath

The Arab countries are a major producer of dates in the world, accounting for ~73% of the world production. A huge amount of lignocellulosic byproducts results from the annual pruning (serving) of date palms. Unfortunately, this wealth of renewable materials is often treated as waste: either open-field burnt or sent to landfills, resulting into a great economic loss and/or environmental pollution. Meanwhile, there is a great potentiality to use these byproducts to manufacture lactic acid: an organic acid highly needed in a wide spectrum of pharmaceutical, leather and food industries, as well as for the production of biodegradable plastics.

An important research [7] has been conducted for the bioconversion of date palm lignocellulosic byproducts (e.g. fronds, petioles and leaf sheath) into high value bioproducts by saccharification and fermentation. The date palm byproducts have been ground and pretreated: either by alkaline pretreatment: 2NNaOH at 30 °C for 48 h; or by acid-steam pretreatment: 1% H_2SO_4, 120 °C for 100 min. The dried material was ground to obtain a particle size ≤ 1 mm. The enzymatic saccharification has been conducted using the bacterial species *Geobacillus stearothermophilus* as a source of cellulases. For lactic acid fermentation, the culture of *Lactobacillus delbrueskii* subsp. Lactis (B.01357)—a homo fermentative lactic acid producer—was utilized in this study.

The results show that the alkaline pretreatment was much more effective for reducing lignin in all researched byproducts as compared with acid-steam pretreatment removing ~79.05%, ~68.4% and ~63.2% of the lignin in fronds, petioles and leaf sheath, respectively. The maximum rate of saccharification was found at a substrate concentration of 4% and enzyme concentration of 3OFPU/g of substrate. The optimum pH and temperature for bioconversion were 5.0 and 50 °C, respectively, after 24 h of incubation, with a yield of 31–36 mg/ml of glucose at a saccharification degree 71%. This percent was increased to 99% by removal of hydrolysate after 24 h by using a two-step hydrolysis. Significant lactic acid production (28.8 mg/ml) has been obtained by separate saccharification and fermentation after 72 h of incubation. As a conclusion, this study proves that date palm lignocellulosic byproducts could be almost completely converted to glucose that can be ultimately converted into lactic acid by employing alkaline pretreatment and separate saccharization and fermentation.

3.8 Lactic Acid Production from Date Waste

A considerable amount of dates production may be inappropriate for marketing due to: over-ripening, improper handling in palm gardens, processing and storage. These quantities are most frequently treated as waste! There are, however, different alternatives for use of wasted dates: in the food product manufacturing of items such as bakery products, ice-cream, caramel products, ethanol, vinegar, oxytetracycline, single-cell protein, glutamic acid, carotenoids and curdlan gum [8]. Dates indeed contain large amounts of glucose at levels of 73–83% (dry basis) making them suitable as a fermentation substrate [3].

In the present work, the production of lactic acid from date waste has been investigated using *Lactobacillus casei* subsp. *rhamnosus* in batch and fed-batch cultures. Waste dates were thoroughly cleaned manually to remove dust and foreign materials. The seeds were manually separated. Tap water was added at a ratio of water to dates: 2 to 1. The mixture was heated to 80 °C for 2 h with continuous stirring and then centrifuged at 20,000 × g for 10 min. Before experimentation, the appropriate amount of date juice was diluted to attain the required concentration of glucose. Two concentrations of the feeding medium (62 and 100 g/L of date juice glucose) were used at different feeding rates (18, 22, 33, 75 and 100 ml/h). The highest volumetric productivity (1.3 g/L.h) and lactic acid yield (1.7 g/g) were got at a feeding rate of 33 ml/h and a date juice glucose concentration 62 g/L in the feeding medium. As a result, most of the date juice glucose has been completely utilized (residual glucose 1 g/L) and a maximum acid production level (89.2 g/L) was obtained. In conclusion, the fed-batch culture could significantly improve the lactic acid production and the suggested approach is simple and easy to apply.

3.9 Lactic Acid Bacteria from Date Palm Sap

This research [9] has been devoted to the study of the distribution of the microbiological groups in date palm fresh sap and the potentiality of use of lactic acid bacteria as a starter culture for the production of fermented beverage. The date palm sap is widely produced as a fresh juice in Southern Tunisia and other locations in the Arab world. There is a wide potentiality to convert date palm sap to other high value-added substances, such as caramel, palm sugar or to use it as an industrial culture to produce ethanol, lactic acid, palm wine and vinegar.

Fresh date palm sap samples were collected in the early morning from a palm grove in Southern Tunisia using the local familiar collection method dominant there. These samples were directly stored in an ice box (4 °C) to avoid their fermentation until reaching the laboratory. The microbiological analysis of the samples has been conducted within a day. Two lactic acid bacteria strains were selected as starter cultures for palm sap fermentation proceeding from their acid production ability.

The results indicate that the date palm sap has a diversity of microflora including yeasts, coliforms and lactic acid bacteria. Ten species of lactic acid bacteria dominated by regular rods and lenticular in shape were tentatively identified. The use of lactic acid strains (autochthonous L. debrueckii KH3 and Leuconostoc mesenteriodes 5B4) gave a fast acidification rate. These results may serve as a springboard for the design of a process of production of lactic acid beverage. But further research is needed for full characterization of the technological properties of the starter culture and prediction of properties of the fermented beverage.

3.10 Insoluble Fibers from Date Wastes

In date palm processing, i.e. in sorting, storage and conditioning, a considerable amount is not accepted and thus treated as waste. Thus, it is extremely important to find economic avenues of utilization of these date wastes. There is a growing demand, among consumers on insoluble fibers for the prevention and treatment of several diseases, such as coronary heart-related diseases, diabetes, constipation, diverticular disease, colonic cancer, etc. The date fibers, derived from date processing industry, could be used to manufacture health-related food products. A study [10] has been conducted to extract insoluble fibers from three potential cultivars of date palms: Barhee, Sultana and Owadi. The insoluble date fiber has been extracted by multistage water extraction of date flesh using microwave heating, followed by freeze-drying and grinding. After the 7th extraction, high-performance liquid chromatography analysis confirmed the absence of sugars in the fiber. The inductively coupled plasma optical emission spectrometry has confirmed that date fibers are rich in potassium, calcium and magnesium (1.5–2.4 g/kg) and low in sodium content. Besides, the date fibers demonstrated a high water and oil holding capacity. Therefore, the date fibers are a good dietary source, rich in mineral contents and can thus be used as an excellent food ingredient.

3.11 Remedy for Skin Wrinkles from Date Palm Kernel

Aging involves many changes in skin properties and the appearance of wrinkles. A hormonal decrease with age is one of the major mechanisms involved in the aged appearance of skin [11]. Consequently, one of the central approaches to reduction in age-related wrinkles has become compensation of age–related hormonal decreases. Studies have shown that date palm kernel extract has a very interesting composition of protein, fat, carbohydrates, many minerals including calcium and magnesium, ten different amino acids and many fatty acids (C12 to C20); studies have also confirmed the presence of phytohormones in the extract.

In this research, the antiaging properties of date palm kernel have been investigated in a vivo study on wrinkles. Ten healthy women volunteers between the ages of 46 and 58 years applied the cream formula with 5% date palm kernel or placebo on the eye area twice a day for 5 weeks. Measurement of skin relief was obtained by image analysis using specific software (Quantirides Monaderm) to calculate standard parameters of wrinkles such as their number, complete surface (mm^2), depth (mm) and total length (mm). After 5 weeks at the end of the test, the improvement in skin wrinkles and appearance was observed in 60% of the volunteers on their date palm kernel-treated side compared with the placebo treated side. In conclusion, this in vivo study demonstrates that date palm kernel extract exhibits a significant antiaging effect and suggests that it is of interest in antiaging skin care products.

3.12 Cosmetic Cream, Liquid Shampoo and Bar Shaving Soap from Date Palm Seed Oil

Date palm seeds are mostly treated as a waste except for small quantities, used in animal feed applications. A research [12] has been conducted with the purpose of finding potential uses for date palm seeds in cosmetic and pharmaceutical applications. Seeds (~10% of fruit weight) from Dekel Noor, Zahidi, Medjool and Halawy cultivars were soaked in water and well-washed from any date palm flesh and sun-dried. Samples from each date variety were ground in a hammer mill and treated with petroleum ether at 40–60 °C. The atomic absorption spectrometry has been used to determine the mineral contents in the date seeds ash. The oils were converted to methyl easters, and the fatty acid methyl easters were analyzed by gas chromatography.

The low protein content (5.28–17.3%), coupled with high fiber content (15.1–17.3%), limits the use of date seeds in feeds. The extracted pale-yellow date seed oils have iodine values (48–58%) similar to palm oil and their saponification values were within the range of those for palm and kernel/coconut oils. The oleic (41–44%) and lauric (19–24%) were most abundant fatty acids found. Thus, as distinct from palm oil (palmitic-oleic) and palm kernel/coconut oils (lauric-myristic), the date seed oil may be regarded as oleic-lauric oil. But from oil extraction industry point of view the date palm seed oil (~8%) is of a little commercial interest. The date seed oil has been successfully used to replace other vegetable oils (coconut/palm oils, coconut oil and coconut oil/tallow) in three types of cosmetic cream, three types of liquid shampoo (called “date seed oil shampoo”) and one type of bar shaving soup. The obtained encouraging marketing comments support the feasibility of establishing a small plant to use ~850 tons of date seeds annually available in Israel at low or no cost rendering an oil crop of ~70 tons of date palm seed oil.

3.13 Natural Wax from Date Palm Leaflets

The world is witnessing a growing increase of demand on natural waxes, motivated by the consumer preference of greener and more sustainable natural products, especially in the areas of cosmetics and health care products. The search for sources for natural waxes is also motivated by the unsustainability of reliance on petroleum as a main source of wax. The sources of natural waxes at the present time are limited and restricted in certain geographical regions of the world (e.g. Candelilla in Mexico and South America and Carnauba in Brazil). Therefore, it is essential to discover new sources of natural waxes [13].

The date palm is one of the most cultivated palm tree varieties in the world exceeding 140 million palms [14]. The date palms should be annually pruned as an important service of the date palm. Each date palm produces during annual pruning a quantity of leaflets $\approx$ 14.6 kg air-dry weight [15]. This annual huge quantity of date palm leaflets (~2 million tons) is most dominantly open-field burnt resulting in tremendous environmental pollution. It has been proven that supercritical carbon dioxide ($ScCo_2$) provides solvent-free extracts thus providing an ideal extraction process for industries, such as pharmaceutical, food and cosmetic industries. Over and above $ScCo_2$ process could be a first step in an integrated biorefinery process leading to a high value-added.

In this study [13], cuticle waxes from date palm leaflets were extracted with $ScCo_2$. Leaflets were removed from the midribs, rinsed with water to remove dust and dried in sunlight for 3 weeks, milled and passed through a 2 mm screen before extraction by $ScCo_2$. The highest yield (3.49%) was obtained at 400 bar and 100 °C, much higher than other agricultural residues, such as stover, rice straw, Miscanthus and sugar cane. Moreover, 97% of the yield could be isolated after 120 min, significantly shorter than Soxhlet systems requiring 4–5 h extraction times. The obtained wax exhibited a melting point, (78 °C), comparable with carnauba wax. The economic study of the process of $ScCo_2$ extraction of wax from date palm leaflets gave a result of € 3.78 kg^{-1} falling within the range of non-petroleum waxes: € 7.8/kg, € 2.4/kg and € 6.2/kg respectively for beeswax, candelilla and carnauba waxes. Therefore, the environment-friendly $ScCo_2$ extraction of wax from date palm leaflets, abundantly and sustainably available represents a great future opportunity for wide industrial production of natural wax for a host of applications including cosmetics, coating and lubricants.

References

1. Al-Senaidy AM, Ismael MA (2011) Purification and characterization of membrane-bound peroxidase from date palm leaves (*Phoenix dactylifera* L.). Saudi J Biol Sci 18(3):293–298. https://doi.org/10.1016/j.sjbs.2011.04.005
2. Aldhaheri A, Alhadrami G, Aboalnaga N, Wasfi I, Elridi M (2004) Chemical composition of date pits and reproductive hormonal status of rats fed date pits. Food Chem 86(1):93–97. https://doi.org/10.1016/j.foodchem.2003.08.022
3. Abou-Zeid A-ZA (1991) Increasing the protein content of palm by-products. Biores Technol 37(3):239–242. https://doi.org/10.1016/0960-8524(91)90190-U
4. Elsanhoty RM, Al-Turki IA, Ramadan MF (2012) Screening of medium components by Plackett-Burman design for carotenoid production using date (*Phoenix dactylifera*) wastes. Ind Crops Prod 36(1):313–320. https://doi.org/10.1016/j.indcrop.2011.10.013
5. Kabirifard A, Fazaeli H, Kafilzadeh F (2012) Comparing the growth rate of four pleurotus fungi on wheat stubble and date palm leaf. J Res Agri Sci 8(1):35–43
6. Acourene S, Djafri K, Ammouche A, Amourache L, Djidda A, Tama M, Taleb B (2011) Utilisation of the date wastes as substrate for the production of baker's yeast and citric acid. Biotechnology (Faisalabad) 10(6):488–497. https://doi.org/10.3923/biotech.2011.488.497
7. Alrumman SA (2016) Enzymatic saccharification and fermentation of cellulosic date palm wastes to glucose and lactic acid. Braz J Microbiol 47(1):110–119. https://doi.org/10.1016/j.bjm.2015.11.015
8. Nancib A, Nancib N, Boubendir A, Boudrant J (2015) The use of date waste for lactic acid production by a fed-batch culture using *Lactobacillus casei* subsp *Rhamnosus*. Braz J Microbiol 46(3):893–902. https://doi.org/10.1590/S1517-838246320131067
9. Ziadi M, Mrsquo hir S, Kbaier N, Hamdi M, Ferchichi A (2011) Microbiological analysis and screening of lactic acid bacteria from Tunisian date palm sap. Afr J Microbiol Res 5(19):2929–2935.https://doi.org/10.5897/AJMR11.325
10. Ahmed J, Almusallam A, Al-Hooti SN (2013) Isolation and characterization of insoluble date (*Phoenix dactylifera* L.) fibers. LWT—Food Sci Technol 50(2):414–419. https://doi.org/10.1016/j.lwt.2012.09.002
11. Bauza E, Dal Farra C, Berghi A, Oberto G, Peyronel D, Domloge N (2002) Date palm kernel extract exhibits antiaging properties and significantly reduces skin wrinkles. Bioscience Ediprint Inc, XXIV(4):131–136
12. Devshony S, Eteshola E, Shani A (1992) Characteristics and some potential applications of date palm (*Phoenix dactylifera* L.) seeds and seed oil. JAOSC 69(6)
13. Al Bulushi K, Attard TM, North M, Hunt AJ (2018) Optimisation and economic evaluation of the supercritical carbon dioxide extraction of waxes from waste date palm (*Phoenix dactylifera*) leaves. J Clean Prod 186:988–996. https://doi.org/10.1016/j.jclepro.2018.03.117
14. Zaid A, Piesik S (2020) The Khalifa awards report
15. EL-Mously H, Darwish EA (2020) Date palm byproducts: History of utilization and technical heritage. In: Date palm fiber composites processing, properties and applications. Springer

Chapter 4
Date Palm Byproducts in Fibers, Textiles and Composites

Abstract Date palm (*Phoenix dactylifera* L.) is a very rich source of cellulosic fibers. Fibers could be extracted from different parts of the palm including midribs, spadix stems, leaflets, mesh and even the trunk at the end of the palm life. The extracted date palm fibers have promising properties when compared to other leaf fibers like sisal, abaca and banana. This chapter sheds the light on the great potential of date palm fibers as a newcomer to the natural fibers' library and then provides a comparative analysis between fibers extracted from the palm midrib, spadix stem and leaf sheath to other commercial vegetable fibers. The chapter also benchmarks the performance of the automotive-grade composite panels reinforced by date palm fibers to other leaf fiber composites. Further, the thermal and acoustical insulation properties of nonwoven fiber batts made from date palm midrib were also analyzed. Finally, the potential applications of date palm fibers were highlighted in the context of the circular bioeconomy of the future. The information presented in this chapter confirms the excellent opportunity of valorizing date palm byproducts in producing sustainable textile grade fibers with competitive properties and high availability.

Keywords Date palm fiber · Textile · Composite · Thermal insulation · Nonwoven · Acoustic · PalmFil®

4.1 Introduction

4.1.1 The Global Textile Industry

The global textile industry has grown tremendously over the past few decades, boosted by population growth, a booming global middle class and rising income levels. In parallel to this growth, there have been lots of concerns related to the sustainability of the textile value chain, and the current fiber mix in the industry is not sustainable. The industry relies heavily on synthetic fibers which are derived from the ever-depleting oil resources and result in the problem of microplastics in the oceans, in addition, to posing serious environmental issues at their end-of-life

H. EL-Mously et al., *Date Palm Byproducts: A Springboard for Circular Bio Economy*, Materials Horizons: From Nature to Nanomaterials,
https://doi.org/10.1007/978-981-99-0475-4_4

and resulting in a significant overall carbon footprint. On the other hand, traditional natural fibers like cotton and linen are facing major challenges, due to their intensive water, chemical and land use, in addition to their vulnerability to climate-driven shocks and stresses, which makes their future sustainability as a textile feedstock highly uncertain [1].

Recently, there has been an industry wide hunt for newer, better, and more sustainable fibers. Research has indicated that biobased alternatives derived from underutilized agricultural residues such as date palm, show great promise. Activating such value chain offers great potential to: decrease extensive crop residues burning and its associated negative environmental and climate impacts; generate new job opportunities, and extra income to agricultural communities and activate a new, scalable and more environmentally sustainable source of fiber for the fast-growing textile industry.

4.1.2 Date Palm Fibers

The term fiber may refer to many types of fibers, which could be textile fiber, wood fiber, paper fiber or dietary fiber. Date palm byproducts are very diverse and can offer all the previously mentioned grades of fibers, yet the focus of this chapter will be limited to the textile grade fibers. Textile fibers often refer to fibers with high aspect ratio (length to diameter ratio), high flexibility and sufficient length to be spun into yarns and further processed into textile products.

Date palm byproducts are very rich source of vegetable (cellulose) fibers, with different sources from which fibers can be extracted, including, midrib, spadix stem, leaflets, leaf sheath (mesh) as well as the trunk at the end of the palm life as shown in Fig. 4.1 [2]. However, every source of those byproducts has a different type of fiber. To be able to valorize such fibers, it is very important to understand the physiochemical, morphological and mechanical properties of such fibers.

Fig. 4.1 Date palm byproducts of pruning, including, midribs, spadix stems, and leaf sheath [2]

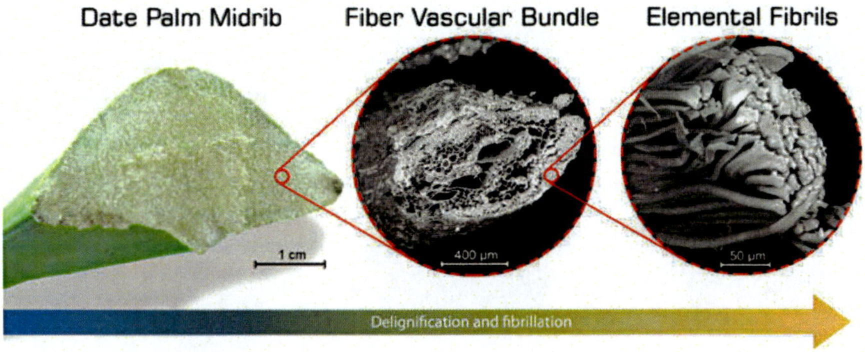

Fig. 4.2 Date palm midrib fiber extraction by effective delignification and fibrillation [4]

There are two types of date palm fibers (DPF); fibers that are present in a fibrous form which require minimal cleaning and opening such as the leaf sheath, and other fibers that are embedded inside the biomass which require an extraction process, such as the midrib and spadix stem fibers. There are two major challenges associated with extracting fibers from such sources. First, the fibers are embedded inside the biomass and surrounded by a matrix of lignin and hemicellulose, which makes it difficult to extract the pure fibers without breaking or damaging them. Second, in most cases the fibers are coarse and hollow (vascular), and they do not have enough flexibility to be processed into any textile form. The fibers break whenever they are bent or twisted. This explains why all previous attempts were limited to crushing or shredding the date palm biomass into particles or chips, and none of these attempts were capable of extracting long textile fibers [3].

There are several approaches used for extracting vegetable fibers, which can be generally classified into mechanical, chemical and biological. While each method can be used separately, yet it is more common that two or more methods are combined to achieve effective extraction. The efficiency of the extraction process is often judged by the ability of the process to remove the lignin–hemicellulose matrix holding the vascular bundles together (delignification), without causing any damage. In addition to its ability to split (fibrillate) the coarse vascular bundles into finer fibrils and eliminating the hollow content, resulting in fine flexible textile fibers as shown in Fig. 4.2 [4].

4.2 Comparative Analysis

4.2.1 Midrib and Spadix Stem Fibers

Date palm midrib and spadix fibers are considered leaf fibers. Generally, leaf fibers are classified as monocotyledons which consist of numerous fiber vascular bundles

Fig. 4.3 Date palm fibers and other leaf fibers used in the analysis [2]

in varying sizes, which are responsible for the moisture transportation. Hence, to introduce date palm midrib and spadix fiber as newcomers to the family of leaf fibers, their properties should be evaluated and compared to other leaf fibers.

Three commercial leaf fibers were used for the comparative analysis: sisal, abaca and banana. Whereas, for the date palm fibers, three types were used, midrib core, midrib skin and spadix fibers as shown in Fig. 4.3 [2].

4.2.1.1 Morphology

By comparing the cross-sectional features, it was found that the leaf fibers are generally in the form of large bundle of fibrils which can have multiple cell walls and a central lumen. Most fiber bundles are circular or oval with single or multiple vascular voids, but they may split into longitudinal segments of irregular shapes during the fiber extraction. For instance, sisal fiber bundle has one single vascular void located at the periphery of the bundle, and during extraction, the bundle may be ruptured into kidney shaped bundle instead of circular. Hence, the final cross-sectional shape of the bundle would depend on the number of vascular voids, their distribution within the bundle and severity of the extraction operation. Another important feature is the cell wall thickness of the fibril with respect to the size of the central lumen. In this case, banana fiber has thinner cell walls compared to sisal and abaca as shown in Fig. 4.4, while date palm (DP) fibers have very thick cell walls and negligible central lumen, which might be due to cell densification resulting from the extraction stage [5].

DP midrib core fiber is similar to oil palm and bagasse fibers, having large vascular voids [6, 7]. However, DP midrib skin fiber was quite different from the core, because it consists of pure fiber bundles unlike the hollow vascular nature of the midrib core fiber as shown in Fig. 4.4. Furthermore, DP spadix fiber had approximately the same geometry and size of the DP core fiber.

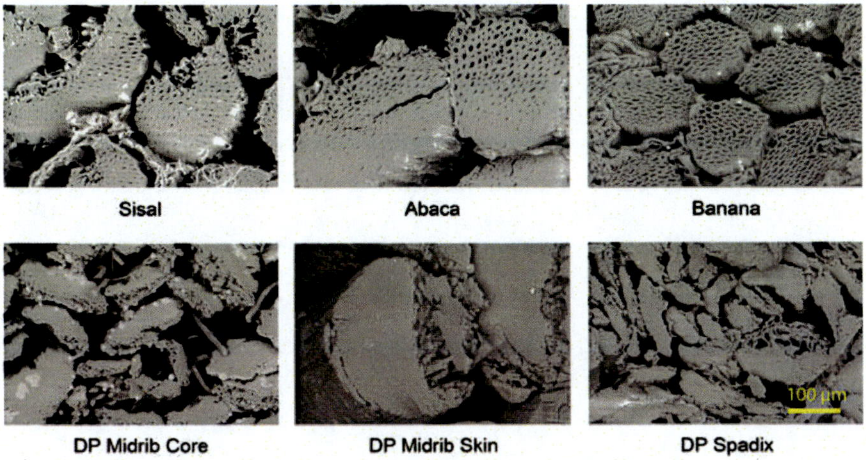

Fig. 4.4 SEM micrographs showing the cross-section of sisal, abaca, banana, DP midrib core and skin fibers and DP spadix fibers [2]

By observing the surface morphology, it was found that the inner vascular void of sisal is lined with parenchyma tissues which is introduced to the surface when the bundle splits during extraction as this was also observed in date palm midrib (DPM) core and skin fibers. The outer surface of banana fiber was lined with bright square beads aligned in the fiber axial direction. This surface feature was also similar to the silica crystals embedded on the surface of DPM core and skin fibers. On the other hand, the DP spadix stem fiber had smoother surface with minimal observed scales to an extent that the elemental fibrils were exposed as shown in Fig. 4.5 [2].

The average cross-sectional areas of the six samples are shown in Fig. 4.6. It was noticed that the banana, DP midrib core and DP spadix stem fibers had nearly average cross-sectional areas ranging from 0.01 to 0.015 mm^2. However, the average cross-sectional area of sisal and abaca fibers had higher area values equals to 0.022 $\pm$ 0.009 mm^2 and 0.031 $\pm$ 0.016 mm^2, respectively. Finally, the cross-sectional area of DPM skin fiber was the highest among the six fibers equals to 0.0498 $\pm$ 0.02 mm^2. It was expected that DP midrib skin fiber would have higher area values because it is composed of pure fiber bundles with large amount of pectic matter that prevents the bundle fibrillation during extraction. However, to better assess the size differences between the six fibers, the equivalent diameters of the fibers were calculated. The equivalent diameter values in micrometer of sisal, abaca, banana, DPM core, DPM skin and DP spadix were found to be 167.7 $\pm$ 9.3, 199 $\pm$ 16, 121.4 $\pm$ 5.4, 131.7 $\pm$ 9.7, 252 $\pm$ 21.4, and 123 $\pm$ 8.5 μm, respectively. The fiber cross-sectional area is an important measure to evaluate the fiber fineness, which has a direct impact on its flexibility and ability to be processed into any textile form. Leaf fibers tend to be coarser than bast fibers; hence, they are less likely to be used in apparel and home furnishing. Yet, finer fibers such as banana, DP midrib core and DP spadix may be good candidates in this context [2].

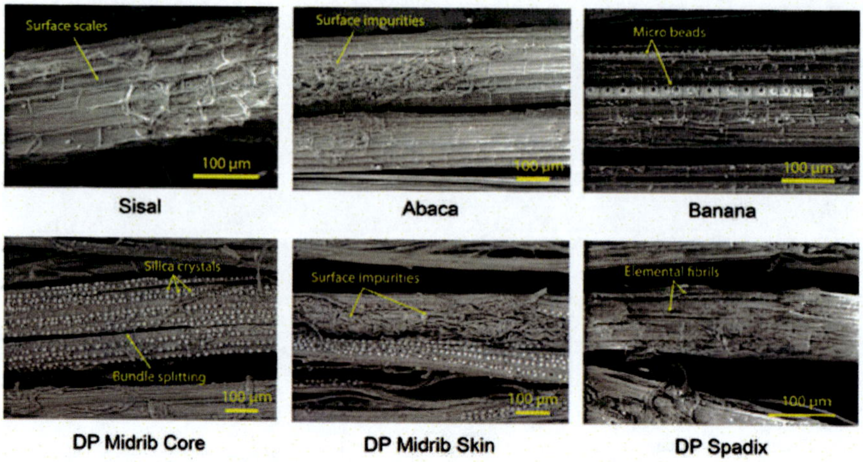

Fig. 4.5 Longitudinal view of sisal, abaca, banana, DP midrib core and skin fibers and DP spadix fibers [2]

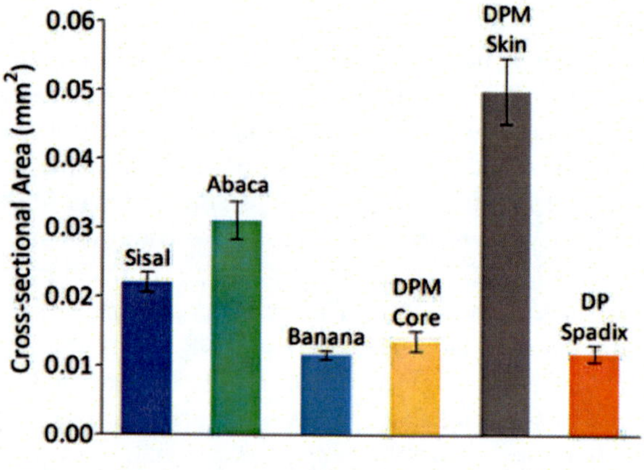

Fig. 4.6 Average cross-sectional areas of sisal, abaca, banana, DP midrib core and skin fibers and DP spadix fibers

4.2.1.2 Chemical Composition

Determination of the fiber chemical composition, including cellulose, lignin and hemicellulose contents, is very important to judge the purity of the fibers and the expected fiber yield. The compositional analysis results are shown in Table 4.1; it was found that abaca fiber had the highest cellulose wt% which was 70.9 ± 0.35% and low hemicellulose wt. 9.88 ± 0.35% which is comparably lower than most of the other fibers. On the other hand, sisal fiber had the lowest cellulose wt% which

Table 4.1 Chemical composition of sisal, abaca, banana, DP midrib core and skin fibers and DP spadix fibers [2]

Fiber	Cellulose (%)	Hemicellulose (%)	Lignin (%)	Ash (%)	Extractives (%)
Sisal	49.5 ± 0.56	6.09 ± 0.004	6.505 ± 0.247	1.23 ± 0.07	36.7
Abaca	70.93 ± 0.35	9.88 ± 0.35	7.455 ± 0.035	0.44 ± 0.05	11.3
Banana	57.18 ± 0.27	14.8 ± 0.24	4.38 ± 0.113	0.39 ± 0.04	23.2
Midrib Core	52.52 ± 0.162	20.41 ± 0.06	8.91 ± 0.23	2.55 ± 0.04	15.6
Midrib Skin	52.67 ± 0.085	20.98 ± 0.26	10.65 ± 0.247	0.95 ± 0.01	14.8
Spadix Stem	58.91 ± 0.24	19.18 ± 0.4	4.96 ± 0.113	0.56 ± 0.05	16.4

was 49.5 ± 0.6%; this is mainly due to the large proportion of extractives that was present in the fiber, yet the amount of hemicellulose was the least 6.1%.

DP spadix stem and banana fibers had similar cellulose and lignin wt% values. The cellulose wt% of DP spadix and banana fibers were 58.9 ± 0.24% and 57.2 ± 0.27%, respectively. While the lignin wt% of banana fibers and DP spadix stem fibers were 4.38 ± 0.11% and 4.96 ± 0.11%, respectively. It should be noted that banana and DP spadix fibers had the lowest lignin wt% among the six leaf fibers. As for the DPM core and skin fibers, they both had nearly equal chemical composition, because they are obtained from the same source, and they only differ in structure rather than chemical composition. The DPM core fiber had cellulose wt. 52.5 ± 0.16% and hemicellulose wt. 20.4 ± 0.61% and lignin wt. 8.9 ± 0.23%. Finally, the ash weight percentage in sisal, abaca, banana, DPM core, DPM skin and DP spadix was found to be 1.23 ± 0.07%, 0.44 ± 0.05%, 0.39 ± 0.04%, 2.55 ± 0.04%, 0.95 ± 0.01% and 0.56 ± 0.05%, respectively. The cellulose wt% of DP fibers were better than the other leaf fibers except for abaca.

4.2.1.3 Thermogravimetric Analysis (TGA)

The thermal properties of natural fibers indicate their thermal stability and their range of processing temperatures and end-use applications. Thermal gravimetric analysis (TGA) is the most commonly used method to analyze the thermal decomposition of natural fibers represented by weight loss as a function of temperature. By analyzing the thermogravimetric behavior of the fibers in Fig. 4.7, it was found that the fibers experienced four weight loss stages: dehydration, hemicellulose degradation, cellulose degradation and lignin degradation. Sisal fiber was thermally stable until reaching a temperature of 216 °C. The first stage, which corresponds to the water loss, was nearly between 30 and 150 °C which was followed by the degradation of hemicellulose at onset temperature of 204 °C. Afterward, cellulose started to degrade nearly at 285 °C with an onset temperature of 327 °C. The degradation of lignin followed the degradation of cellulose which occurred nearly at 445.6 °C and continued till 527.4 °C. Sisal fibers had 5% residues. The dehydration of Abaca fiber started from 30 °C till 176.4 °C. The fibers remained thermally stable till nearly

245 °C. The second degradation stage was due to the degradation of hemicellulose which had an onset temperature of nearly 200 °C. Subsequently, this stage was followed by the degradation of cellulose at 330 °C with nearly 50% drop in weight. The degradation of lignin was maximum at 492.77 °C. However, in case of banana, the fiber cannot be considered thermally stable above 200 °C. The results of this work were similar to the work of Guimarães et al. [6]. Their banana fibers also showed thermal stability between 100 and 200 °C and large weight loss at onset temperature 300 °C. The hemicellulose in banana fiber degraded at nearly 190 °C. The third stage due to cellulose degradation is associated with the major weight loss of nearly 52%. Finally, lignin in banana fiber was at its maximum at 422 °C with 25% weight drop [6].

As for DP fibers, the DPM skin and DP spadix thermal behavior was almost the same regarding the start and end of thermal degradation stages and the weight loss

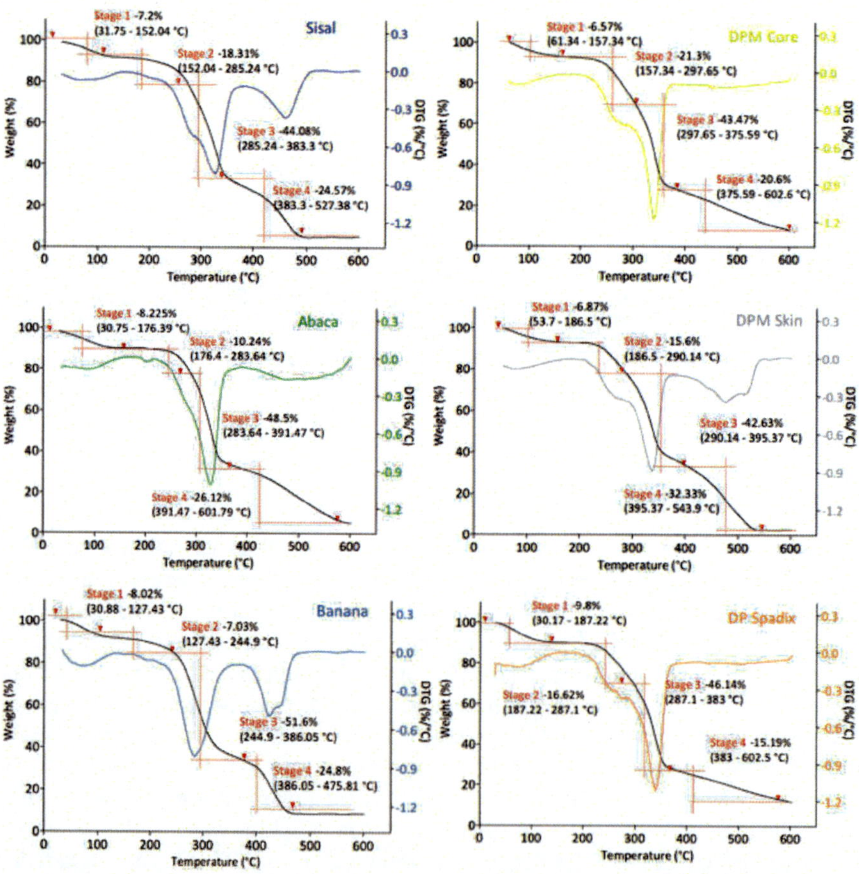

Fig. 4.7 TGA and DTG curves of sisal, abaca, banana, DP midrib core and skin fibers and DP spadix fibers

percentages. They were both thermally stable till almost 250 °C. The two samples experienced weight loss at first due to dehydration then hemicellulose degradation started at 190 °C and continued till 290 °C. The weight loss due to cellulose degradation was the largest and was nearly 43% in case of DPM skin and 46% in case of DP spadix stem fibers. At 390 °C, the lignin in the samples started degrading and corresponded to 32.33% mass drop in case of DP skin and 15.19% drop in case of DP spadix stem fibers. As for DPM core fibers, the fibers were thermally stable till nearly 250 °C. The hemicellulose and cellulose degradation stages had onset temperatures of nearly 243 °C and 342 °C, respectively, and corresponded to 21.3% and 43.5% weight loss, respectively. Finally, the lignin degradation occurred nearly at 410 °C with 20% weight loss. The thermal properties of DP fibers were comparable to that of the other leaf fibers. DP fibers were found to have better thermal stability than sisal, abaca and banana fibers. This is considered an advantage since DP fibers can be used when high processing temperature is required unlike the other leaf fibers.

4.2.1.4 X-Ray Diffraction (XRD) Analysis

The X-ray diffraction (XRD) analysis is a method that can be used to analyze the mechanical properties of a fiber by determining its degree of crystallinity. Date palm fibers are composed of crystalline cellulose, in addition to amorphous hemicellulose and lignin. The XRD patterns of the six fibers are shown in Fig. 4.8. All fibers had the highest peak around $2\theta = 22.6°$, average intensity peak around $2\theta = 16.5°$ and a peak of low intensity around $2\theta = 35°$, which were assigned respectively to the (200), (110) and (004) crystalline diffraction planes of cellulose [4]. The XRD patterns

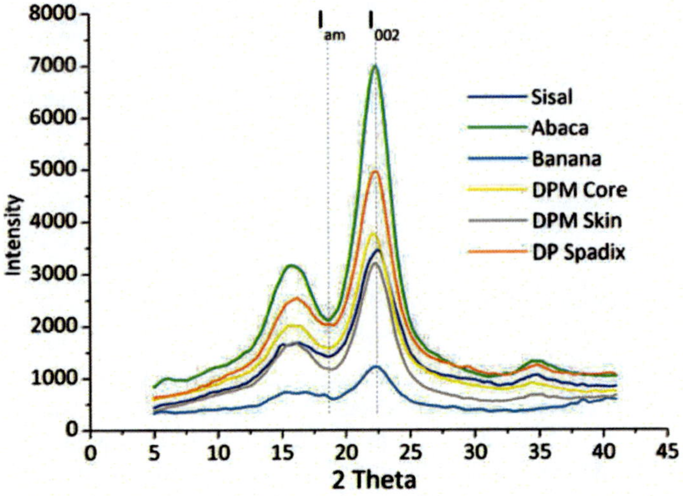

Fig. 4.8 XRD patterns of sisal, abaca, banana, DP midrib core and skin fibers and DP spadix fibers [2]

confirm that the fibers are of type cellulose Iβ [8]. The CrI of sisal, abaca, banana, DPM core, DPM skin and DP spadix fibers were 59.49, 69.77, 52.36, 58.5, 64.02 and 59.93, respectively. The crystallinity index of abaca fiber was the highest among the other fibers followed by DPM skin fibers. The crystallinity index of sisal, DPM core and DP spadix fibers was all within same range. Banana fibers had the lowest crystallinity index [2].

4.2.1.5 Single Fiber Tensile Test (SFTT)

The typical stress–strain curves of the six fibers are shown in Fig. 4.9a. Figure 4.9b which demonstrates a comparison between the tensile strength and modulus of elasticity of the six leaf fibers. It was noticed that among the six fibers, abaca (946.7 ± 170 MPa) had the highest strength followed by sisal (777.9 ± 212 Mpa), DPM core (583.5 ± 280 Mpa) and DP spadix stem fibers (601.7 ± 167 Mpa). The tensile strength of banana and DPM skin fibers was 377.55 ± 149 Mpa and 393 ± 144.7 Mpa, respectively. On the other hand, the modulus of elasticity of the fibers did not follow the same trend of the tensile strength. DPM core and sisal fibers had the highest modulus; 22.5 ± 8.83 Gpa and 19.6 ± 8.77 Gpa, respectively. As for the other fibers, the modulus of elasticity values of DPM skin, banana, abaca and DP spadix was found to be 13.9 ± 3.9 Gpa, 12.2 ± 5.34 Gpa, 11.4 ± 1.3Gpa and 11.67 ± 1.95 Gpa, respectively. Concerning the strain at break results, the average strain values are shown in Fig. 4.9d. It was found that abaca fibers had the highest strain values among the six fibers of 8.26 ± 0.57%. Following was the strain at break of DP spadix and sisal fibers 5.09 ± 0.66% and 4.3 ± 1.2%, respectively. Finally, the strain of banana, DPM skin and DPM core fibers was found to be 3.1 ± 0.115%, 2.788 ± 0.37% and 2.58 ± 0.32%, respectively. In conclusion, the mechanical properties of DP fibers are comparable and sometimes better than the other fibers. In the application where high stiffness is desired, DP midrib core fibers can be used as reinforcement. On the other hand, when high toughness is favored, DP spadix stem fibers can be used.

The results of the previous analysis showed that the properties of DP fibers are comparable to other commonly used leaf fibers proving that DP is a very promising source of vegetable fibers.

4.2.2 Leaf Sheath Fiber (Coir)

Unlike coconut coir, which is obtained from the husk surrounding the coconut fruit, date palm coir is obtained from the sheath layer surrounding the date palm trunk. This sheath layer is often referred to as leaf sheath, mesh or coir, and it originates from the tender tissues covering the new date palm leaves as they grow out. The sheath layer remains attached to the trunk after the leaves grow, and it turns into a coarse-brownish woven-like mesh, which can be separated from the trunk during

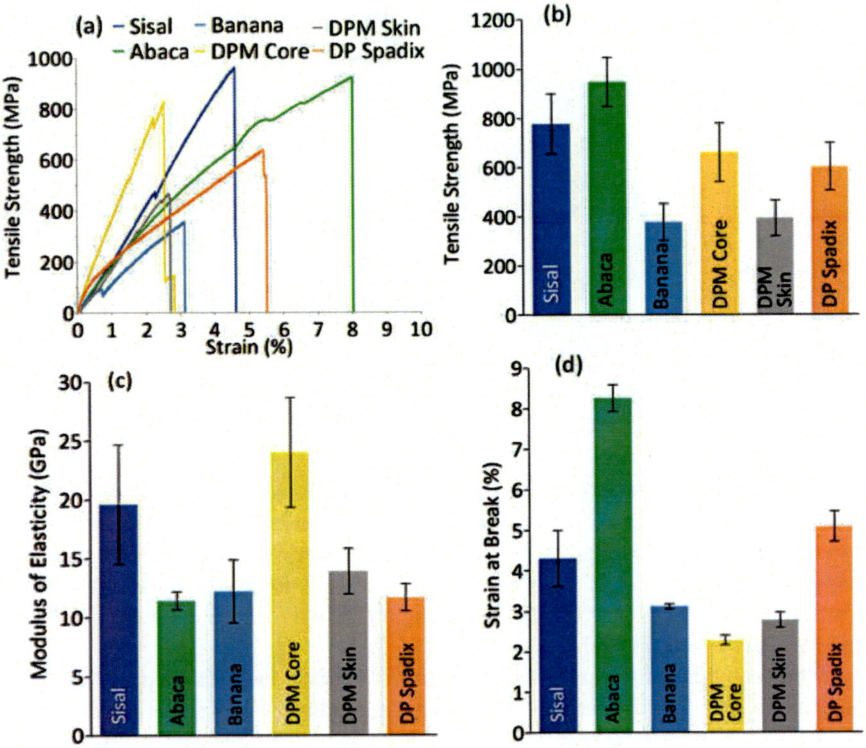

Fig. 4.9 **a** Typical tensile stress–strain curves and average values of **b** tensile strength, **c** modulus of elasticity and **d** stain at break of the six leaf fibers

the annual pruning [9, 10]. Figure 4.10 illustrates the similarity in appearance of the coconut coir and the date palm coir fibers.

Fig. 4.10 Visual appearance of coconut coir fiber after extraction (left) and date palm coir before fiber separation (right)

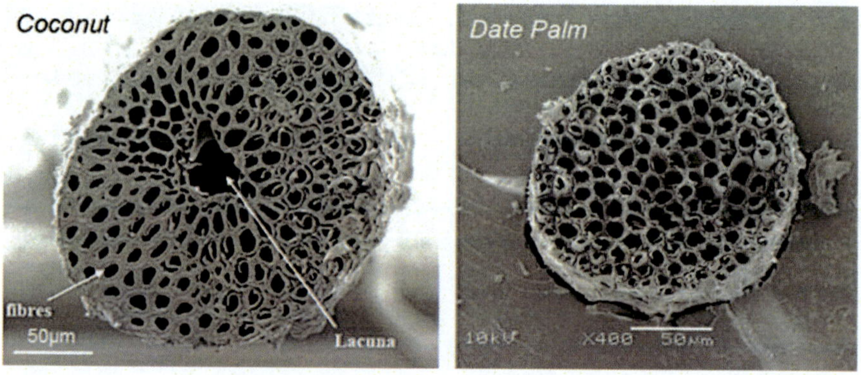

Fig. 4.11 Cross-sectional shape of coconut coir (left) and date palm coir (right) [11, 12]

4.2.2.1 Physical Properties

The fiber physical properties generally refer to the density, diameter, length, cross-sectional shape and surface morphology. However, the physical properties of natural fibers suffer from high variability depending on the source, pretreatment conditions and extraction method [10].

The density of the structural long coconut coir fiber was measured by Tran et al. [11], and it was found to be 0.9 g/cm^3. Similar method was used by Al-khanbashy et al. [12] to measure the density of long date palm coir fiber, and it was found to be 0.917 g/cm^3. This low density is attributed to the large void content and internal porous structure of both the coconut and date palm coir.

The fiber diameter of coconut coir is in the range of 100–450 μm, and length is up to 300 mm [13], while the fiber diameter for date palm coir ranges 100–2000 μm and length ranges 50–300 mm [3, 14].

The microstructural features of coir fibers are observed using scanning electron microscopy (SEM). The cross-sectional shape of both coconut and date palm coir is circular and has a cellular structure made up of hollow fibrils bonded together by a primary layer as shown in Fig. 4.11. However, the coconut coir has a large central lumen known as a lacuna which is not observed in date palm coir [11, 12]. Further, by looking at the surface morphology of both types of coir after cleaning with 2.5% NaOH to remove surface impurities, it is evident that the fiber surface is rough with micropores aligned in the fiber axial direction as shown in Fig. 4.12 [3, 12].

4.2.2.2 Chemical Properties

Table. 4.2 shows the chemical compositional analysis of coconut and date palm coir from previous research work. It was noticed that the cellulose content of both coconut coir and date palm coir was in very close range 41.14–50.6%. However, coconut coir

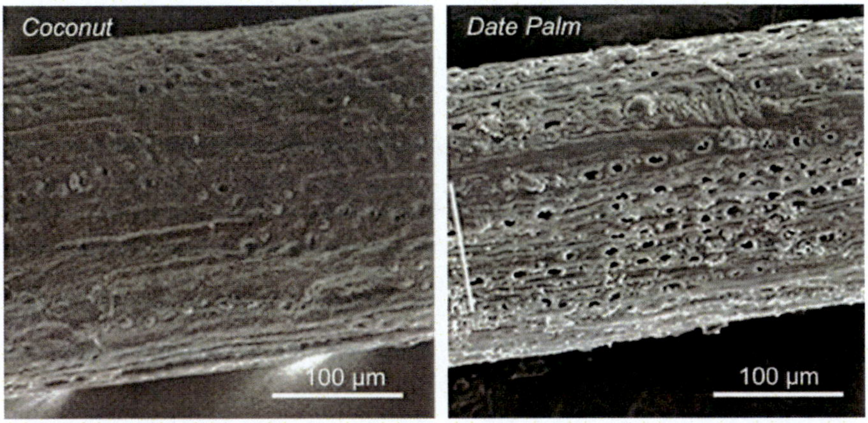

Fig. 4.12 Surface morphology of 2.5% NaOH treated coconut coir (left) and date palm coir (right) [3, 12]

Table 4.2 Comparison between the chemical composition of coconut coir and date palm coir fiber

	Cellulose (%)	Lignin (%)	Hemicellulose (%)	Pectin (%)	Source
Coconut coir	42.44	45.4	0.25	3	[16]
	43.44	45.84	0.25	3	[17]
	43.4	45.8	0.25	3	[18]
	41.14	35.25	15–17		[19]
Date palm coir	50.6 ± 1.3	31.9 ± 1.3	8.1 ± 0.3	–	[20]
	47.5	39.86	12.64	–	[21]
	48	24	19	–	[22]
	43 ± 2	35 ± 5	8 ± 2	–	[23]

has higher lignin content and lower hemicellulose content when compared to those of date palm coir. The low lignin content of date palm coir could be due to the fact that it is present in nature as fibrous mesh, requiring only minor cleaning and opening to remove the surface impurities [15–23].

4.2.2.3 Thermal Properties

The initial weight loss stage in the TGA curve corresponds to the release of bound water and other volatile extractives. The second stage corresponds to the decomposition of thermally sensitive hemicellulose, whereas the third stage corresponds to the decomposition of cellulose. Further weight loss stages correspond to the decomposition of lignin which has high thermal stability. The TGA curves differs depending on the fiber treatment conditions due to the change in the weight fraction of the

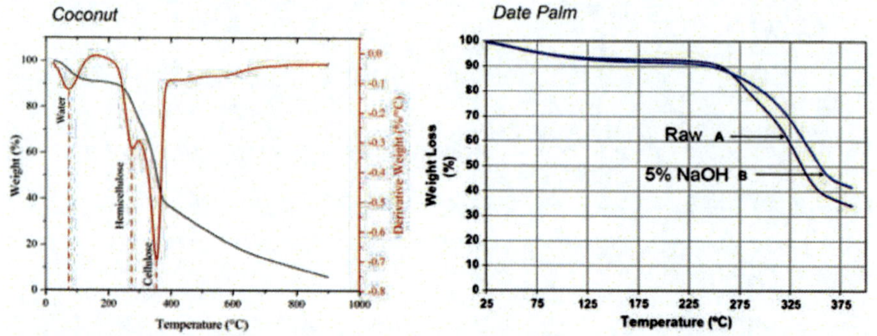

Fig. 4.13 Typical TGA curves of raw and soda treated coconut coir (left) and date palm coir (right) [24, 25]

chemical constituents within the fiber. In case of both coconut and date palm coir fibers, the thermal decomposition of hemicellulose is in the range of 225–300 °C, followed by the decomposition of cellulose 300–350 °C, which is then followed by the decomposition of lignin above 350 °C as shown in Fig. 4.13 [22, 24, 25]. The thermal stability of the fiber is governed by the amount of thermally sensitive hemicellulose, and the lower the weight fraction of hemicellulose, the higher the thermal stability would be, and generally, both types of coir are stable up to 225 °C.

4.2.2.4 Mechanical Properties

Both coconut coir and date palm coir are characterized by high ductility and toughness and the ability to deform plastically, unlike most vegetable fibers which are considered perfectly elastic. This high extensibility in the fiber under tensile loading is attributed to the cellular structure of the fiber shown in Fig. 4.11 which results in much lower resistance to lateral contraction and hence lower stiffness in axial direction during loading. The typical stress–strain curves of both coconut coir and date palm coir are shown in Fig. 4.14 [11, 12]. Both fibers have tensile strength in the range of 250 MPa and initial elastic modulus around 5 GPa, yet coconut coir has higher failure strain that can reach up to 40% which is four time greater than date palm coir. The tensile properties of the fibers may vary depending on many factors, including, extraction method, treatment conditions, cultivar, geography, climate, etc. Comparison between tensile properties of coconut coir and date palm coir is presented in Table. 4.3.

Figure 4.15 shows the typical XRD spectra of coconut and date palm coir fibers, the graphs show peaks (2θ) around 16.5°, 22° and 34.6° which represent the typical cellulose I structure, and these peaks are due to (110), (200) and (004), respectively. The calculated crystallinity index (CI%) using Segal method for both NaOH treated coconut coir and date palm coir was very close 39.5% and 38.5%, respectively [28, 29].

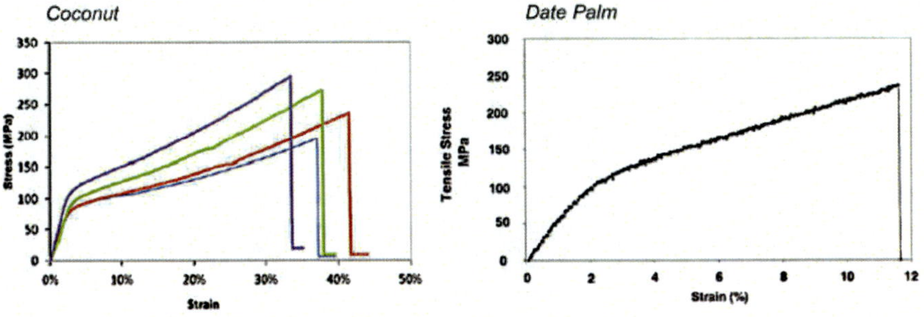

Fig. 4.14 Typical stress–strain curve of coconut coir (left) and date palm coir (right) [7, 8]

Table 4.3 Comparison between the tensile properties of coconut coir and date palm coir fiber

	Strength (MPa)	Elastic modulus (GPa)	Failure strain (%)	Source
Coconut coir	210–250	4.6–4.9	33–43	[11]
	131–175	4–6	15–40	[26]
	186–345	4–7	–	[27]
Date palm coir	275	0.8	–	[24]
	170–275	5–12	5–10	[12]
	233 ± 27.1	7.15 ± 2	10.3 ± 1.6	[28]
	178.2 ± 58	5.35 ± 1.9	11.4 ± 3.7	[22]

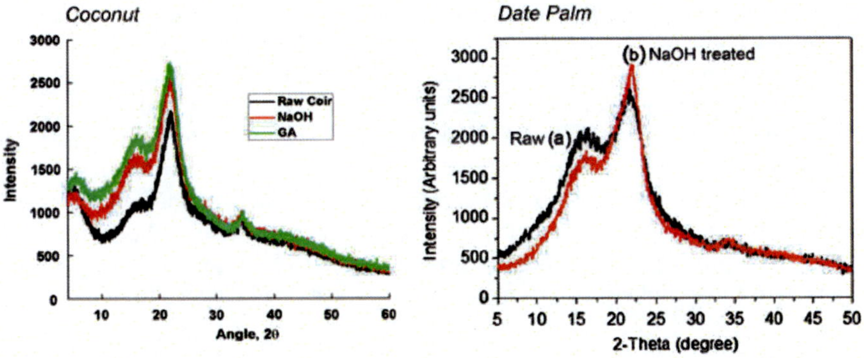

Fig. 4.15 Typical XRD spectra of coconut coir (left) and date palm coir (right) [28, 29]

4.3 Date Palm Composites Comparative Analysis

The main objective of this part is to compare date palm midrib and spadix fiber composites to other commercial leaf fiber composites, such as, sisal, abaca, and banana. For this purpose, six different nonwoven composite samples were fabricated,

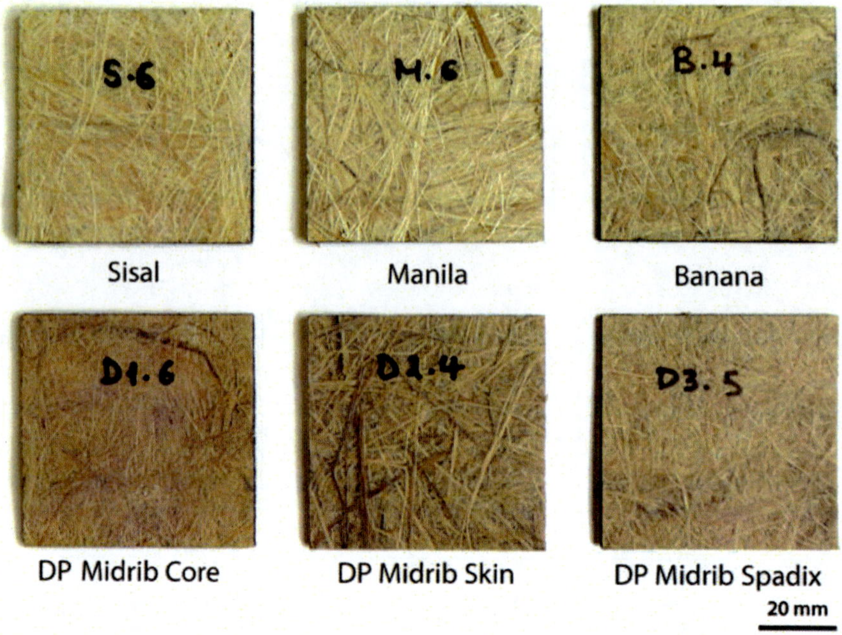

Fig. 4.16 Composite samples reinforced with date palm fibers and other leaf fibers

each of which were made of 50% wt. leaf fiber and 50% wt. PP fiber. This blend ratio was selected with reference to the automotive standard for natural fiber composites interior components, such as the door panels, trunk liners and parcel shelves. The samples were made by thoroughly blending the leaf fibers with PP fibers to ensure uniform distribution. Afterward, the blended fibers were then dry laid to create a nonwoven web with area 30 × 30 cm. The nonwoven web was then dried at 80 °C for 4 h to eliminate the moisture content. In order to make the composite samples, the nonwoven web was placed between the plates of a hot press and preheated for 1 min under 3 MPa pressure, then hot pressed under 15 MPa for 15 min at 183 °C. Figure 4.16 shows the visual appearance of the 6 fabricated composite samples.

4.3.1 Density and Void Content

The physical properties of the composite materials, such as volumetric density and void fraction, are very important parameters in evaluating their performance. Though, it is very common in nonwoven natural fiber composites to have void fraction in the range of 30%, which provides good thermal and acoustical insulation, even if this will be at the expense of the mechanical performance. Table 4.4 lists the densities

Table 4.4 Thickness, void content, areal and volumetric densities of the six composites

Fiber	Thickness (mm)	Areal density (gm/m^2)	Volumetric density (Kg/m^3)	Void content (%)
Sisal	2.7 ± 0.25	1815	680.6 ± 70	36.9 ± 2.2
Abaca	2.6 ± 0.14	1815	690.3 ± 36	40.3 ± 1.02
Banana	1.98 ± 0.20	1488	750.9 ± 87	29.3 ± 2.2
DPM core	2.07 ± 0.06	1694	818 ± 28	24.7 ± 0.1
DPM skin	2.12 ± 0.07	1730	815.8 ± 28	22.1 ± 0.45
DP spadix	1.9 ± 0.12	1597	829.7 ± 59	25.7 ± 0.36

and void fraction of all six composites. Generally, the void fraction in composites is initiating from two sources; first the voids in the fiber microstructure (lumens and vascular cavities) and second the voids between the fiber and the matrix due to poor wetting and formation of dry spots at the cross-over points between the fibers. The void created between the fiber and the matrix is highly influenced by the flexibility of the fiber and its ability to deform and fill the gaps in the structure, which highly depends on the fiber bending rigidity, which is influenced by the fiber diameter.

The volumetric densities of date palm composites were in the range 818–829 kg/m^3 and the void fraction 22.1–25.7%, which is considered the least void fraction compared to the other leaf fibers.

4.3.2 *Tensile Properties*

The composite tensile behavior is considered the major performance parameter. Generally, the factors that affect the composite tensile properties are the fiber stiffness, fiber weight fraction, presence of defects and interfacial shear strength between the fiber and the matrix. Figure 4.17a illustrates the typical tensile stress–strain curves of the composite samples. Figure 4.17b shows the average tensile strength of the composite samples with respect to their fiber tensile strength. It was evident that the composite's strength followed the same pattern of the fiber's strength, which was expected, since the fiber strength is a determinant factor. Date palm fiber composite strength from the midrib skin and core as well as spadix stem was in the range of 20.3–23.56 MPa, which was comparable to that of sisal (24.5 ± 2.78 MPa) with no significant statistical difference and slightly lower than that of abaca (32.13 ± 4.02 MPa), whereas the failure strains of the date palm composites were in the range 2.09–2.71% which is considerably lower than abaca. As for the modulus of elasticity results, the date palm composites ranged from 2.23 to 2.93 GPa, with date palm midrib core composites having higher modulus than any other fibers including abaca and sisal. The tensile properties results show that date palm composites are comparable to other composites reinforced with commercially available leaf fibers and in some cases superior to those fibers.

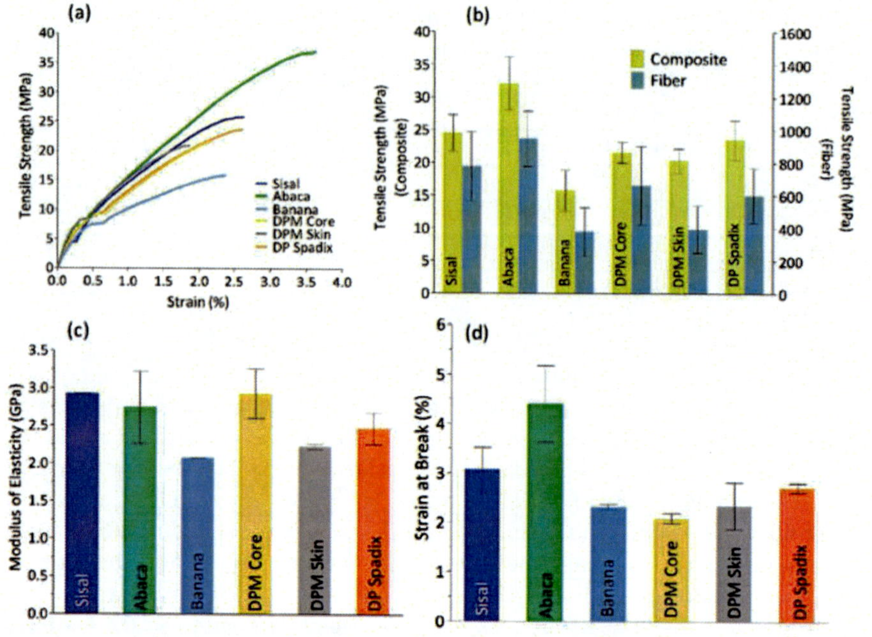

Fig. 4.17 **a** Typical tensile stress–strain curves and average values of **b** tensile strength, **c** modulus of elasticity and **d** strain at break of the six samples

By looking at the fracture surface of the composite specimens after tensile testing in Fig. 4.18, it was obvious that the failure of all specimens was dominated by fiber-matrix debonding, and pullout followed by fiber fracture. This failure mode has been exacerbated by the presence of dry spots due to the high void fraction in the composites. The dry spots and matrix discontinuity can be easily observed in the sisal and abaca micrographs in Fig. 4.18.

4.3.3 Flexural Properties

The main factor that affects the composite bending strength is the in-plane modulus of elasticity. Consequently, it was expected that the flexural strength of all composites follows the same trend of the tensile modulus. However, the fiber diameter in the composite also contributes to the flexural strength; the coarser the fibers, the higher the bending resistance will be. Figure 4.19a shows the typical flexural stress–strain curves of all the composite samples, whereas Fig. 4.19b compares the average values of the flexural strength and flexural chord modulus. Date palm composites had flexural strength 33–36.8 MPa, which were in the same range of abaca and sisal fibers with no statistical significant difference. On the other hand, the date palm

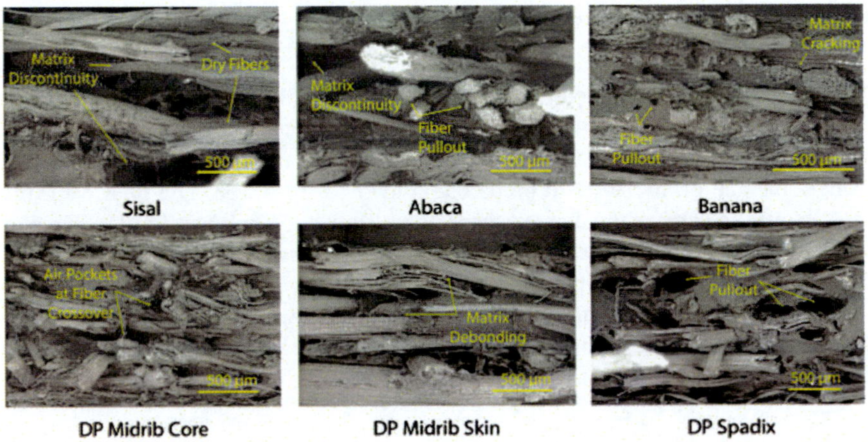

Fig. 4.18 SEM micrographs of the composites fractured surfaces

composite chord modulus was in the range of 202–207 GPa, which were also in the same range of sisal and abaca. The date palm midrib skin consistently showed higher bending resistance, which is due to the large fiber size of the midrib skin fiber bundles. The results show that date palm fiber composites have bending performance similar to other commercial leaf fiber composites, with potential to be used in similar applications and provide high durability.

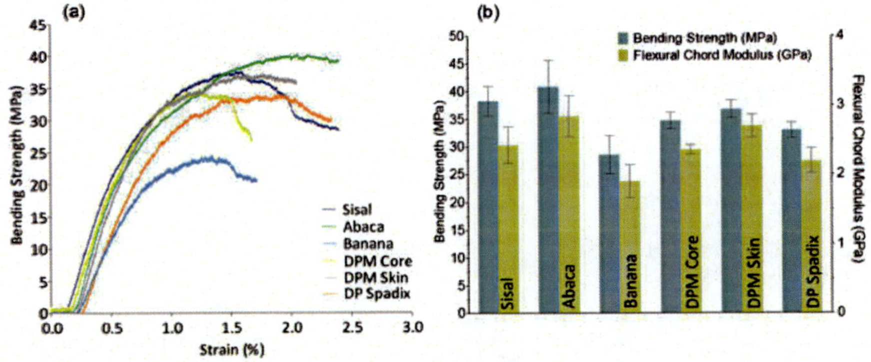

Fig. 4.19 **a** Typical bending stress–strain curves and **b** average values of bending strength and modulus of rupture of the six fibers

4.4 Date Palm Fibers Thermal and Acoustical Insulation

This part demonstrates the potential of date palm midrib fiber as an insulation material for building and construction to provide both thermal and acoustical insulation. For this purpose, 300 mm long fibers were first cut into 70 mm length and then carded using a manual drum carder to prepare a nonwoven fiber web. The fiber web was then stacked to achieve the desired shape and weight and slightly misted with water and confined in a mold to achieve the desired thickness and consequently density defined in the experimental plan. After drying the fiber web, good adhesion between layers was achieved due to interfiber friction as well as formation of hydrogen bonds. Figure 4.20 depicts the prepared nonwoven insulation batt.

4.4.1 Thermal Conductivity

Determining the thermal conductivity of date palm midrib fiber nonwoven batt was conducted using GUNT WL 376 thermal conductivity of materials experimental unit, according to DIN 52616 and DIN 52612 standards for "Determination of the thermal conductivity using heat flow plate equipment." The sample dimensions were 30 × 30 × 1.25 cm and had density of 150 kg/m^3.

Figure 4.21 demonstrates the results of the thermal conductivity test. Thermal conductivity value at the steady state of the date palm midrib nonwoven fiber batt was 0.0496 W/mK, which is considered within the range of thermal insulation materials in construction, as its thermal conductivity lies in the range 0.02–0.06 W/mK [30]. Comparing the thermal conductivity of DPM sample with other natural and synthetic insulators of equivalent densities, demonstrated in Fig. 4.22, showed that DPM occupies an intermediate ranking in between other insulators, scoring lower conductivities than straw bales, cotton stalks, oil palm, kenaf and sugar cane.

Fig. 4.20 Date palm midrib fiber nonwoven batt used in analyzing the thermal and acoustical properties

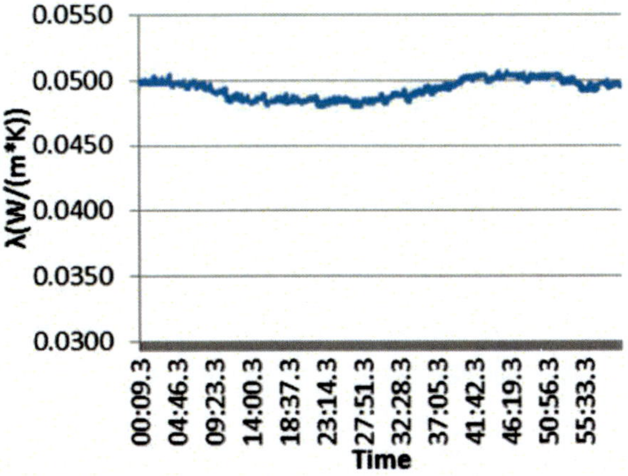

Fig. 4.21 Thermal conductivity measurements of DPM nonwoven batt $\lambda = 0.0496$ W/mK

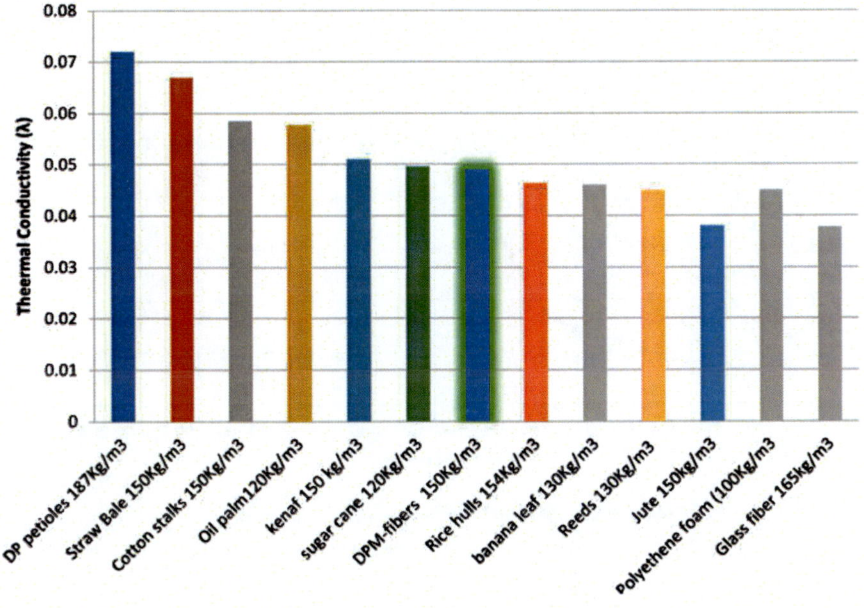

Fig. 4.22 Comparison of thermal conductivity of DPM fibers and other synthetic and natural insulators of equivalent density [31–36]

4.4.2 Acoustic Absorption

Several samples of date palm midrib nonwoven fiber batt were prepared in varying densities and thicknesses. Impedance tube was used to determine the normal acoustic absorption coefficient (AAC). Normal AAC is measured for sound waves falling perpendicularly onto the material surface using transfer function method in a standard impedance tube method; normal AAC can be defined as: "the ratio of absorbed energy to incident sound energy. Figure 4.23 shows the AAC values of the DPM samples in comparison with conventional synthetic porous absorbers [37–39]. The results show that mineral wool and propylat provide better absorption at low frequencies than natural fibers, but this performance is still surpassed by DPM sample. After the dip in mid-range frequencies, DPM samples return to highly comparable to propylat, acoustic foam, mineral wool, rock wool and polyurethane PUR foam. These results indicate that DPM fibers can provide a sustainable substitute for synthetic fibers at low (63–315 Hz) and high frequencies (1000–4000 Hz).

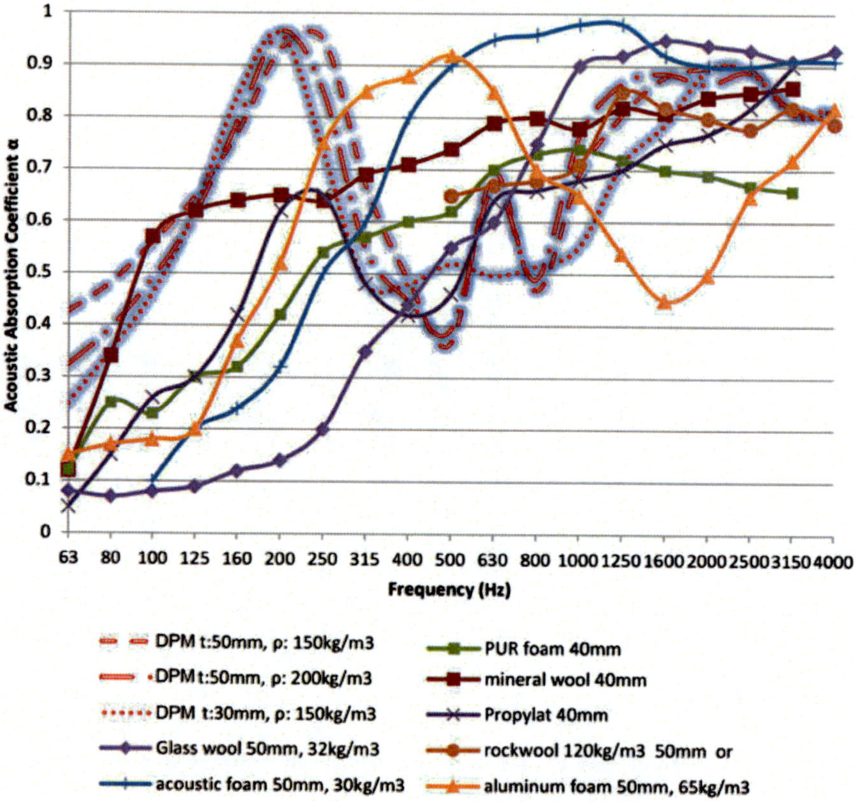

Fig. 4.23 Comparison between AAC values with synthetic porous absorbers

Table 4.5 Potential applications and markets for date palm fibers [40]

	Application	Market size in 2019 (US$)	CAGR 2010–2019 (%)
Emerging applications	Natural fiber composites	1.2 bil	12
	Thermal insulation of vegetable fibers	107 mil	8.4
	Biodegradable tableware of cellulose fiber	1.2 bil	6.5
Established applications	Carpets of vegetable fibers	550 mil	0.9
	Carpet backings of vegetable fibers	344 mil	3.9
	Gypsum plaster reinforced with vegetable fibers	431 mil	0.8
	Pulp and paper from fibrous cellulose & linter	724 mil	0.5
Traditional applications	Burlap sacks of vegetable fibers	249 mil	− 1
	Ropes and twines of vegetable fibers	488 mil	2.7

4.5 Potential Applications of Date Palm Fibers

The unique features and characteristics of date palm fibers make them excellent sustainable material base for a wide spectrum of industries, including emerging applications such as natural fiber composites, thermal insulation panels and biodegradable tableware. Moreover, well-established markets such as carpets/ backing, gypsum plaster panels and non-wood paper. In addition to, traditional applications such as burlap sacks and ropes. Table 4.5 lists major potential applications and their market values.

4.6 Conclusion

The results discussed in this chapter are very promising and highlights the points of strength of date palm fibers and their composites. Date palm fibers are still in their embryonic stage, and with further research and development the fibers can reach new horizons to help support the regional transition toward the circular bioeconomy of the future. Date palm fibers represent an opportunity to all date palm lovers to change the prevailing look of palms as part of the past and envision them as a whole resource for the circular bioeconomy of the future. This represents a chance for the younger generations to unite with their traditions in pride and continue the heritage of using the byproducts of pruning in modern industrial applications. Moreover, it is a chance to redirect the palm byproducts from the waste stream into the value stream.

References

1. Midani M, Hassanin A, Hamouda T, Elseify L (2021) Reversing the shift back to natural fibers. Int Fiber J 1
2. Elseify LA, Midani M, El-badawy AA, Awad S, Jawaid M (2022) Comparative study of long date palm (*Phoenix Dactylifera* L.) midrib and spadix fibers with other commercial leaf fibers. Cellulose. https://doi.org/10.1007/s10570-022-04972-1
3. Elseify LA, Midani M, Shihata LA, El-Mously H (2019) Review on cellulosic fibers extracted from date palms (*Phoenix dactylifera* L.) and their applications. Cellulose 26(4):2209–2232. https://doi.org/10.1007/s10570-019-02259-6
4. Elseify LA, Midani M, Hassanin A, Hamouda T, Khiari R (2020) Long textile fibers from the midrib of date palm: physiochemical, morphological, and mechanical properties. Ind Crops Prod 151:112466. https://doi.org/10.1016/j.indcrop.2020.112466
5. Li Z, Chen C, Xie H, Yao Y, Zhang X, Brozena A, Li J, Ding Y, Zhao X, Hong M, Qiao H, Smith LM, Pan X, Briber R, Shi SQ, Hu L (2021) Sustainable high-strength macrofibres extracted from natural bamboo. Nat Sustain. https://doi.org/10.1038/s41893-021-00831-2
6. Guimarães JL, Frollini E, da Silva CG, Wypych F, Satyanarayana KG (2009) Characterization of banana, sugarcane bagasse and sponge gourd fibers of Brazil. Ind Crops Prod 30:407–415. https://doi.org/10.1016/j.indcrop.2009.07.013
7. Aulya Syamani F, Suryani A (2015) Changes in oil palm frond fiber morphology, cellulose crystallinity and chemical functional groups during cellulose extraction phases. Chem Mater Res 7:113–121
8. French AD (2014) Idealized powder diffraction patterns for cellulose polymorphs, pp 885–896. https://doi.org/10.1007/s10570-013-0030-4
9. Barreveld WH (1993) Date palm products. FAO Agricultural Services Bulletin No. 101. https://www.fao.org/docrep/t0681e/t0681e00.htm#con
10. Midani M, Al-Othman O, Saba N (2020) Date palm fiber composites: processing, properties and applications. Springer Nature, Singapore. https://doi.org/10.1007/978-981-15-9339-0
11. Tran LQN, Minh TN, Fuentes CA, Chi TT, Vuure AWV, Verpoest I (2015) Investigation of microstructure and tensile properties of porous natural coir fibre for use in composite materials. Ind Crops Prod 65:437–445. https://doi.org/10.1016/j.indcrop.2014.10.064
12. Al-Khanbashi A, Al-Kaabi K, Hammami A (2005) Date palm fibers as polymeric matrix reinforcement: fiber characterization. Polym Compos 26:486–497. https://doi.org/10.1002/pc.20118
13. Adeniyi AG, Onifade DV, Ighalo JO, Adeoye AS (2019) A review of coir fiber reinforced polymer composites. Compos B Eng 176:107305. https://doi.org/10.1016/j.compositesb.2019.107305
14. Al-Rifaie WN, Al-Niami M (2016) Mechanical performance of date palm fibre-reinforced gypsums. Innov Infrastruct Solut 1:18. https://doi.org/10.1007/s41062-016-0022-y
15. Midani M, Elseify LA, Hamouda T, Hassanin A (2022) Chapter 15: comparison of coconut coir and date palm coir (sheath fiber) and their composites. In: Jawaid M (ed) Coir fiber and its composites. Elsevier. https://doi.org/10.1016/B978-0-443-15186-6.00070-9
16. Khan A, Ahmad MA, Joshi S, Al Said SA (2014) Abrasive wear behavior of chemically treated coir fibre filled epoxy polymer composites. Am J Mech Eng Autom 1(1):1–5

17. Haque M, Rahman R, Islam N, Huque M, Hasan M (2010) Mechanical properties of polypropylene composites reinforced with chemically treated coir and abaca fiber. J Reinf Plast Compos 29(15):2253–2261
18. Verma D, Gope P, Shandilya A, Gupta A, Maheshwari M (2013) Coir fibre reinforcement and application in polymer composites: A. Environ Sci 4(2):263–276
19. Siakeng R, Jawaid M, Ariffin H, Salit MS (2018) Effects of surface treatments on tensile, thermal and fibre-matrix bond strength of coir and pineapple leaf fibres with poly lactic acid. J Bionics Eng 15(6):1035–1046
20. Saadaoui N, Rouilly A, Fares K, Rigal L (2013) Characterization of date palm lignocellulosic by-products and self bonded composite materials obtained thereof. Mater Des 50:302–308. https://doi.org/10.1016/j.matdes.2013.03.011
21. Nasser R, Salem M, Hiziroglu S et al (2016) Chemical analysis of different parts of date palm (*Phoenix dactylifera* L.) using ultimate, proximate and thermo-gravimetric techniques for energy production. Energies 9:374. https://doi.org/10.3390/en9050374
22. Taha I, Steuernagel L, Ziegmann G (2006) Chemical modification of date palm mesh fibres for reinforcement of polymeric materials. Part I: examination of different cleaning methods. Polym Polym Compos 14:767
23. Mekhermeche A, Kriker A, Dahmani S (2016) Contribution to the study of thermal properties of clay bricks reinforced by date palm fiber. AIP Conf Proc 030004:030004. https://doi.org/10.1063/1.4959400
24. Alawar A, Hamed AM, Al-Kaabi K (2009) Characterization of treated date palm tree fiber as composite reinforcement. Compos Part B Eng 40:601–606. https://doi.org/10.1016/j.compositesb.2009.04.018
25. Mittal M, Chaudhary R (2019) Experimental investigation on the thermal behavior of untreated and alkali-treated pineapple leaf and coconut husk fibers. Int J Appl Sci Eng 76(1):01–16
26. Malkapuram R, Kumar V, Negi YS (2009) Recent development in natural fiber reinforced polypropylene composites. J Reinf Plast Compos 28(10):1169–1189
27. Defoirdt N, Biswas S, De Vriese L, Acker JV, Ahsan Q, Gorbatikh L, Vuure AV, Verpoest I (2010) Assessment of the tensile properties of coir, bamboo and jute fibre. Compos A 41(5):588–595
28. Abdal-hay A, Suardana NPG, Jung DY et al (2012) Effect of diameters and alkali treatment on the tensile properties of date palm fiber reinforced epoxy composites. Int J Precis Eng Manuf 13:1199–1206. https://doi.org/10.1007/s12541-012-0159-3
29. Manjula R, Raju NV, Chakradhar RPS et al (2018) Influence of chemical treatment on thermal decomposition and crystallite size of coir fiber. Int J Thermophys 39:3. https://doi.org/10.1007/s10765-017-2324-5
30. Sarkar DK (2015) Steam generators. In: Thermal power plant. Elsevier, pp 39–89. https://doi.org/10.1016/B978-0-12-801575-9.00002-0
31. Asdrubali F, D'Alessandro F, Schiavoni S (2015) A review of unconventional sustainable building insulation materials. Sustain Mater Technol 4:1–17. https://doi.org/10.1016/j.susmat.2015.05.002
32. Agoudjil B, Benchabane A, Boudenne A, Ibos L, Fois M (2011) Renewable materials to reduce building heat loss: characterization of date palm wood. Energy Build 43(2–3):491–497. https://doi.org/10.1016/j.enbuild.2010.10.014
33. Manohar K, Adeyanju A (2016) A comparison of banana fiber thermal insulation with conventional building thermal insulation. BJAST 17(3):1–9. https://doi.org/10.9734/BJAST/2016/29070
34. Manohar K, Ramlakhan D, Kochhar G, Haldar S (2006) Biodegradable fibrous thermal insulation. J Braz Soc Mech Sci Eng 28(1). https://doi.org/10.1590/S1678-58782006000100005
35. Xu J, Sugawara R, Widyorini R, Han G, Kawai S (2004) Manufacture and properties of low density binderless particleboard from kenaf core. J Wood Sci 50(1):62–67. https://doi.org/10.1007/s10086-003-0522-1

36. Goodhew S, Griffiths R (2005) Sustainable earth walls to meet the building regulations. Energy Build 37(5):451–459. https://doi.org/10.1016/j.enbuild.2004.08.005
37. Naguib EE (2009) Traditional crafts encyclopedia: date palm products handicrafts, 1st ed., vol 4. Asala Organization For Traditional Crafts, Cairo
38. Zunaidi NH, Tan WH, Majid MSA, Lim EA (2017) Effect of physical properties of natural fibre on the sound absorption coefficient. J Phys: Conf Ser 908:012023. https://doi.org/10.1088/1742-6596/908/1/012023
39. Gumanová V, Sobotová L, Dzuro T, Badida M, Moravec M (2022) Experimental survey of the sound absorption performance of natural fibres in comparison with conventional insulating materials. Sustainability 14(7):4258. https://doi.org/10.3390/su14074258
40. International Trade Center. Trade Map—trade statistics for international business development. Retrieved from: https://www.trademap.org/Index.aspx

Chapter 5
Date Palm Byproducts for Cellulose and Cellulose Derivatives Production

Abstract Nanofibrillated cellulose and cellulose nanocrystals were extracted from date palm midribs and successfully used in the reinforcement of nanocomposites (latex). Microfibrillated cellulose and oxidized microfibrillated cellulose were extracted from date palm midribs and successfully used for the improvement of paper sheet properties. Enzymatic treatment was used for the isolation of microfibrillated cellulose from date palm fruit stalks dominantly treated as waste. Microcrystalline cellulose having wide applications in pharmaceutical, food and cosmetics applications has been isolated from date palm stalks with a yield of 35.4%. Its high crystallinity (79.4%) endowing it with high rigidity opens a great potentiality for its use as a reinforcing material in nanocomposites. Oxidized nanocellulose has been successfully extracted from date palm leaf sheath fibers and used as a packaging additive for better packaging properties. Cellulose whiskers were successfully extracted from date palm midribs and leaflets and characterized. Only cellulose whiskers extracted from midribs were used in the reinforcement of nanocomposites (rubber) due to their higher geometrical characteristics (higher aspect ratio). Microfibrillated cellulose was successfully extracted from date palm fruit stalks using enzymatic treatment. This proves the economic and developmental potentiality of realizing such a high value-added product from date palm fruit stalks dominantly treated as waste.

Keywords Date palm byproducts · Cellulosic nanocrystals · Cellulose nanocrystals · Microfibrillated cellulose · Oxidized microfibrillated cellulose · Microcrystalline cellulose · Cellulosed whiskers · Nanocomposites

5.1 Nanofibrillated Cellulose and Cellulose Nanocrystals from Date Palm Rachis for the Reinforcement of Nanocomposites

In a research work [1], cellulosic nanocrystals and nanofibrillated cellulose have been extracted from Alfa and date palm rachis and used in the reinforcement of latex. The objective of the research was to study the influence of morphology and the

H. EL-Mously et al., *Date Palm Byproducts: A Springboard for Circular Bio Economy*, Materials Horizons: From Nature to Nanomaterials,
https://doi.org/10.1007/978-981-99-0475-4_5

size of the cellulose fibrils on the mechanical and optical properties of the produced nanocomposites.

Original fibers of length 2–3 cm were first obtained from Alfa (Alfa tenacissma) and date palm rachis, harvested in Tunisia. They were then treated in a digester with NaOH solution (16–17 vol%) at a temperature of 165 °C for 1–2 h. A high-pressure homogenizer was used to prepare nanofibrillated cellulose. The obtained product was a high viscosity translucent gel. The cellulose nanocrystals were obtained from alfa and date palm rachis using classical sulfuric acid hydrolysis. In order to obtain nanocomposite films, the nanofibirillated cellulose and cellulose nanocrystal suspensions were mixed with latex; the cellulose content ranged from 0 to 15%. After stirring for 1 h, the mixture was poured in a Teflon mold. The product was a transparent to translucent film (depending on the nanofiller content) with a thickness of 300–400 μm. In addition, rigid films of thickness 80–100 μm were obtained by casting nanofibrillated cellulose and nanocrystal solvent of 1 wt% in a Teflon mold.

The dynamic mechanical properties of the test samples have been investigated and a strong correlation between the reinforcing capability of the nanofiller and its aspect ratio has been found. This can be explained by the formation of a stiff percolating cellulose network in case of higher aspect ratio particles. It was also found that the use of smaller diameter nanofiller has a detrimental influence on the transparency of the polymeric matrix. This can be explained by the percolation effect of the nanofiller. Thin and high aspect ratio nanoparticles can form a continuous network negatively impacting the transparency of matrix by promoting light scattering.

5.2 Extraction of Microfibrillated Cellulose and Oxidized Microfibrillated Cellulose from Date Palm Rachis for the Improvement of Paper Sheet Properties

There is a growing interest in the use of cellulosic nanomaterials for the improvement of properties of paper. Within this context, a research [2] has been conducted to compare the improvement of paper sheet properties as a result of adding different ratios of microfibrillated cellulose and oxidized microfibrillated cellulose, extracted from date palm rachis, with the corresponding improvement as a result of beating the fibers.

Bleached Kraft bagasse pulp was obtained from a company in Egypt, whereas the wood pulp was obtained from a company in Sweden. Palm rachis was obtained from local field, cleaned and chopped into 2–3 cm long pieces. The date palm rachis was then treated with 15% NaOH at 150 °C for 3 h and bleached. The date palm rachis pulp was treated with xylanase enzymes to facilitate isolation of microfibrillated cellulose.

The extracted microfibrillated cellulose and the oxidized microfibrillated cellulose were added at ratios from 2.5 to 20% to paper sheets, prepared from unbeaten softwood and bagasse pulps. The results proved that the addition of microfibrillated cellulose and oxidized microfibrillated cellulose has brought about an increase in density, wet and dry tensile strength, tear resistance and a decrease in air permeability of paper sheets prepared from unbeaten softwood or bagasse. Meanwhile, usual beating of softwood fibers was found much more active in improving the strength properties of softwood paper. Conversely, the addition of microfibrillated cellulose and oxidized microfibrillated cellulose to the bagasse paper sheets was much more efficient in the improvement of properties of paper sheets than beating of bagasse pulp. In addition, it was proven that the use of oxidized microfibrillated cellulose gave better improvement in tensile strength (wet and dry) with unbeaten softwood or bagasse fibers as compared with the addition of microfibrillated cellulose.

5.3 Enzyme-Assisted Isolation of Microfibrillated Cellulose from Date Palm Fruit Stalks

The date palm fruit stalks are generally treated as a waste in the date palm growing countries. One of the rare exceptions is the use of a small portion of them in the manufacture of streets cleaning brooms in Egypt. This means that ~17% of products of pruning of date palms [3] is wasted on the world level. This points to the economic and developmental signification of a research [2], conducted with the aim of extraction of microfibrillated cellulose from date palm stalks using ultrafine grinding. To facilitate the extraction of microfibrillated cellulose and to improve its strength properties enzymatic pretreatment has been used.

The date palm fruit stalks were sourced from palm plantations in Giza, Egypt. They were cleaned with water and cut to pieces 2–3 cm. The pulp was made by alkali treatment using 15% NaOH at 150 °C for 3 h. Bleaching was conducted using sodium chlorite/acetic acid mixture at 80 °C for 1 h. The bleached pulp included 71.5% α-cellulose and 18.4% pentosans. To prepare the citric buffer (pH 5.3), sodium citrate and citric acid were used. The enzymatic pretreatment of date palm pulp was conducted using xylanase enzymes. The reaction mixture was kept under gentle stirring at 50 °C for 4 h. Then the temperature was raised to 90 °C to deactivate the enzymes. The pulp was filtered and washed with distilled water. A high-shear mixer was used to disintegrate the pulp suspension (2% consistency). A high-shear ultrafine friction grinder was used to treat the fiber suspension by passing them through the device up to 60 times. Then the microfibrillated cellulose was centrifuged at 10,000 rpm to reduce its water content and kept wet in the fridge till used. A 0.5% (by weight) microfibrillated suspension containing 1 g was used to make test sheets by vacuum filtration. These microfibrillated cellulose test sheets were air-dried and conditioned at 50% relative humidity for 48 h at 25 °C before testing. Tensile strength tests (wet and dry), as well as air permeability and water absorption tests (5 specimens each), have been conducted and the results averages calculated.

The results show that the enzymatic pretreatment has given microfibrillated cellulose sheets with higher density than those obtained from untreated pulp. This can be explained by the partial removal of hemicellulose leading to the increase of compressibility of microfibrillated cellulose. The maximum increase of density was obtained at an enzyme dose of 0.02 g/g. The enzymatic pretreatment of date palm fruit stalks fibers has also led to a profound increase of the tensile strength properties: strength value and moduli, wet and dry, as well as the water absorption characteristics of the microfibrillated cellulose sheets.

5.4 Microcrystalline Cellulose from Bunch Stalk of Date Palm: Isolation and Characterization

An important research [4] has been devoted to the extraction of microcrystalline cellulose (MCC) from date palm fruit bunch stalks: a date palm byproduct usually treated as waste. The microcrystalline cellulose (MCC) is a biodegradable material that finds wide industrial applications as water retention and a suspension stabilizer in pharmaceuticals, food and cosmetics. A lot of research has been conducted to isolate (MCC) from different sources, such as rice husk, jute, oil palm empty fruit bunch fibers, etc.

The date palm fruit stalks were obtained from Riyadh, Saudi Arabia. The retting technique has been used to extract fibers from the fruit bunch stalks. The fibers were washed to remove impurities and dried in an over for 24 h at 60 °C. They were then ground and sieved to a size of 10 mm. The extraction of MCC has been conducted by bleaching of fibers using 10(w/w) sodium hypochlorite (NaClO) solution for 1 h at a temperature 70–80 °C. The ratio of fiber to NaClO was 1:60 (g/ml). The solution was acidified by acetic acid to reach pH4. After washing several times with distilled water until reaching white-yellowish color, the bleached fibers were filtrated and dried in an oven for 24 h at 60 °C. Then the bleached fibers were treated with 80% (w/w) sodium hydroxide (NaOH) solution based on a 1:50 (g/ml) fiber-to-NaOH ratio for 30 min at room temperature. Then the alkali-treated bleached pulp was filtrated, washed and oven-dried for 24 h at 60 °C. The alkali-treated bleached pulp was then hydrolyzed using 2.5 mol/L of hydraulic acid for 30 min at 85 °C based on a 1:30 (g/ml) solid-to-liquor ratio [5, 6] with constant stirring, followed by cooling at ambient temperature. To obtain MCC, the cooled hydrolyzed alkali-treated bleached pulp was washed again until a pH7 was achieved. The obtained MCC was then put in a vacuum oven for 5 h at a temperature of 70–80 °C.

In order to characterize MCC FTIR was conducted, as well as SEM, EDX and partial size analysis, XRD was used to identify the samples crystallinity. The thermal stability of samples was done. The yield of MCC in this work was with a total of 35.4%. The highest crystallinity was 79.4%. Therefore, it can be concluded that in this work a successful attempt has been made to isolate microcrystalline cellulose from date palm fruit bunch stalks: a waste resource in date-producing countries. This may have a strong economic and developmental impact on these countries. The production of MCC was quite easy and practical. Its high crystallinity (79.4%) endowing it with high rigidity opens a great potentiality for its use as a reinforcing material in nanocomposites.

5.5 Extraction of Oxidized Nanocellulose from Date Palm Leaf Sheath Fibers to Obtain a Packaging Additive for Better Packaging Properties

It is widely acknowledged that nanocelluloses enjoy specific strength properties higher than those of commercially available metallic and polymeric products, in addition to their being toxicologically and ecologically harmless.

The date palm leaf sheaths (DPS) are produced during the annual process of pruning of date palms. With the decay of the traditional handicrafts, associated with DPS, the huge amounts of DPS became redundant and treated as a waste. A pioneer research [7] has been conducted to obtain oxidized nanocellulose from date palm leaf sheaths. The ultimate objective of this research was to obtain a packaging additive to improve the packaging properties of chitosan films.

Date palm sheath fibers (DPS) were obtained from a date palm grove in Sohag Province, Egypt. They were manually washed with water and dried at 50 °C for 72 h. Samples of DPS of dimensions 5 × 5 cm^2 were cut and soaked for 3 h in 18% NaOH (w/w) at liquor ratio 1:7 and then autoclaved in 2.5 L high-pressure stainless steel reactor at 140 °C for 1.5 h. The sodium chlorite bleaching process was used in treating DPS pulp: 3.5% sodium chlorite at pH3–4 adjusted by 10% acetic acid and 5% consistency at 70 °C for 1 h for each step. This was followed by washing with distilled water to remove the lignin totally and hemicellulose partially. Then the mercerization treatment was conducted using 17.5% NaOH (w/w) for 3 h at 20 °C. The cellulosic materials resulting from bleaching and mercerization were filtered by cotton bags and washed with water till neutrality. For the preparation of homogenous oxidized nanocellulose (ONCs): the unbleached pulp and mercerized fibers were milled and 1 g from each was added to 100 ml of concentration from (0.5 to 1.5 M) ammonium per sulfate solution. The mixture was heated at 50–80 °C for 12–20 h for unbleached pulp and for 8-14 h for mercerized fibers under a continuous mechanical stirring to produce a suspension of ONC1 and ONC2. The suspension was centrifuged at 12,000 rpm for 10 min. and then washed with distilled water. This was repeated until the solution conductivity was ~5 μs cm^{-1} (pH4). Sodium form of ONCs was prepared by adding 1 M NaOH until the suspension reached pH7. Then washing/centrifugation was conducted. The chitosan solution (CS) was prepared by mechanical stirring of 2% chitosan (wt/wt) in 1% acetic acid solution (v/v) at 50 °C for 3 h. Films were prepared by mixing chitosan solution with 7% (wt/wt_{ch}) ONC1 and ONC2 fractions. The mixtures were homogenized at 13,500 rpm for 5 min, followed by adding 30% glycerol (wt/wt$_{ch}$). The homogenous mixtures were cast into PTFE plates.

The chemical compositions of different treated DPS fibers were determined. Zeta-potential and average size analyses of the obtained ONCs were carried out. FTIR spectra of the different fiber samples and ONCs were recorded. The XKD, TGA and morphological analyses have been performed. The mechanical and water vapor permeability tests for bionanocomposite films have been conducted.

It can be concluded from the research results that it is possible to obtain from date palm sheath fibers homogenous oxidized nanocellulose suspensions from unbleached pulp and mercerized cellulose using ammonium sulfate treatment. The optimum conditions to prepare ONCs were found as follows: 1.25 M for 16 h and 1 M for 10 h, respectively, at 60 °C and 1:100 liquor ratio for both products. The research results show that the cellulose polymorphism has a positive influence on the mechanical, water vapor repellence and thermal properties of bionanocomposite. It was also proven that chitosan/ONC1 bionanocomposite enjoys higher mechanical and water vapor repellence properties than those loaded with ONC1, whereas the latter exhibits thermal stability higher than chitosan/ ONC1 bionanocomposite.

5.6 Cellulose Whiskers from Date Palm Rachis and Leaflets for the Reinforcement of Nanocomposites

The present research [8] has been devoted to the extraction of cellulose whiskers from the rachis and leaflets of date palms, their characterization and their use for the reinforcement of natural rubber. The samples of rachis and leaflets were first turned into powder, treated with toluene–ethanol (62:38) for 24 h and then subjected to alkaline treatment NaOH 2%, 80 °C for 2 h twice and then bleached by $NaClO_2$, pH 4.8, 70 °C for 1 h twice to obtain bleached pulp. The colloidal suspensions of whiskers in water were obtained by treatment with H_2SO_4 65% for 45 min. Only cellulose whiskers extracted from the rachis were used due to their higher geometrical characteristics. The nanocomposites films were prepared from rubber, reinforced by cellulose whiskers with a content 0 to 15 wt%. Stirring of the mixture was conducted using a magnetic stirrer for 8 h. Then water evaporation was conducted to determine the crystallinity index. The transmission electron micrographs of cellulose whiskers were taken, and the atomic force microscopy observation was made and the differential scanning calorimetry was performed, as well as the dynamic mechanical analysis was carried out.

The research results show that the ratio of the rachis and leaflets in a date palm leaf is 53.4 wt% and 46.6 wt%, respectively. The protein content is 2 wt% and 6 wt% for the leaflets and rachis, respectively, suggesting the superiority of the rachis in feed applications. The cellulose content was found 44 wt% for rachis and 33.5 wt% for the leaflets, whereas the lignin content was found 27 wt% for leaflets and 14%

for the rachis, respectively. The crystallinity index was found 55% for the rachis and 50% for the leaflets, respectively.

Concerning the cellulose whiskers extracted from the rachis and leaflets, they were found 260, 6.1 nm for the rachis and 180, 6.1 nm, respectively. This proves that the rachis whiskers aspect ratio (43) is superior to that of the leaflets (30). This suggests that the extraction of cellulose whiskers from the rachis is more profitable than from leaflets (higher aspect ratio) and with a yield 15% of the dried ratios. Therefore, only the cellulose whiskers extracted from the rachis were used for the reinforcement of nanocomposites. The nanocomposites samples with low whisker content showed a ductile behavior, whereas those with high content showed a quasi-brittle behavior. The elongation at break decreased with increasing the nanocrystal content. With the increase of the cellulose nanocrystal content, the tensile strength and modulus increased drastically (e.g. the tensile modulus increased from 0.5 MPa for pure rubber samples to 187 MPa for samples with 15 wt% nanoparticles). The research results in open great potentialities for the valorization of such a neglected resource as the date palm leaves.

5.7 Enzyme-Assisted Isolation of Microfibrillated Cellulose from Date Palm Fruit Stalks

The date palm (Phoenix dactylifera) is an essential element of flora in the Arab world possessing ~111 million date palms [9]. The annual pruning of date palms produces huge amounts of byproducts including leaves and fruit stalks. Except for a small amount that is being used in manufacture of street brooms, these fruit stalks are mostly treated as waste. A genuine research work [10] has been conducted with the objective of the extraction and characterization of microfibrillated cellulose (MFC) from date palm fruit stalks using ultrafine grinding. In order to facilitate the extraction of MFC and improve the strength properties of the isolated MFC, the enzymatic treatment has been used.

The date palm fruit stalks were obtained from a palm grove in Giza, Egypt. After cleaning with water, the stalks were cut into pieces of the length 2–3 cm and alkali-treated using 15% NaOH (w/w based on oven-dried stalks) at 150 °C for 3 h. The chemical composition of the bleached pulp using sodium chlorite/acetic acid mixture at 80 °C for 1 h was: α-cellulose 71.5%, pentosans 18.4%, degree of polymerization 1264 and ash 0.64%. For the preparation of citric buffer (pH 5.3), sodium citric and citric acid were used. The bleached pulp was treated with xylanase enzymes at a concentration of 0.01, 0.02 and 0.04 g/g of pulp. The mixture was maintained under gentle stirring at 50 °C for 4 h. Then the temperature was raised to 90 °C to deactivate the enzymes, and the pulp was filtered and washed with distilled water. The pulp suspension (2% consistency) was disintegrated using a high-shear mixer and then treated in a high-shear ultrafine friction grinder by passing through the device up to 60 times. Then MFC was centrifuged at 10,000 rpm to reduce the water content and kept wet in the fridge.

For the characterization of MFC, the atomic force microscopy was conducted. The diffraction patterns were obtained and the crystallinity index was calculated. The surface charge was determined, and morfi analysis was used to measure the morphological properties of the pulp and MFC suspensions. The MFC sheets were prepared by vacuum filtration of 0.5% by weight suspension of MFC. The wet and dry tensile strength testing has been conducted on MFC sheets samples of 10 cm length and 1 cm width. The air permeability test was also conducted.

The results of this research emphasize the success of extraction of microfibrillated cellulose from date palm fruit fibers. The MFC sheets extracted from enzymatically treated fibers enjoyed lower porosity, more compact and tight structure and higher wet and dry tensile strength than those produced from non-treated fibers. The application of the results of this research may have significant economic and developmental impact in date-producing countries since they offer the potentiality of realizing a high value-added product from such a neglected resource as the date palm fruit stalks.

References

1. Boufi S, Kaddami H, Dufresne A (2014) Mechanical performance and transparency of nanocellulose reinforced polymer nanocomposites: mechanical performance and transparency of nanocellulose .... Macromol Mater Eng 299(5):560–568. https://doi.org/10.1002/mame.201300232
2. Hassan ML, Bras J, Mauret E, Fadel SM, Hassan EA, El-Wakil NA (2015) Palm rachis microfibrillated cellulose and oxidized-microfibrillated cellulose for improving paper sheets properties of unbeaten softwood and bagasse pulps. Ind Crops Prod 64:9–15. https://doi.org/10.1016/j.indcrop.2014.11.004
3. El-Mously H, Saber M (2019) Medium density fiberboards from date palm residues a strategic industry in the Arab world. By-products of palm trees and their applications
4. Alotabi MD, Alshammari BA, Saba N, Alothman OY, Kian LK, Khan A, Jawaid M (2020) Microcrystalline cellulose from fruit bunch stalk of date palm: isolation and characterization. J Polym Environ 28(6):1766–1775. https://doi.org/10.1007/s10924-020-01725-8
5. Chuayjuljit S, Su-uthai S, Charuchinda S (2010) Poly(vinyl chloride) film filled with microcrystalline cellulose prepared from cotton fabric waste: properties and biodegradability study. Waste Manage Res: J Sustain Circ Econ 28(2):109–117. https://doi.org/10.1177/0734242X09339324
6. Chauhan YP, Sapkal RS, Sapkal VS, Zamre GS (2009) Microcrystalline cellulose from cotton rags (waste from garment and hosiery industries). Int J Chem Sci:681–688
7. Adel A, El-Shafei A, Ibrahim A, Al-Shemy M (2018) Extraction of oxidized nanocellulose from date palm (*Phoenix Dactylifera* L.) sheath fibers: influence of CI and CII polymorphs on the properties of chitosan/bionanocomposite films. Ind Crops Prod 124:155–165. https://doi.org/10.1016/j.indcrop.2018.07.073

8. Bendahou A, Habibi Y, Kaddami H, Dufresne A (2009) Physico-chemical characterization of palm from <I>Phoenix Dactylifera</I>–L, preparation of cellulose whiskers and natural rubber-based nanocomposites. J Biobased Mater Bioenergy 3(1):81–90. https://doi.org/10.1166/jbmb.2009.1011
9. Ayana RA (2017) Palm by-products. Al Rajhi Endowment Administration (in Arabic)
10. Hassan ML, Bras J, Hassan EA, Silard C, Mauret E (2014) Enzyme-assisted isolation of microfibrillated cellulose from date palm fruit stalks. Ind Crops Prod 55:102–108. https://doi.org/10.1016/j.indcrop.2014.01.055

Chapter 6
Date Palm Byproducts as Timber and Wood Substitutes

Abstract The rationale behind the research topic: lumber-like products from date palm midribs, is to test the potentiality of reliance on such a locally sustainable resource as date palm midribs to manufacture lumber-like products as a measure to decrease the country's wood imports, which are expected to reach ~22.8 billion US$ in 2050! The research results proved the potentiality of manufacture of palm midrib blocks enjoying mechanical properties (e.g. modulus of rupture, modulus of elasticity, maximum compressive strength, nail pull-through test results and hardness) comparable with those for spruce and beech woods.

The rationale behind the research topic: organic products from date palm midribs and leaflets is to test the technical and developmental feasibility of reliance on the traditional artisans and the tools they possess in the manufacture of new designs of products and thus opening new markets locally, nationally and worldwide. Besides, this suggested approach means beginning the first cycle of production from the top of cascade of utilization of palm midribs and leaflets thus giving wide chances for subsequent life cycles. It is expected that the green products markets and ecotourism will be most interested in organic products, made from date palm midribs and leaflets. The following organic products from palm midribs and leaflets have been designed and manufactured:

- Room divider
- Armchair
- Library
- Chair
- Partition
- Basket
- Photo frame
- Window unit
- Sweeper stick

The rationale behind the research topic: Mashrabiah products from date palm midribs, is that Mashrabiah (Arabesque) is an important feature of Islamic house architecture in Egypt and the whole Arab world. Mashrabiah (Arabesque) in houses helps in preserving the privacy of house dwellers and in ameliorating the harshness of sun rays

H. EL-Mously et al., *Date Palm Byproducts: A Springboard for Circular Bio Economy*, Materials Horizons: From Nature to Nanomaterials,
https://doi.org/10.1007/978-981-99-0475-4_6

especially in summer, while introducing light and allowing the house residents to have a look from windows on the outside world without being exposed to the public! The drastic increase in prices of imported beach wood has led to the shrinkage of demand on Mashrabiah (Arabesque) products. Thus, the discovery of the date palm midrib enjoying physical and mechanical properties comparable with those for imported woods (spruce, pine and beech) opened before us a new realm of development: going to remote villages where the date palm plantations are extensive, teaching the poor there (especially women) the crafts of turning palm midrib pieces on lathes and thus reviving the traditional skills of Mashrabiah (Arabesque) manufacture. Thus, a village (Gededia village) in the New valley governorate in Egypt was chosen as a site for the project. A new multi-purpose lathe was specially designed and manufactured to suit work in houses of beneficiaries. A training center has been established to secure palm midrib pieces for the beneficiaries. Thus, Mashrabiah (Arabesque) products were manufactured by beneficiaries with high quality.

How modern pieces of furniture were manufactured from date palm midribs? Four designers joined our team of rediscovery of the date palm byproducts. They put their bet on the uniqueness of the date palm midrib and its specific beauty features as compared with imported wood species. Their efforts have led to the design and manufacture (in village conditions) of new pieces of furniture to satisfy contemporary needs of high and high-middle class citizens in Egypt. Thus, a new market for products from date palm midribs has been opened to satisfy local contemporary needs in Egypt.

Within the framework of the research on flooring and parquet products from date palm midribs, different chemical treatments have been applied on samples of date palm midribs with the objective of improvement of dimensional stability properties, static bending properties and abrasion resistance to be used in flooring and parquet products.

A research has been conducted on eco-friendly laminated strand lumber from date palm midribs. The research results have proven that the laminated strand lumber manufactured from date palm midribs enjoys similar or superior strength properties compared to solid lumber and engineered products from wood or other lignocellulosic materials for building sector.

The rationale behind the research on blockboards with core layer from date palm midribs was that we found that the local blockboard industry in Egypt is not economically competitive, because it totally relies on imported wood, the price of which is continuously increasing. Meanwhile, our research endeavors have shown that the date palm midribs being a sustainable locally available lignocellulosic resource enjoys physical and mechanical properties comparable with those for imported wood (e.g. spruce and beech). Here came the spark of rediscovery: could we replace the inner wooden core of the blockboard with palm midribs and thus save ~80% of the wood being imported for the manufacture of blockboards in Egypt? The research results have proven that the date palm midrib core blockboard enjoys excellent quality as compared with the wooden-core blockboard according to DIN standards and can be thus used as a substitute for spruce-core blockboards in furniture, wall and ceiling paneling, containers etc.

A research has been conducted on the use of date palm petioles as a sandwich core. The petioles constitute the bases of the date palm leaves and are most dominantly treated as waste. But the petiole is very anisotropic with the best mechanical properties in the longitudinal direction, having a limited density and thus a high fatigue life. Proceeding from the above research results and the almost zero-price of the date palm petioles, they are a good candidate for the development of core for sandwich panels.

Given the decline in plywood manufacturing worldwide due to limited large log supply, the oriented strand board—being typically manufactured from fast-growing small tree species—is one of the world's most commonly used engineered wood-base panel products in residential construction. In a pioneer study, date palm midribs from barhi, saqui, khalas and sukkari cultivars were used in making untreated strands to manufacture oriented strand boards with acceptable physical and mechanical properties.

Within the framework of a research conducted on the use of date palm products of pruning in the manufacture of MDF, samples of products of pruning were collected with quantities proportional to those obtained during the traditional pruning activities in Bahariah oases in Egypt. These samples were sent to the laboratories of Naga Hammady Company of Fiber Boards (MDF) in Egypt. The results of tests of physical and chemical properties, conducted according to EN322, EN317, EN120 and mechanical properties, conducted according to EN310; EN319; EN311 prove that it is possible to manufacture MDF boards from date palm products of pruning satisfying the international standards with respect to their physical, chemical and mechanical properties. This opens the potentiality—worldwide—to establish MDF industrial projects in locations having extensive date palm plantations.

Within the framework of research on the potentiality of production of particleboards from date palm midribs, two semi-industrial experiments have been conducted. The first industrial experimental was conducted in October 1993 on 1.15 tons of air-dried date palm midribs in the Nasr Company for particleboards in Egypt according to the Egyptian standard 906/1991 for particleboards. The results of tests of the modulus of rupture were 30.3 N/mm^2 satisfying the requirements of the above-mentioned standard. The second industrial experiment was conducted in August 1994 on 60 tons of air-dried date palm midribs in the Modern Arabian Company for Industry of Wood. The experiment on 100% date palm midrib boards gave the following results:

- Density: 0.844 gm/cm^3
- Modulus of rupture: 21.9 N/mm^2
- Face strength: 1.07 N/mm^2
- Internal bond: 0.9 N/mm^2

A research has been conducted on the potentiality of production of particleboards from date palm midribs. Date palm leaves were collected from Barhi, Saqie and Sukkari cultivars in Riyadh, Saudi Arabia. To manufacture particleboard panels, dried particles were blended with urea–formaldehyde resin 10% (oven dry particle weight), 1% liquid paroffin wax, as well as ammonium chloride as a hardener (2% based on the hard resin weight). From the results of tests, it can be concluded that the date palm midribs can be used for the manufacture of particleboards. Another research has been conducted on the potentiality of use of fast-growing tree species (e.g. A. Saligna, C. erectus L and M. azedarach L.) and date palm midribs for the manufacture of particleboards. As a conclusion of this research, all the tested species can be used for the manufacture of particleboards of density 750 kg/m^3. A research has been conducted to evaluate the potentiality of use of the date palm midribs for the production of particleboards. Date palm midribs were collected from local areas in Quena governorate, south of Egypt, cut into small strips and then converted into small chips 25 × 25 × 5 mm. Three-layer particleboards 400 × 400 × 12 mm with density 680 kg/m^3 were produced at a pressure 35 kg/cm^2. The results of tests proved that the date palm midrib panels have satisfied the requirements of load-bearing boards for use in dry conditions type (p4) of the European standard (EN 314.2010). This opens a wide potentiality of use of such an available renewable resource in manufacture of a value-added product as particleboards. Another study has been conducted to test the potentiality of use of date palm midribs considered as waste and vemiculite (as an inorganic filler) in the manufacture of particleboards. Midribs were obtained by defoliation of date palm leaves obtained from Kerman region in Iran, cut to suitable lengths (30–40 cm) and air-dried. A hammer mill was then used to cut them into pieces of 15 mm length, 2 mm width and 0.4 mm thickness. A sample of vermiculate commercially available in Iran was used in micro and nanosize. Experimental panels were manufactured with resin content 10%, hardener content 2%, press closing rate 6 mm/s, press pressure 35 kg/cm^2, press temperature 175 °C, board thickness 15 mm and target density 0.75 g/cm^3. The experimental values were number of layers (single- and 3-layer), the size of vermiculate particles (micro and nano). The research results have led to the conclusion that date palm midribs are potentially feasible for the manufacture of particleboards for indoor applications, to absorb noise, preserve the temperature of indoor living spaces as a substitute for wooden boards.

An important study has been conducted to evaluate the potentiality of use of date palm trunks and midribs in the manufacture of particleboards as a substitute for imported wood. Date palm trunks and midribs were sourced from a local plantation in south Algeria, reduced to a particle size 1–2 cm, oven-dried at 100 °C to reach a MC of 3%. To manufacture the particleboard specimens, the chips were placed in a drum blender and sprayed with phenol formaldehyde or melamine–urea–formaldehyde for 1 min. The resin content was 10% based on dry particle content. The panels were produced at a density 0.7 g/cm^3. The test panels reached after trimming 320 × 270 × 14 mm^3. The total press time was 7.5 min and temperature 195 °C. Proceeding from the research results, it can be concluded that the manufacture of particleboards from date palm trunks and midribs is technically feasible. It is worth noting that the particleboards produced from the trunk enjoy significantly higher MOR, MOE and 1B values as compared with those for midribs. This opens a great potential for use of a new material resource (old unproductive date palm trunks) for a high value-added product such as particleboards.

Research has also been conducted on self-bonded particleboards from date palm products of pruning. The rationale behind the idea of self-bonding particleboards is the dispensing with the synthetic resins, the presence of which hinders the recycling of the products at the disposal stage. To conduct this study samples of date palm leaflets, midribs, petioles and fibrillum were sourced from Marrakesh province in Morocco, hammer milled, sieved by 3-mm sieve and oven-dried at 105 °C for 12 h.

The self-bonding was successful: all samples were cohesive. The fibrillum boards enjoyed the highest MOR and MOE properties (12.9 N/mm^2 and 1257 N/mm^2, respectively). This can be explained by the high lignin content in the fibrillum (the lignin plays an important role as a binder forming a coherent thermosetting matrix during board manufacture gluing particles togethers). The panels of leaflets and midribs had the second highest MOR values of 8.4 and 8.5 N/mm^2, respectively. This can be explained by the highest amount of hemicelluloses in midribs and highest content of extractives in leaflets, beside its average content of hemicelluloses.

Keywords Date palm byproducts · Timber · Wood substitutes · Lumber · Like products · Organic products · Mashrabiah · Arabesque · Palm midribs · Furniture · Flooring and parquet · Petiole · Sandwich core · Oriented strand board · MDF · Blockboard · Particleboard · Vermiculite · Palm trunk · Self-bonded · Leaflets · Petioles · Fibrillum

6.1 Lumber-Like Products from Date Palm Midribs

Egypt is located in an arid zone with a very poor forest coverage. This situation has made Egypt rely on importation for the satisfaction of the country's need of wood and wood products. Egypt's imports of wood and wood products in 2014 have approached $\approx$ 2 billion US$ [1]. Taking into consideration the expected increase of population, the future estimates of the cost of wood imports and wood products will reach ~22.8 billion US$ in 2050! This represents an unacceptable burden on the future generations. Proceeding from this situation, a research has been conducted [2] with the objective of verifying the potentiality of manufacture of lumber substitutes from the sustainably locally available date palm midribs being a product of pruning of the date palms. The results of this research (Fig. 6.1) show that the date palm midrib blocks enjoy mechanical properties (e.g. modulus of rupture, modulus of elasticity, maximum compressive strength, nail pull-through and hardness test results), comparable with those for spruce and beech woods. This opens a great potentiality for dispensing with importation of wood and wood products and realization of endogenous development in provinces having extensive palm plantations in Egypt.

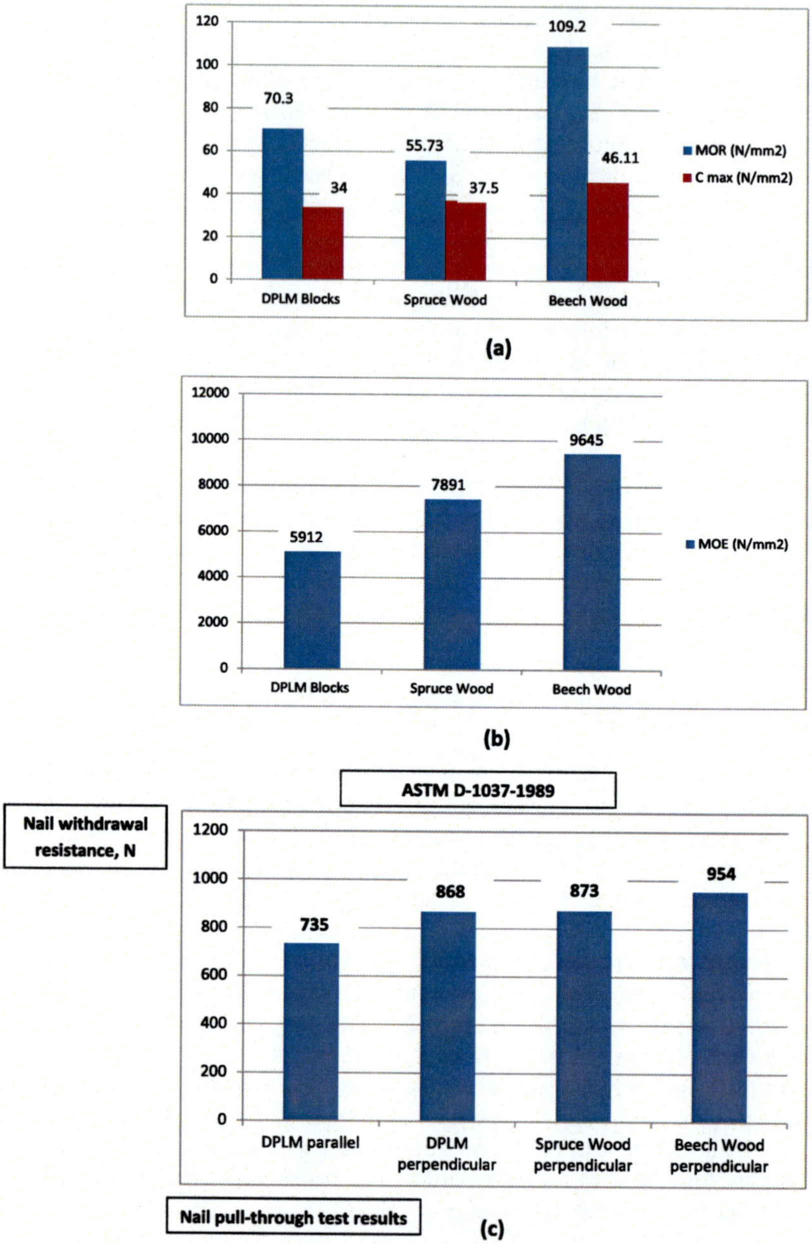

Fig. 6.1 Comparison between the mechanical properties of lumber-like blocks from date palm midribs and species of wood: **a** bending strength MOR and compressive strength Cmax, **b** modulus of elasticity MOE, **c** nail withdrawal resistance N and **d** hardness H [2]

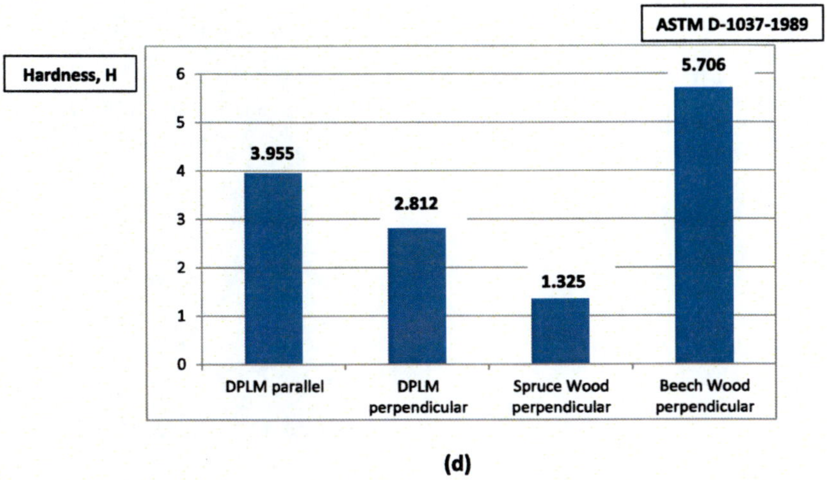

Fig. 6.1 (continued)

6.2 Organic Products from Date Palm Midribs

What are the arguments behind this line of research?

(1) The organic products from date palm midribs mean that you are making use to the maximum of the inherent structural properties of the date palm midribs. The anatomical study of the date palm midrib (Fig. 6.2) points to the increase of the intensity of the fibrovascular bundles and fibers and consequently the strength, as we approach the peripheral zone, i.e. the strength of the date palm midrib in its natural state will be certainly higher than that of the core, found by research to compete with the imported wood.

(2) The manufacture of organic products from date palm midribs means that you are beginning the first life cycle from the top of cascade of utilization (Fig. 6.3). This gives the chance for subsequent life cycles for example in Arabesque, particleboard and fiberboard, etc. This is a good example of application of the full utilization principle [3].

(3) This application will provide the chance to make use of distinguishing physical properties of date palm midribs. For example, the epidermal layer of the date palm midrib is not porous, which makes it water resistant, in addition to its being covered by a layer of natural wax protecting it from dehydration in the conditions of aridity of the Arab region, where palms grow. This means that the organic products from date palm midribs will enjoy a high water resistance without using paints or chemicals, as well as good insulation properties.

(4) From a life cycle perspective [4], the organic products from date palm midribs enjoy considerable environmental advantages, since the net energy requirement for their manufacture is minimal as compared with other fields of manufacture

Fig. 6.2 Diagrammatic illustration of a transverse section of the date palm midrib showing the peripheral (1), transitional (2) and core zone (3) (25X) [10]

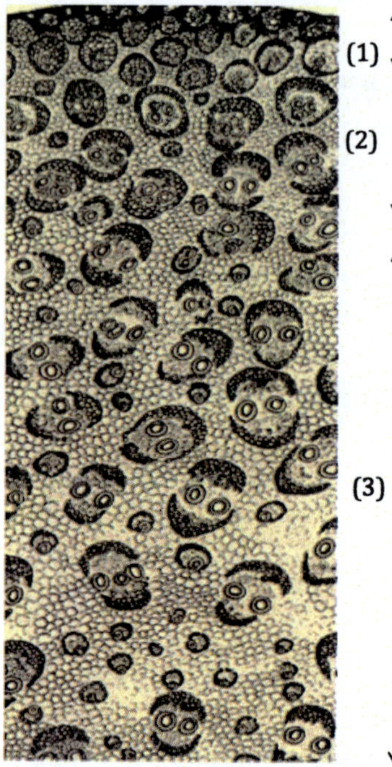

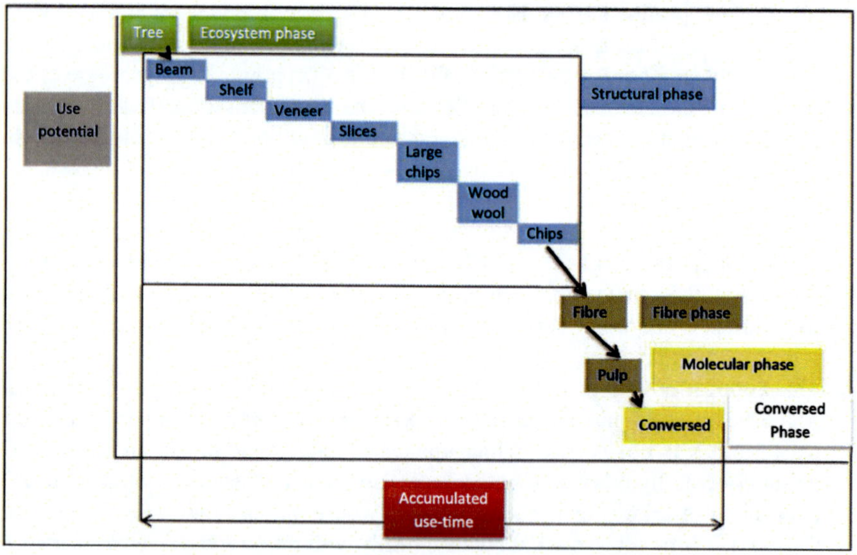

Fig. 6.3 Cascade of utilization of wood from a life cycle perspective [11]

of date palm midribs. In addition, these products are totally biodegradable; i.e. they will not represent any problem during the disposal stage, since they could be used in fodder (for poultry [5], and livestock) [6], or in fuel as a CO_2-neutral material [7].

(5) The organic products from date palm midribs represent an eloquent example of use of the ecological specificity of the Arab region as a strategic comparative advantage. The date palm midrib carries the imprint of the local ecosystem. Therefore, the organic products will not disguise the distinguishing ecological identity of the Arabian region, which guarantees competitiveness in export (compare for example these products with those manufactured from imported wood or plastics!) and in ecotourism projects relying on the use of ecologically specific elements of flora and fauna in designs of the lifestyle and modes of consumption in the ecologes.

(6) It is expected that the organic products from date palm midribs and date palm residues, as well as other ecologically specific elements of flora and fauna, will make use of the chance that the green market offers in the international level. There are strata of environmentally conscious and socially aware consumers, who prefer those organic products, produced in the south in socially just contexts [8]. The success of the trade outlets, such as Fair Trade Net [8], opens the potentiality of manufacture of organic products from date palm midribs and leaflets, as well as products of pruning of fruit trees in general. Besides, the General Preferential System of the EU gives preference for products, produced in the south in an environment-friendly way. In addition, the date palm midrib products are not in need of a certificate of origin, i.e. that they were manufactured from wood, obtained from sustainably managed forest. I may suggest in marketing of these products the slogan: tree-free products [9].

(7) The organic products from date palm midribs and leaflets mean that we could realize a qualitative shift in development of uses of date palm midribs and leaflets simply by the innovation of new designs of products. This means that we could rely on the date palm midrib and leaflets artisans who manufacture crates and baskets in the villages in the production of these organic products using their very procedures and traditional tools via their training on the production of new designs of products from date palm midribs and leaflets. Therefore, the organic products from date palm midribs and leaflets open wide potentiality for the endogenous development of Arabian villages utilizing—and building upon—the locally accumulated traditional knowledge and experience. This in turn gives high guarantee of success of development and makes development more sustainable.

The following organic products have been designed and manufactured from date palm midribs and leaflets:

- Room divider: three pieces (Fig. 6.4).
- Armchair (Fig. 6.5).
- Library.
- Chair.

Fig. 6.4 Room divider: a sample of organic products from date palm midribs and leaflets (Designed by Prof. Dr. Adel Y. M., Ain Shams Univ.)

- Partition.
- Basket.
- Photo frame.
- Window unit.
- Sweeper stick.

Ecotourism is one of the most appropriate areas to make use of the organic products from date palm midribs and leaflets and other date palm byproducts.

Fig. 6.5 Armchair: a sample of organic products from date palm midribs

6.3 Mashrabiah Products from Date Palm Midribs

How did the idea of use of date palm midribs in Mashrabiah (Arabesque) come to our minds? The Mashrabiah (Arabesque) was being intensively used in houses in Egypt and the Arab World. It helps in preserving the privacy of dwellers and in ameliorating the harshness of sun rays, while allowing the house residents at the same time to have a look from windows on the outside world without being exposed to the public!. The drastic increase in prices of imported beech wood has led to the shrinkage of demand on Mashrabiah (Arabesque) products. We were conducting machinability tests on date palm midrib samples on a metal-cutting center lathe. The date palm midrib piece reached a diameter of ~2 mm without breakage. This means that date palm midrib is strong enough to replace wood! Thus, the idea came to us (again the rediscovery of material resources): why would not we go to remote villages possessing palm plantations and teach the populace there how to make Arabesque products that may replace wooden Arabesque products? The idea looked fantastic: reviving the traditional skills of Arabesque manufacture to support the Mashrabiah as a distinguishing component of our traditional way of life, while combating poverty and providing sustainable labor opportunities in remote villages, especially for women who could work on lathes at their houses.

The Gedeida village in El-Dakhla oases was chosen as a site of the project, which was launched by the Center for Development of Small-Scale Industries, Ain Shams Univ. by a grant from GTZ. A new multi-purpose machine has been especially designed for this project, tailored to the needs of beneficiaries to work at home (Fig. 6.6). The machine could perform turning, drilling, saw cutting of date palm midribs, in addition to turning solid pieces of local wood species. A training center has been established to secure date palm midribs for beneficiaries, prepare date palm midrib pieces and train beneficiaries (Fig. 6.6), as well as a permanent exhibition to help in marketing of products. Figure. 6.7 illustrates samples of arabesque products from date palm midribs.

Fig. 6.6 Training of women on manufacture of Arabesque products from date palm midribs

Pens holder from date palm midribs

Handkerchief case cover from date palm midribs.

Fig. 6.7 Mashrabiah (Arabesque) products from date palm midribs

6.4 Furniture Pieces from Date Palm Midribs

Four designers joined our team of rediscovery of the date palm byproducts. They came with their long history of design of products and interaction with consumers from the high and high-middle classes in Egypt. They did not perceive the date palm midrib just as a locally available cheap material. They perceived the date palm midrib as a material with a long history of utilization and a symbol of the year in the old kingdom in Egypt (Fig. 6.8a) [12]. They thus put their bet on the uniqueness of the date palm midrib and its specific beauty as compared with imported wood species. Their efforts have led to the design and manufacture (in village conditions) of new pieces of furniture to satisfy contemporary needs of high and high-middle class citizens in Egypt [12].

Thus, a new market has been opened for the use of the date palm midribs in furniture. Figure 6.8b gives examples of these products.

6.5 Flooring and Parquet Products from Date Palm Midribs

Different treatments [13] were applied to improve physical and mechanical properties of date palm midribs and namely, treatment with chemicals: water-soluble phenol formaldehyde resin (PF_w), alcohol-soluble phenol formaldehyde resin (PF_a), melamine–formaldehyde resin (MF), linseed oil (LD), methyl methacrylate (MMA), polystyrene (PS) and polyester (P_{est}), with different concentrations (concentration of

Thus a new market has been opened for the use of the date palm midribs in furniture. Fig.8b gives examples of these products.

A table from date palm midribs

A console table from date palm midribs

A chair from date palm midribs

Fig. 6.8 **a** Palm midrib is a symbol of eternity, **b** examples for products from date palm midribs treatment material in solution), on weight to weight basis in suitable solvent fluids for the PF_w, the PF_a and the MF: 20%, 30%, 40%, 50%, for the MMA and the LS: 100%, 90%, 80%, 70%, for the P_{est}: 10%, 15%, 20%, 30% and for the PS: 30%, 40%, 50%, 60%) in a vacuum process with suitable preconditioning before impregnation, curing and post curing after impregnation.

The studied properties included dimensional stability properties: water absorption ($WA_{2,24}$), volumetric swelling coefficient ($S_{2,24}$), anti-shrink efficiency ($ASE_{2,24}$) after 2 and 24 h water soak as well as static bending properties: modulus of rupture (MOR) and modulus of elasticity (MOE), shear stress parallel to grain (S_s) and abrasion resistance (A_r).

The results of this study showed significant differences between the behavior of the treated specimens compared with the untreated control specimens. The analysis of the results points to an appropriate value of concentration to achieve the maximum retention levels of impregnation media reaching 25% with MF, LS, PF_w and P_{est}, while attaining only 15% with other polymers.

Most of the treatments showed recognized influence on dimensional stability. For example, the treatment with MF has resulted into decrease of volumetric swelling coefficient from 20.2% to 1.5% and from 71.6% to 9.6% after 2 and 24 h water submersion test, respectively, as compared to the control, and water absorption decreased by weight from 37.2% to 4.9% and from 165.1% to 32.5% after 2 and 24 h water submersion test, respectively, as compared to the control.

The impregnation by PS, P_{est}, LS and PF_a increased the MOR values by up to 35% for both P_{est} and PS, 30% and 23% for alcohol-soluble phenol formaldehyde and linseed oil, respectively. The MOE values increased by up to 45% for PS, 35% for polyester and 33% for alcohol-soluble phenol formaldehyde.

The treatment with LS, MMA, P_{est} and MF increased the abrasion resistance by up to 61% for methyl methacrylate, 43% for linseed oil, 23% for polyester and 17% for melamine–formaldehyde, while no treatment improved the shear strength of the DPLM.

Thus, selective treatments could be chosen for the improvement of physical and mechanical properties of date palm midribs to be used in flooring and parquet products manufacture.

6.6 Eco-Friendly Laminated Strand Lumber from Date Palm Midribs

This study [14] has been conducted with the objective of use of the date palm midribs, as a low value biowaste, in the manufacture of a high value eco-friendly wood substitute.

The Taguchi design of experiments was used to analyze the effect of raw material and product parameters on the mechanical properties of laminated strand lumber, manufactured from the date palm midribs. As a result of use of Taguchi design: only eight experiments have been carried out instead of 96.

The study results indicate that the laminated strand lumber, manufactured from date palm midribs, exhibits similar or superior strength properties compared to solid lumber and engineered products from wood or other lignocellulosic materials for building sector. The maximum values for the mechanical properties of laminated strand lumber from date palm midribs were obtained from a combination of 20 mm product thickness, 10% (w/w) resin content, and 4 mm strand thickness and 850 kg/m^3 product density. Product thickness with 81.3% contribution and strand thickness with an 80% contribution have the highest effects on the flatwise stiffness and compression strength perpendicular to grain, respectively.

6.7 Core Layer of Blockboard from Date Palm Midribs

Why we chose this topic of research? The blockboard is highly needed in Egypt for the manufacture of high-quality furniture, as well as wall and ceiling paneling, interior fittings, etc. The local blockboard industry could satisfy only 15% of the local demand, the rest being imported. We found that this industry is not economically competitive, because it totally relies on imported wood, the price of which is continuously increasing due to the success of environmental movements worldwide putting increasing pressure on the cutting of wood from forests. The research, conducted by the faculty of engineering, Ain Shams Univ. in collaboration with the Academy of Scientific Research and Technology, has proven that the date palm midrib enjoys mechanical properties comparable with those for imported wood (e.g. spruce and beech) (Fig. 6.9). If we take a blockboard of thickness 16 mm, the wood core layer will be ~13 mm. Here comes the spark of rediscovery of the palm midrib: could we replace the inner wooden core of the blockboard with palm midribs and thus save ~80% of the wood being imported?. Thus a master thesis has been registered with the headline: "A Study on the Appropriate Conditions of the Press-cycle for the Manufacture of Date Palm-Midrib Core Blockboards" [15]. The appropriate conditions of the press cycle were determined and found as follows:

Pressure: 0.8 N/mm^2, temperature 120 °C.
Press time: 5 min [15].

Side by side with conducting research the Center for Development of Small-Scale Industries, Ain Shams University launched a project for the establishment of a blockboard production pilot unit in El-Kharga, the New Valley Governorate by a grant from GTZ. The date palm midrib core blockboard specimens were sent to Munich Institute for wood Research in Aachen University. The results of tests indicate that the date palm midrib core blockboard has excellent quality according to DIN standards (DIN 52,375, 52,364, 53,374, 53,255, 52,371 and 68,705) and can

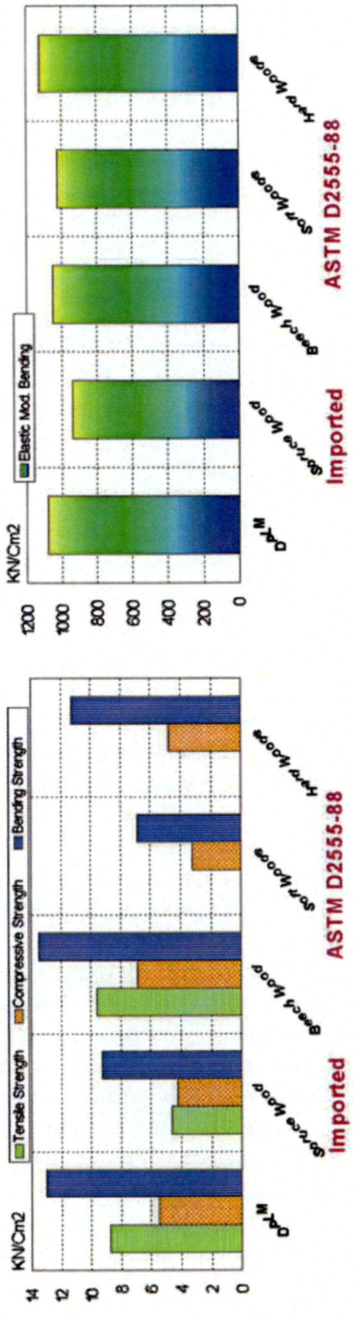

Fig. 6.9 A comparison between the mechanical properties of date palm midribs, spruce and beech wood, as well as soft and hardwoods according to ASTM 2555-88 [16]

be used as a substitute for spruce-core blockboards: in furniture, wall and ceiling paneling, containers, equipment, etc. (Appendix 1).

The production of the blockboard pilot unit has been utilized by the UNICEF for the manufacture of furniture for 100 community schools in Asiut, Souhag and Kena governorates in 1995: The UNICEF was satisfied with the results (Appendix 2).

6.8 Use of Date Palm Petioles as a Sandwich Core

A research work [17] has been devoted to the use of date palm petioles as a sandwich panel core. The petioles constitute the bases of the date palm leaves. Not only in Algeria, but also in all Arab countries, petioles are treated as waste. This represents a great economic loss of such a renewable resource as the date petioles. In this research, morphology and experimental static and fatigue behavior of petioles were investigated. The northern (near to the Mediterranean sea) and southern (semi-arid region) petioles have been tested in X, Y and Z directions.

From a mechanical point of view, the petiole can be considered a unidirectional composite material with a fiber volume ratio of 18% to 32%. The petiole density varies from 150 kgm^{-3} in the north to 210 kgm^{-3} in the south. It is assumed that the cause of variation in density is the semi-arid climate south of Algeria. The reinforcement has been analyzed by X-ray tomography and by SEM analysis. It appears as a complex fibrovascular ultrastructure mainly oriented in the direction of the date palm leaf. Two types of macroscopic fibers have been identified. The fiber bundles, typically 180 μm diameter, constituted by thin-wall elementary fibers and possessing one vascular hole for some of them; and the fibrovascular bundles, typically 560 μm diameter, mainly constituted by honeycomb material and some elementary fibers, possessing two vascular holes. The matrix is constituted by the parenchyma. SEM analysis has shown that the transition from reinforcement to matrix is very gradual. The mechanical properties of the macroscopic fibers—$E_{FB} = 16.9 \pm 4.8$ GPa, $G_{FB} = 287 \pm 101$ MPa, $E_{FB} = 2.55 \pm 0.82\%$—are not very high comparing with other plant fibers, but they clearly play a role in the reinforcement for the petiole. The tensile properties of the tested petioles are $E = 0.465 \pm 0.15$ MPa, $\sigma^U = 6 \pm 2.5$ GPa and $\varepsilon^U = 1.75 \pm 0.42\%$ for the best petiole (southern) and the best orientation (longitudinal); and they are three times lower for transversal direction. The tested southern petiole wood is more rigid and resistant than the northern one and its density is higher.

The complexity of fracture modes of the petioles is consistent with the observed structure for the main components (fiber bundles, fibrovascular bundles and parenchyma). Concerning the parenchyma, the main failure modes are cell wall tearing and spheroidal cell detachment. As far as the macroscopic fibers are concerned, pull-out has been observed (particularly for fiber bundles), but fiber fracture just at the fracture surface has been also observed (mainly for fibrovascular bundles). This reveals an inherent superiority of the mechanical properties of the fiber bundles compared to the fibrovascular bundles.

Big differences of behavior have been noticed between the longitudinal and transversal directions of the petiole in all tests. This is easy to explain, since the natural load of the palm leaf is a bending load. Therefore, the petiole is very anisotropic with the best properties in the longitudinal direction. The date palm petiole has a limited density and thus a high fatigue life. Proceeding from the above research results and the neglected price of the date palm petioles, they are a good candidate for the development of core for sandwich panels.

6.9 Oriented Strand Board from Date Palm Midribs

The oriented strand board (OSB)—a structural composite panel—is one of the world's most commonly used engineered wood-based panel products in residential construction. Given the decline in plywood manufacturing in many countries due to limited large log supply and escalating environmental concerns, it is expected that the OSB production will continue to increase world-wide, especially because OSB is typically manufactured from fast-growing small tree species. The OSB is traditionally manufactured as a multi-layered panel made from strands of wood, with approximate dimensions of 15–25 mm width, 75–150 mm length, and 0.3–0.7 mm thickness, bonded together with an exterior adhesive under pressure and heat. One of the most important arguments of expected future expansion of OSB—as compared to other wood panels—is its reliance on low-quality trees, as well as waste from lumber, in addition to non-wood-based agricultural products (e.g. bamboo and products of pruning of fruit trees).

The date palms (~105 million palms in the world) producing annually ~2 million tons of products of pruning may serve as a material base for the manufacture of OSB, especially in the Arab countries, very poor in forest coverage, and possessing ~62 million date palms [18].

In a pioneer study [19], fronds of four date palm cultivars barhi, saqui, khalas and sukkari were collected from date palm farms in AL-Kharj, located 100 km east of Riydh, Saudi Arabia. The leaflets were mechanically cut from the midribs. Next the midribs were cut into sections of 10–15 cm length and then converted to strands with thickness 0.7 mm and width ranging from 1.25 to 2 cm. Strands: either without treatment or treated by soaking in water at 80 °C temperature for 20 h and then rinsed were oven-dried at a temperature of 90 °C for several days to a moisture content of 3%. Both the washed and unwashed strands were sprayed with phenol formaldehyde adhesive at a rate of 10%, based on the dry weight of the raw material for 15–20 min in a rotary drum-type blender equipped with a pneumatic spray gun. One percent wax and hardener were also added to the panels. Using a forming box 50 cm × 50 cm with spacers, strands were laid parallel in one direction, prepressed for 3 min before being compressed at 170 °C, pressure 4 MPa for 6 min. The mats of strands were thus compressed to a nominal thickness of 13 mm at 2 density levels, 0.65 and 0.75 g/cm^3. All panels were conditioned for 2 weeks in a climate cabinet 65% humidity and 20 °C before conducting the physical and mechanical tests.

Bending characteristics, internal bond strength, thickness swelling, water absorption and linear expansion along and across the grain orientations of the samples were tested. The results of tests show that the internal bond strength values are satisfactory. It was found that the samples, manufactured from water-soaked strands had lower mechanical and physical properties as compared with those made from unwashed strands. Thus, it could be concluded that untreated strands of date palm midribs could be used as a material for the manufacture of orientated strand board. But further research is needed to open the full potential of use of date palm midribs in the manufacture of OSB.

6.10 Use of the Date Palm Products of Pruning in the Manufacture of MDF

Within the framework of the project of Care for the Palms in El-Bahriah Oases,[1] samples of the products of pruning of date palm were collected with ratios, equal to the masses of each of these products with respect to the whole mass of products of pruning.

Item	Mass, kg (air-dried)
Palm midribs	15
Palm leaflets	14.6
Spadix stems	9
Leaf sheaths fibers	1.56
Petioles	14
Total	54.16

The samples were sent to the laboratories of Naga Hammady Company of Fiber Boards. The results of tests (Appendix 3) are as follows:

Physical and chemical properties

- Humidity (5.2%, which falls within the limits of EN 322 standards: 4–11%);
- Water absorption (12.7%, which is less than the corresponding value in EN 317: 15);
- Formalin emission (22.54 mg/100 mg, which is less than the corresponding value in EN 120: 30).

[1] This project has been conducted by the Faculty of Engineering, Ain Shams Univ. in collaboration with the Ministry of Environment during the period from January to October, 2016, the project leader was Prof. Dr. Hamed El Mously.

Mechanical properties

- Modulus of rupture (24.4 N/mm^2, which is higher than the corresponding value in EN 310: 20);
- Modulus of elasticity (2911 N/mm^2, which is higher than the corresponding value in EN 310: 2200);
- Internal bond (0.9 Nmm^2, which is much higher than the corresponding value in EN 319: 0.55);
- Surface strength (1.35 N, which is higher than the corresponding value in EN 311: 1.2).

This means that it is possible to manufacture MDF boards from date palm products of pruning satisfying the international standards with respect to their physical, chemical and mechanical properties. This opens the potentiality—worldwide—to establish industrial projects in locations having extensive date palm plantations.

6.11 Particleboard from Date Palm Midribs

6.11.1 Particleboard from Date Palm Midribs: Semi-Industrial Experiments

In October, 1993 the factory of the Nasr Company for Particleboards[2] and Resins in El Mansourah has been operated using date palm midribs as a base material. An amount of 1.15 ton of date palm midribs was used to produce particleboards of size 2240 × 1220 × 16 mm. Samples of the factory production were tested according to the Egyptian standard 906/1991 for particleboards. The results of tests showed that the average value of the modulus of rupture (MOR) for these specimens was 20.3 N/mm^2, which satisfies the requirements of the above-mentioned standard. In August, 1994 the factory of the Modern Arabian Company for Industry of Wood. (MATIN)[3] was operated by about 60 tons of date palm midribs brought from Siwa oasis to manufacture three-layer particleboards with melamine-impregnated paper veneer of dimensions 4.3 × 1.83 m and thickness 8 mm, whereby 20 tons were used to produce 100% date palm midribs boards and 40 tons using a blend of date palm midribs and casuarina wood (50% each). The experiment gave positive results, where the properties of the 100% date palm midrib boards were as follows:

- Density: 0.844 gm/cm^3.
- Modulus of rupture: 21.9 N/mm^2.
- Face strength: 1.07 N/mm^2.
- Internal bond: 0.9 N/mm^2.

[2] Report of the El Nasr Company for Particleboard's and Resins, dated 4/12/1993.

[3] Report of the MATIN Company on 24/8/1994.

6.11.2 Particleboards from Midribs of Different Date Palm Cultivars

The date palm (Phoenix dactylifera) is an important fruit crop with 120 million trees in the world: 62 million of them are located in the Arab countries [20]. These palms are annually pruned resulting in huge quantities of products of pruning, which are mostly treated as waste: either open-field burnt or sent to landfills. Therefore, new avenues of utilization of these renewable materials should be used [21]. Date palm midribs having rich fiber content have been investigated as a raw material for manufacture of particleboards [19, 22–24]. In addition, the basic physical and mechanical properties of experimental particleboards manufactured from date palm midribs have been evaluated and studied [23].

In Saudi Arabia, the main date palm cultivars are Barhi, Saqie and Sukkari. A study [25] had been devoted to determine the physical and mechanical properties of particleboard samples, made from the midribs of the aforementioned cultivars.

Barhi, Saqie and Sukkari, date palm leaves have been collected from date palm farms in Al-Kharj, located 100 km east of Riyadh, Saudi Arabia. The leaflets were mechanically stripped, and the midribs were cut into pieces with 150 mm length, before being converted into particles using a ring flaker. A shaker-type screen was used to classify the obtained raw material into different particle size classes. Half of the obtained particles were soaked in a hot water container at a fixed temperature 80 °C for 16 h to eliminate the extractive and sugar contents. Then these particles were washed with distilled water at room temperature and dried in a laboratory oven dryer to 3% moisture content along with the untreated fibers.

To manufacture the panels for testing, the dried particles were blended with urea–formaldehyde resin using a pneumatic spray gun. Based on the oven dry particle weight, a 10% UF resin and 1% liquid paraffin wax were added, as well as ammonium chloride as a hardener (2% based on the hard resin weight). The experimental panels have been manufactured from hot water extracted and non-extracted, and fine and coarse particles under two target panel densities of 650 and 750 kg/m^3. The bending properties and internal bond strength, as well as dimensional stability in the form of thickness swelling, water absorption and liner expansion was tested. Proceeding from the findings of this study, the panels manufactured from high density level and Saqie cultivar, as well as fine particles, had better performance regarding their mechanical properties. It can be thus concluded that the date palm midribs can be used for the manufacture of particleboards.

6.11.3 Suitability of Some Fast-Growing Trees and Date Palm Midribs for Particleboard Production

The demand for composite wood products is increasing substantially throughout the world, probably due to the increasing gap between the demand on wood and wood supply [26]. The particleboard forms about ~57% of the total production of wood-based panels and is growing constantly at a rate of 2–5% annually [27]. According to the Food and Agriculture Organization of the United Nations (2010), the world consumption of particleboards has increased from 56. 2 × $10^6 m^3$ in 1998 to 104 × $10^6 m^3$ in 2006 (almost doubled). Deforestation, forest degradation and continuous increase of demand for wood-based panels have led to a growing shortage of materials in this sector. One of the most effective responses to this challenge is the establishment of fast-growing tree plantations and the use of agricultural lignocellulosic residues as a substitute of traditional wood materials.

In one of the outstanding researches [28], fast-growing tree species: A. saligna (rendering 65–80 tons/ha yield after 3 years), C. erectus L. (button mangrove) (rendering a yield of 67 tons/ha at the age of 3 years), M. azedarach L. (known as Chinaberry), as well as date palm midribs were used as prospective materials for the manufacture of particleboards.

The wood materials have been sourced from King Saud university experimental station, cut into 1-m logs, air-dried, converted to thin 0.5–1.0 cm disks and then converted to chips and sieved with openings 2.4, 0.85, 0.43 and 0.25 mm. The particles were then oven-dried at 90 °C for 40–48 h to reach 3% MC. Particles were then blended with urea–formaldehyde resin at 10% using UF 50% solid content, 1% wax, 1% ammonium chloride as a hardener (w/w of resin). Samples 30 × 30 cm were pressed at a temperature of 160 °C (palm boards at 120 °C) and a pressure 30 kg/cm^2 using a press cycle 8 min including 2 min closing time. Two target densities: HD and LD 750 and 650 kg/m^3 were used; 13 mm thickness was targeted.

The modulus of rupture (MOR) mean values for the manfuctured boards ranged from 13.34 to 6.7 MPa for date palm midrib boards pressed at high density (HD) and C. erectus boards pressed at low density (LD), respectively. Modulus of elasticity (MOE) mean values ranged from 2674 to 1149 MPa for M. azedarach boards at HD and A. saligno boards at LD, respectively. Referring to the American National Standards Institute, all boards pressed at HD passed the minimum requirements for both MOR and MOE except for A. saligna boards. The LD boards did not pass the standard, expect for M. azedarach boards. As far as the internal bond is concerned, the mean values of all boards were higher than the requirements, made by the standards. However, the boards did not satisfy the linear expansion requirements for general uses. As far as the thickness swelling is concerned, only the date palm boards compressed at LD passed the English Standard requirements for both 2-h and 24-h tests. As a conclusion of this research all the tested species can be used for the manufacture of particleboards of density 750 kg/cm^3. The dimensional stability properties could be improved by specific treatments, such as coating the board surface with malemine-inpregnated paper or laminates.

6.11.4 Properties of Particleboard Based on Date Palm Midribs as a Renewable Egyptian Lignocellulosic Material

A research has been conducted [29] to evaluate the potentiality of use of the date palm midribs for the production of particleboards. The cell walls of any plant are composed of cellulose, hemicellulose, lignin and extractives. The cellulose is a high molecular weight linear homopolymer of repeated units of cellobiose (two anhydrous glucose rings joined via a β1,4-glycosidic linkage [30]). Hemicellulose is a linear and branched heterogeneous polymer typically made up of different sugars. Lignins are amorphous, highly complex, mainly aromatic polymers of encrusting substance. Lignins can be classified in several ways, but they are usually divided according to their structural elements [31]. Extractives can be defined as chemical compounds that are extractable with various neutral solvents [32].

Date palm midribs were collected from local areas in Quena governorate, south of Egypt and cleaned from leaflets and spines, cut into small strips and then converted into small chips 25 × 25 × 5 mm using an electric fiber mill. For comparison bagasse samples were collected from Nag-Hammady Fiberboard Company in Dishna, Quena governorate. For particle classification a screen shaker with 6 layers were used. The particles that remained on 2 and 1 mm sieves were used in the core layer, while particles on 0.5 and 0.25 mm sieves were used in the surface layer. Then the particles were dried to 3% MC. Three-layer particleboards 400 × 400 × 12 mm with density 680 kg/m^3 were produced at a pressure 35 kg/cm^2.

The values of date palm midribs fiber lengths, diameters and cell wall thickness were found in the range of the hardwoods values. As far as the date palm manufactured board samples are concerned the physical properties (density and thickness swelling) and mechanical properties: bending strength (MOR), modulus of elasticity (MOE) and internal bond (IB) were measured. The results of tests prove that the date palm midrib panels have satisfied the requirements of load-bearing boards for use in dry conditions type (P4) of the European standard (EN 314:2010). This opens a wide potentiality of use of such an available renewable resource as date palm midribs in the manufacture of a value-added product as particleboards.

6.11.5 Physical Properties of Particleboards Panels, Manufactured from Date Palm Midribs

A research [33] has been devoted to the study of the physical properties of chipboard panels, manufactured from date palm midribs. It is generally agreed that chipboard with a density lower than that of the employed wood furnish is unacceptably inferior [34] and that an improvement in binder efficiency is realized when wood of lower specific gravities are used to manufacture the particleboards [35].

The density samples were taken from the bending specimens at the time of testing. The test pieces for density determination were 50 × 50 mm square having the thickness of the test boards. The thickness measurements were taken at four different positions near to the edges of the test pieces. Every piece was weighed to an accuracy of ± 0.1 g. The thickness measurements and length measurements were carried out to an accuracy of 0.05 mm and 0.1 mm, respectively. The volume of the test piece has been calculated to the nearest 0.1 cm^3. The mass per unit volume was computed to the nearest 0.01 g/cm^3.

For the conduction of water absorption, thickness swelling and equilibrium moisture content tests, the test samples were 25 ± 0.1 × 25 ± 0.1 mm size. The test specimens were maintained at 65 ± 11% relative humidity and 20 ± 3 °C temperature for 3–7 days. After conditioning, the specimens were weighed to an accuracy of not less than ± 0.2%. Length and thickness were measured to an accuracy of not less than ± 0.3%. The thickness was measured to an accuracy of ± 0.3% at four points midway, along each side (ASTM D1037-78) or at the point of intersection of the diagonals of the square face (TS 180 UDK 675.815). The specimens were submerged horizontally under 25 mm depth of distilled water, maintained at 20 ± 1 °C. After 2 h submersion, the specimens were suspended to drain for 10 min, the excess surface water removed and the specimens weighted and thickness determined. An XL30 ESEM model environmental scanning electron microscope was used to study the microstructural features of the date palm midrib before and after compaction.

The scanning electron microscopy revealed that the date palm midrib is composed of numerous vascular bundles. These vascular bundles are embedded in a matrix of ground tissue consisting of parenchymatous cells. In the middle of the vascular bundle, there is a tubular hole about 80 μm diameter extending along the full length of the date palm midrib and transferring water and soluble salts from the root of the tree to the leaves.

As far as density is concerned, the increase of temperature increases the panel density. Chips containing moisture undergo plastic deformation: the interparticle porosity and surface irregularities get eliminated leading to densification. This effect is more pronounced in cases of smaller-size chips. The smaller the size of chips the higher is the packing factor. The densification effect is also associated with the resin, which has a higher density than the lignocellulosic (e.g. date palm midribs) materials. The higher the percentage of the UF resin, the higher is the densification effect. Proceeding from the fact that the optimum pressure is that which consolidates the mat to the required thickness without causing damage to the internal structure in the form of cell wall disruption, or fiber rupture within each particle or particle fragmentation, the pressure effect on the board density from medium-size and course-size particles is very small.

Concerning the thickness swelling, the increase of press temperature leads to the increase of swelling, especially in the 180–200 °C range. The decrease of the particle size leads to more thickness swelling. This can be explained by the increase of particles surface area per unit volume coming in contact with water. But the increase of the compaction pressure leads to a significant reduction of thickness swelling. This can be explained by the fact that higher pressures lead to the reduction of macroscopic porosity, which provides primarily the ports for water entry into the interior across the section of the specimens. The influence of varying the press cycle time on the thickness swelling is controvertial. In the first phase (polymerization) the increase of the press cycle time leads to decrease of the thickness swelling especially for fine particles. This can be explained by the fact that extended heating leads to efficient formulation of polymer network reducing porosity. In the second phase (polymer degradation) the increase of the press cycle time leads to the increase of thickness swelling. This can be explained by polymer degradation causing the UF polymer film to rupture and disintegrate as a result of residual stress build up and thermo-oxidative deterioration in the polymer film.

Concerning water absorption, the press temperature plays a major role. At 180 °C the lowest water is attained due to full polymerization at this temperature minimizing the water ingress into the material. Beyond this temperature polymer degradation leads to the increase of water absorption through damaged sites in the polymer film. Concerning the effect of pressure, the increase of pressure leads initially to compacting leading to the reduction of water absorption. This is explained by the fact that effective compaction reduced the interparticle porosity resulting in decreased water entry into the specimens. But the further increase of pressure beyond a certain value promotes absorption due to increased permanent damage of individual particles facilitating water absorption. The press cycle influences the water absorption via compaction and UF polymerization. Increasing the pressing time firstly decreases absorption due to reduced porosity resulting from efficient compaction. But increasing further the press cycle time leads to the increase of water absorption due to marked UF degradation resulting into the generation of cracks and crazes in the UF film readily admitting water into the composite board.

As a conclusion, the present research shows that monolayer chip board panels of excellent physical properties can be manufactured from date palm midribs.

6.11.6 Mechanical and Acoustical Properties of Particleboards Made with Date Palm Midribs and Vermiculite

In one of the studies [36] date palm midribs and vermiculite have been used in the manufacture of particleboards. The particleboards are the most popular lignocellulosic composite panels, because they basically rely on relatively low-quality wooden resources, or other available lignocellulosic resources, such as bamboo or agricultural residues [37]. In order to manufacture particleboards,, particles of these materials are subjected to heat and pressure with addition of an organic binder [5, 38] Agroresidues in different parts of the world: sunflower stalks [39], wheat cereal straws [40], rice straw [41], bagasse [42], rapeseed straw [43], oil palm byproducts [44] and cotton carpel [45] have been successfully used in the manufacture of particleboards. There are at least 30 plants around the world using these agrowaste in the production of particleboards.

The main objective of this study is to test the potentiality of use of date palm midribs (considered as an agro-waste) and vermiculite (VER) (as inorganic filler) in the manufacture of particleboards. Both the date palm midribs and VER are abundantly available in Iran. VER has layer cation exchange capacity when compared to morillonite [46].

Midribs were obtained by defoliation of date palm leaves, obtained from Kerman region in Iran, cut to suitable lengths (30–40 cm) and air-dried. A hammer mill was then used to cut them into pieces of 15 mm length, 2 mm width and 0.4 mm thickness. These particles were then oven-dried at 103 °C to reach 4% MC. Urea–formaldehyde adhesive with a solid content 63%, density 1.28 g cm^{-3}, viscosity of 45 cp, gelation time of 67 s, pH of 7.5 was applied. Ammonium chloride (NH_4CL) solution (solid content: 20%) was used as a hardener. A sample of vermiculite (VER) commercially available was used in micro and nanosize. This sample was ground in a planetary mill at 1000 rpm using balls from SiN_4 of diameter 10 mm for 4 h. Experimental panels were manufactured with resin content (10%), hardener content (2%) press closing rate (6 mm/s), press pressure (35 kg/cm^2), press temperature (175 °C), board thickness (15 mm) and target density (0.75 g/cm^3). The experimental values were: number of layers (single-and 3-layer), the size of VER particles (micro and nano) and VER content (0, 10 and 20% by weight). The normal dimensions of the boards were 420 × 420 mm^2. Stop bars were used to achieve constant thickness and boards were trimmed to a final size of 400 × 400 mm^2.

The analysis of the research results shows that none of the tested boards could meet the minimum MOR requirements of the EN standards of particleboards for general use. The increase in VER content has significantly reduced the MOR and MOE of the tested boards. This is in agreement with results, reported by Li et al. [38] using VER in composite. This can be explained by the weak interfacial interaction between VER and urea–formaldehyde resin. It is interesting to note that MOE of boards containing micro size of VER particles was 84% and 50% higher than the corresponding boards (with 10% and 20%) containing VER with nanosize. Similarly, the addition of VER has led to the decrease of the internal bond (IB). This can be explained by the reduced bonding ability. Nevertheless, all the boards containing micro size of VER particles and made with 3 layers exceeded EN standard. Concerning the water absorption, the research results show that it increases with the VER content, both for micro and nanoscale particles. This can be explained by poor adhesion between the matrix and lignocellulosic material due to the presence of more gaps in the interfacial region. Many studies have supported this observation [17, 38]. Concerning the sound absorption coefficient (SAC), it was found that SAC of boards filled with VER were higher in the middle and high frequency range than the control specimens. Nano-VER has a higher sound absorption coeffient than micro VER with the same thickness and board layers. Thus, it can be concluded that the date palm midribs are potentially feasible for the manufacture of particleboards for indoor applications, to absorb noise, preserve the temperature of indoor living spaces as a substitute for wooden boards.

6.11.7 Particleboards from Date Palm Trunks and Midribs

An important research [47] has been conducted with the objective of testing the suitability of use of the date palm trunks and rachises for the manufacture of particleboards. The date palm is an important plant in the arid and semi-arid regions in the Arab countries. Beside its role in realizing food security the date palms help in creating a microclimate ameliorating the damage from sand storms and wind erosion, and thus facilitating the growth of fruit trees beneath them and the growing of field crops [48].

In Algeria—one of the main date palm growers in the Arab region—the date palm grove has witnessed a significant expansion of 69% rising from 101,000 ha in 2000 to 169,361 in 2009 with a total of 18.7 million palms. The activity of pruning of the date palms in Algeria produces 25 million tons of biomass annually. This treasure of renewable materials is unfortunately treated as waste: either open-field burnt or sent to landfills. This biomass includes date palm trunks (30% of Algerian palms are above production age). Over 150 million tons of solid wood is needed for wood-based industries in Algeria. Therefore to decrease the burden on the country's balance of payment the present study aimed to evaluate the potentiality of use of date palm trunks and rachises in the manufacture of particleboards as a substitute for wood products.

To conduct this study, date palm trunks and rachises were sourced from a local plantation in southern Algeria, reduced to a particle size 1–2 cm, oven-dried at 100 °C to reach a MC of 3%. Specimens of trunks and rachises were prepared according to TAPPI T257Cmo2 (2002) standard for the determination of their chemical composition. Holocellulose and cellulose contents were determined according to wise and Murphy method (1946). Lignin content was determined according to Klason method (TAPPI 222 om-02) (2002). Extractives were determined according to TAPP1 T204 cm-97 (1997).

To manufacture the particleboard panels the chips were placed in a drum blender and sprayed with phenol formaldehyde (PH) or melamine–urea–formaldehyde (MUF) for 1 min. The resin content was 10% based on dry particle content. The panels were produced at a density 0.70 g/cm^3. The test panels were 350 × 300 × 14 mm^3, reaching after trimming 320 × 270 × 14 mm^3. The total press time was 7.5 min and temperature 195 °C.

Concerning the chemical analysis, the cellulose content in the trunk (43.7%) was found higher than in the midrib (35.87%). The date palm trunk had the lowest extractive content of 1.22 wt%, while the midrib had a high lignin content (26.89 wt%). These values are in agreement with previous research [6, 32, 49, 50].

Taking EN312 (2005) as a reference, the minimum requirements for MOR for particleboards are 11.5 and 13.0 N/mm^2 for general uses and interior fitments including furniture, respectively, while the minimum MOE for interior fitments is 1600 N/mm^2. For internal bond strength, the minimum requirement is 0.024 N/mm^2 for general purpose, 0.35 N/mm^2 for interior fitments and 0.50 N/mm^2 for heavy-duty load-bearing boards. Comparing the test results with the aforementioned values, it is clear that the particleboards produced from trunk satisfied all the MOR, MOE and IB standards. The particleboards produced from midribs satisfied the MOR standards using both PH and MUF, as well as MOE standards using pH, but not boards using MUF. Concerning IB, the palm midrib board satisfied all the IB standards. It is interesting to note that the palm trunk particleboards produced from trunk had significantly higher MOR, MOE and IB values, compared with those for palm midribs. This opens a great potentiality for use of a new resource (old unproductive palm trunks) for a high value-added product such as particleboards.

Concerning the thickness swelling, the EN312 15% after 24 h was not met except for the case of use of midribs with PH. This standard can be met by using a suitable water repellent wax within the resin formulation.

As a conclusion, this study proves that the manufacture of particleboards from date palm trunks and midribs is technically feasible. This in turn provides a solution to the problem of shortage of materials for the particleboards industry as well as the release of pressure on cutting of forest resources.

6.11.8 Self-Bonded Particleboards from Date Palm Leaflets, Midribs, Petioles and Fibrillum

A pioneer study [51] has been conducted with the objective of characterizing four date palm byproducts and evaluation of the potentiality of their use in the manufacture of self-bonded particleboards. The shortage of wood supplies has motivated researchers to the use of other lignocellulosic materials, such as the agricultural residues for the manufacture of fiberboards and particleboards using synthetic resins [2]. However, due to environmental concerns, there is a growing trend among researchers to use the above-mentioned materials in the production of self-bonding boards [43, 52–54] and thus dispensing with the synthetic resins, the presence of which hinders the recycling of the products at the disposal stage. The date palm plantations are widely spread with approximately 100 million date palms in the world [18]. These palms are annually pruned, producing large amounts of byproducts being a material base for promising industrial development. Thus, the objective of the present study has been formulated as the characterization of four date palm byproducts: leaflets, midribs, petioles and fibrillum and the evaluation of the potentiality of their use in the manufacture of self-bonding particleboards.

Samples of date palm leaflets, midribs, petioles and fibrillum were sourced from Marrakesh province (Morocco), hammer milled and passed by a 3-mm sieve and oven-dried at 105 °C for 12 h.

The morphological analysis was conducted including the bulk density using a densitap machine and particle size analysis using the American standards sieve series ASTM: E11-09. Binocular microscopy was also used to visualize the state of fibers after grinding. The chemical composition was determined in terms of ash, neutral detergent fiber (NDF), cellulose, hemicellulose, lignin, proteins, wax and hot water extractives. Particleboard test samples of size 150 mm × 150 mm were prepared without using any water or synthetic resin in a press under 100 kg/cm^2 pressure and 180 °C for 2 min. The target density was 1.2 g/cm^3 and thickness 4 mm. Before testing the particleboard, specimens were conditioned in a climatic room for 14 days at 60% RH and a temperature of 25 °C.

Concerning the characterizing of date palm byproducts, the fibrillum, leaflets, rachises and petioles were found to have a high amount of NDF (65–91%), which is normally required for self-bonding adhesion [19]. The fibrillum is distinguished with highest cellulose (50.6%) and lignin (31.9%) content, respectively. It has been reported that lignin has an important role in bonding [55]. Leaflets, midribs and petioles had high contents of hemicellulose and cellulose: similar values were found in other studies [1, 3]. The midrib had the highest hemicellulose content (31.4%). The leaflets had the highest protein content (9.7%) pointing to the suitability for use in fodder. The bulk density values were the least for petioles (0.199 g/cm^3), followed by the fibrillum (0.241), midrib (0.289) and leaflets (0.316).

The self-bonding was successful: all samples were cohesive. Regarding the mechanical properties of test boards, the fibrillum boards enjoyed the highest MOR and MOE properties (12.9 N/mm^2 and 1257 N/mm^2, respectively). This can be explained by the high lignin content in the fibrillum (the lignin plays an important role as a binder forming a coherent thermosetting matrix during board manufacture gluing particles together [49]). The panels of leaflets and midribs had the second highest MOR values of 8.4 and 8.5 N/mm^2, respectively. This can be explained by the highest amount of hemicelluloses in midribs and highest content of extractives in leaflets, beside its average content of hemicelluloses.

It has been reported [8] that the highest hemicellulose content in oil palm byproducts enhances the cross-linking of particles during the pressing stage. It can be also stated that the high strength of leaflets boards is primarily related to high density of leaflets (0.316 g/cm^3). The particleboards made from petioles did not exhibit satisfactory results as far as MOR is concerned (6.1 N/mm^2). This can be explained by their finest particles (0.199 g/cm^3 bulk density). Concerning the internal bond all samples did not meet the minimum requirement of Japanese industrial standard A5908, type-8 (0.15 N/mm^2). These findings could be a result of lack of pretreatment before hot pressing and the use of short pressing time cycle in this study. However, the relatively high IB of boards, manufactured from leaflets can be explained by the high density of the raw material. Concerning the water resistance properties, all panels showed high water absorption ranging from 101 to 304%. To improve these properties some additional processes can be used, such as preheating treatment, chemical or steam treatment could be applied [40, 54, 56]. As a conclusion of this study, the most promising byproducts are the fibrillum and leaflets. It appears necessary to optimize the pressing conditions to take advantage for example of fibrillum high lignin content through longer pressing times and to evaluate the influence of pretreatment on both morphological and chemical properties of the fibers and board properties.

Appendix 1

A Summary of the Report of Munich Institute for Wood Research On Palm Midrib Blockboards Produced in El-Kharga Factory, The New Valley Governorate

Institut für Holzforschung der Universität München
80797 München, Winzererstraße 45
Institutsleiter Prof.Dr.Dr.habil. G.Wegener
Telefon (089) 306 309 0; FAX (089) 306 309 11

Bewertung von "Date Palm Leaves' Midribs Blockboard"

nach deutschen Sperrholznormen

Gutachtliche Stellungnahme

München, 4.9.1996

Institutsleiter
Prof. Dr. G. Wegener

Verfasser
Dipl.-Holzwirt F. Tröger

Fritz Tröger, Institute for Wood Research, Munich University

Summary

Early in 1996, the Institute for Wood Research of the University of Munich was commissioned by the GTZ office in Cairo to assess the quality of blockboard, developed as part of the GTZ project program, in orientation tests using German standards.

The institute's assessment report, in its introductory section, gives a description of the cross-sectional structure of blockboard as defined in German standards. Blockboard is considered a high-quality, very well reputed wood based panel material. For cost reasons alone, applications are restricted to fields with above average quality specifications, e.g. in high-quality interior decorating (wall and ceiling panels), high-quality furniture and equipment manufacture.

Six blockboard samples measuring 1,000 mm x 500 mm x 13 mm constituted the test material supplied by the Egyptian plant El-Kharga. "Date Palm Leaves' Midribs Blockboard", as this panel with a centre layer of palm leaves' midribs is officially called, has a cross-sectional structure according to the principles of cross-banded lumber veneered board. The block core consists of slats cut from palm leaves' midribs, with a cross section of 10 mm x 10 mm. Surface veneers from poplar wood, 1.5 mm thick, are glued onto both sides of the block core using urea formaldehyde resin.

Tests involved the investigation of the following material properties: Moisture content (DIN 52 375), thickness swelling (DIN 52 364), density (DIN 53 374), internal bond strength (DIN 53 255), bending strength (DIN 52 371) and specifications regarding surface veneer and core layer (DIN 68 705, part 2).

Overall, the investigation of the material properties of Date Palm Leaves' Midribs Blockboard led to consistently positive results. Special attention is drawn to the following board characteristics:

1. The surfaces of the poplar wood veneers were evaluated class 1 and class 2 on top and underneath respectively. Some veneer surfaces proved excessively rough. This should be brought to the attention of the supplier.

2. The core of the blockboard is the centre layer which consists of slats made of the midribs of palm leaves. Emphasis is placed on the fact that this centre layer of palm leaves midribs fulfilled most specifications for centre layers made from wood. In future care should be taken to cut palm leaves' midribs in such a manner that slats with angles of 90° are achieved as well as being full edged. Grading for first and second choice could be considered.

3. The results of the internal bond test could not be better. The leverage-break tests gave excellent results regarding the gluing of surface veneers and core layer (No. 1 ranking according to DIN 53 255).

4. For surface veneers with the grain oriented in parallel to the longitudinal axis of the specimens bending strength was in the range of 37 to 43 N/mm^2. As expected, the bending strength range for veneers with grain perpendicular to the longitudinal axis of specimens was, at 27 to 30 N/mm^2, considerably lower.

In summarizing the results, it can be confirmed that Palm Leaves' Midrib Blockboard can be considered a valuable wood-based materials panel. On account of its good mechanical properties this panel is suitable for a great number of applications such as furniture manufacture, interior fitting, container and equipment manufacture as well as wall and ceiling panelling.

Appendix 2

A Certificate from the UNICEF Concerning The Project of Utilization of Palm Midrib Blockboards in the Manufacture of Community Schools Furniture in Asiut, Sohag And Kena Governorates In 1995

unicef

منظمة الأمم المتحدة للأطفال

United Nations Children's Fund
Fonds des Nations Unies pour l'enfance
Fondo de las Naciones Unidas para la Infancia
8 Adnan Omer Sidky Street, Dokki – Cairo – Egypt
Tel.: 700815 – 3616346 – 710578 – 708540 – 708541
Telex: 93164 ICEF UN
Cable: UNICEF Cairo Fax: 3605664

٨ ش عدنان عمر صدقي الدقي - القاهرة - مصر
تليفون : ٧٠٠٨١٥ - ٣٦١٦٣٤٦ - ٧١٠٥٧٨ - ٧٠٨٥٤٠ - ٧٠٨٥٤١
تلكس : ٩٣١٦٤ ICEF UN
عنوان تلغرافي : UNICEF Cairo فاكس : ٣٦٠٥٦٦٤

8 October 95

To Whom It May Concern

This is to certify that the cooperative agreement for the production of Date palm mid rib block board between UNICEF and The Center For Development Of Small Scale Industries, Faculty of Engineering - Ain Shams University and the Governorate of the New Valley was satisfactorily completed. The production was of good quality and delivered on time.

UNICEF has been very happy to support this initiative as it is in line with the development of appropriate local technology. Moreover this type of production is environmentally friendly and contributes to the creation of employment opportunities at the local level.

M. Baquer Namazi
UNICEF Representative

Appendix 3

Results of testing of samples Of MDF, made from the products Of pruning of date palms, Collected from El-Bahariah Oases

NAG-HAMADY FIBER BOARD CO.
(N.F.B)
E-mail : nhfiboco@intouch.com

NFB

شركة نجع حمادى للفيبر بورد
ش.م.م

دشنا في: ٢٢/٨/٢٠١٦

تقرير فنى عن

الاختبارات الفيزيائية والميكانيكية لعينات الالواح المصنعة معمليا من المنتجات الثانوية لنخيل البلح

١- الاختبارات الفيزيائية والكيميائية

الاختبار	السمك	الكثافة	الرطوبة	التشرب	الامتصاص	انبعات الفرمالين
الوحدة	مم	كجم/م٣	%	%	%	مجم/١٠٠جم
النتيجة	١٢,٠٠	٧٥٢,٠٠	٥,٢	١٢,٧	٥٨,٠٤	٢٢,٥٤
الانحراف المعياري	٠,٠٤	٢٤,٢٨	٠١١	١,٣	٢,٥	٠,٨٧
الطريقة القياسية	EN323	EN323	EN322	EN317	BS1142	EN120
المواصفات القياسية	-	--	٤-١١	١٥		≥٣٠

٢- الاختبارات الميكانيكية

الاختبار	السمك	معامل الكسر MOR	معامل المرونة MOE	الرابطة الداخلية IB	قوة شد السطح Surface soundness
الوحدة	مم	نيوتن/مم٢	نيوتن/مم٢	نيوتن/مم٢	نيوتن
النتيجة	١٢	٢٤,٤٠	٢٩١١	٠,٩	١,٣٥
الانحراف المعياري	٠٠,٠٤	٣,٤٩	٢٨٠,٥٢	٠,٢	٠,٣٠
الطريقة القياسية	EN323	EN310	EN310	EN319	EN311
المواصفات القياسية		٢٠,٠٠	٢٢٠٠	٠,٥٥	١,٢

مدير عام المعامل
كيميائي / عبدالحميد محمد محمد

مدير عام البحوث
د./عبدالباسط عبدالحميد ادم

Cairo Office : 17 Gawad Hosni St., Cairo - Egypt
Tel.: +2/02/23922109 +2/02/23902995 Fax: +2/02/23926511
P.O.Box : 432 Mohamed Farid
Factory : Desna - Qena
Tel.: +2/096/6743281 - +2/096/6743002 Fax : +2/096/6743003
website : www.geocities.com/nhfiboco/company-profile html

فرع القاهرة : ١٧ ش جواد حسنى - القاهرة - مصر
تليفون : ٢٣٩٢٢١٠٩ - ٢٣٩٠٢٩٩٥ - فاكس : ٢٣٩٢٦٥١١
ص.ب : ٤٣٢ محمد فريد
المصنع : دشنا - محافظة قنا
تليفون : ٠٩٦/٦٧٤٣٢٨١ - ٠٩٦/٦٧٤٣٠٠٢ - فاكس : ٠٩٦/٦٧٤٣٠٠٣

References

1. Bendahou A, Dufresne A, Kaddami H, Habibi Y (2007) Isolation and structural characterization of hemicelluloses from palm of Phoenix dactylifera L. Carbohyd Polym 68(3):601–608. https://doi.org/10.1016/j.carbpol.2006.10.016
2. Wang D, Sun XS (2002) Low density particleboard from wheat straw and corn pith. Ind Crops Prod 15(1):43–50. https://doi.org/10.1016/S0926-6690(01)00094-2
3. Sbiai A, Maazouz A, Fleury E, Souterneau H, Kaddami H (2010) Short date palm tree fibers / polyepoxy composites prepared using RTM process: effect of tempo mediated oxidation of the fibers. BioResources 5(5):672–689
4. Ibrahim SA, Ali NGM, El-Faham AI, Mohamed M, Khalifa EM (2013) Effect of using date palm waste on performance, carcass characteristics and economic efficiency of Gemaza growing chick 16:309–318. http://esnaf-eg.com/GuidanceNewsPaper
5. Helen L, Gertsakis J (2001) Design and environment. Greenleaf Publishing
6. El-Morsy MMS (1980) Studies on the rachises of the Egyptian date palm leaves for hardboard production. Fibre Sci Technol 13(4):317–321. https://doi.org/10.1016/0015-0568(80)90018-4
7. Brain MLEOD (1999) Panel source international. Dawn in the grain forest. In: Proceedings of the meeting of the Eastern Canadian section of the forest products society, Winnipeg, Manitoba
8. EL-Mously H (1991). A research project on the use of palm leaf as an industrial raw material, the first report. Faculty of Engineering, Ain Shams University, in cooperation with the Academy of Scientific Research and Technology (in Arabic)
9. Azizi K, Tabarsa T, Ashori A (2011) Performance characterizations of particleboards made with wheat straw and waste veneer splinters. Compos B Eng 42(7):2085–2089. https://doi.org/10.1016/j.compositesb.2011.04.002
10. Mazhari Mousavi SM, Hosseini SZ, Resalati H, Mahdavi S, Rasooly Garmaroody E (2013) Papermaking potential of rapeseed straw, a new agricultural-based fiber source. J Clean Prod 52:420–424. https://doi.org/10.1016/j.jclepro.2013.02.016
11. El-Kinawy, Mohamed S (n.d.) Treatments of palm midrib to improve the mechanical properties compared to imported woods
12. Drake PA (1997) The composite panel industry: a global market assessment. In: Proceedings of 31st international particleboard/composite materials symposium. Washington State University, Pullman, pp 29–39
13. Central agency for mobilization and statistics, USDA; Foreign agriculture service, global agriculture information network, Egypt (2015) Wood Sector Report
14. Shafie M, Zarea-Hosseinabadi H (2019) Eco-friendly laminated strand lumber from date palm rachis: analysis of mechanical properties by Taguchi design of experiment. Drvna Industrija 70(4):359–367. https://doi.org/10.5552/drvind.2019.1843
15. Kinawy MSA (1997) Study of the appropriate conditions of the press-cycle for the manufacture of palm midrib-core blockboards [Master]. The Faculty of Engineering, Ain Shams University
16. Akgül M, Çamlibel O (2008) Manufacture of medium density fiberboard (MDF) panels from rhododendron (R. ponticum L.) biomass. Build Environ 43(4):438–443. https://doi.org/10.1016/j.buildenv.2007.01.003
17. Ashori A (2010) Hybrid composites from waste Materials. J Polym Environ 18(1):65–70. https://doi.org/10.1007/s10924-009-0155-6
18. Al-Kaabi K, Al-Khanbashi A, Hammami A (2005) Date palm fibers as polymeric matrix reinforcement: DPF/polyester composite properties. Polym Compos 26(5):604–613. https://doi.org/10.1002/pc.20130
19. Hegazy S, Ahmed K, Hiziroglu S (2015) Oriented strand board production from water-treated date palm fronds. BioResources 10(1):448–456
20. Al-Sulaiman FA (2003) Date palm fibre reinforced composite as a new insulating material. Int J Energy Res 27(14):1293–1297. https://doi.org/10.1002/er.957
21. Dabhi BK, Vyas RV, Shelat HN (2014) Use of banana waste for the production of cellulolytic enzymes under solid substrate fermentation using bacterial consortium. Int J Curr Microbiol App Sci 3(1):337–346

22. Crops for sustainable enterprise, design for sustainable development. European Foundation for the improvement of living and working conditions (2000) Wyattville Read, Longhlinstown Co
23. Iskanderani FI (2008) Physical properties of particleboard panels manufactured from Phoenix Dactylifera-L (date palm) mid-rib chips using ureaformaldehyde binder. Int J Polym Mater Polym Biomater 57(10):979–995. https://doi.org/10.1080/00914030802177404
24. Iskanderani FI (2008) Influence of process variables on the bending strength of particleboard produced from Arabian date palm mid-rib chips. Int J Polym Mater Polym Biomater 58(1):49–60. https://doi.org/10.1080/00914030802461964
25. Hegazy SS, Aref IM, Al-Mefarrej H, El-Juhany LI (2008) Effect of spacing on the biomass production and allocation in Conocarpus erectus L. trees grown in Riyadh, Saudi Arabia. Saudi J Biol Sci 15(2):315–322
26. Sellers T (2000) Growing markets for engineered products spurs research. Wood Technol 40–43
27. Cengiz Guler, Ibrahim Bekta, & Hulya Kalaycioglu (2006) The experimental particle board manufacture from sunflower stalks (Helianthus annuus L.)and Calabrian pine (Pinus brutia Ten.). Forest Prod J 56(4), Article 4
28. Hegazy SS, Aref IM (2010) Suitability of some fast-growing trees and date palm fronds for particleboard production. For Prod J 60(7):599–604. https://doi.org/10.13073/0015-7473-60.7.599
29. Sain M, Panthapulakkal S (2006) Bioprocess preparation of wheat straw fibers and their characterization. Ind Crops Prod 23(1):1–8. https://doi.org/10.1016/j.indcrop.2005.01.006
30. Zayed SE, Adam AA, Hassan EA, Elkady MA (2014) Properties of particleboard based on date palm fronds as renewable Egyptian lignocellulosic materials. Int J Innov Sci Res 9(2):326–334
31. Tayssier AM (1996) An investigation into the conditions of manufacture of lumber-like blocks from date palm leaves' midribs [Master]. The Faculty of Engineering, Ain Shams University
32. Hegazy S, Ahmed K (2015) Effect of date palm cultivar, particle size, panel density and hot water extraction on particleboards manufactured from date palm fronds. Agriculture 5(2):267–285. https://doi.org/10.3390/agriculture5020267
33. El-Mously H, El-Morshedy MM, Megahed MM, Abd El-Hai Y (1993) Evaluation of particleboard made of palm leaves midribs as compared with flaxboard. In: Proceedings of the 4th intermationa conference on production engineering and design for development, Cairo, Egypt
34. El-Mously H, Saber M (2019) Medium density fiberboards from date palm residues a strategic industry in the arab world. By-Products of Palm Trees and Their Applications
35. Li X, Cai Z, Winandy JE, Basta AH (2010) Selected properties of particleboard panels manufactured from rice straws of different geometries. Biores Technol 101(12):4662–4666. https://doi.org/10.1016/j.biortech.2010.01.053
36. Monica Ek, Gellerstedt G, Henriksson G (2009) Pulp and paper chemistry and technology. In: Wood Chem Wood Biotechnol 1
37. van Dam JEG, van den Oever MJA, Teunissen W, Keijsers ERP, Peralta AG (2004) Process for production of high density/high performance binderless boards from whole coconut husk. Ind Crops Prod 19(3):207–216. https://doi.org/10.1016/j.indcrop.2003.10.003
38. Li X, Lei B, Lin Z, Huang L, Tan S, Cai X (2013) The utilization of organic vermiculite to reinforce wood–plastic composites with higher flexural and tensile properties. Ind Crops Prod 51:310–316. https://doi.org/10.1016/j.indcrop.2013.09.019
39. Benzidane R, Sereir Z, Bennegadi ML, Doumalin P, Poilâne C (2018) Morphology, static and fatigue behavior of a natural UD composite: the date palm petiole 'wood.' Compos Struct 203:110–123. https://doi.org/10.1016/j.compstruct.2018.06.122
40. Hashim R, Wan Nadhari WNA,Sulaiman O, Sato M, Hiziroglu S, Kawamura F, Sugimoto T, Guan Seng T, Tanaka R (n.d.) Properties of binderless particleboard panels manufactured from oil palm biomass. BioResources
41. EL-Mously H (2001) Renewable material resources as environmentally friendly engineering materials. Thirteenth Congress of Mechanical Engineering (in Arabic)
42. Cai Z, Wu Q, Lee JN, Hiziroglu S (2004) Influence of board density, mat construction, and chip type on performance of particleboard made from eastern red cedar. For Prod J 54:226–232

43. Álvarez C, Rojano B, Almaza O, Rojas OJ, Gañán P (2011) Self-Bonding boards from plantain fiber bundles after enzymatic treatment: adhesion improvement of lignocellulosic products by enzymatic pre-treatment. J Polym Environ 19(1):182–188. https://doi.org/10.1007/s10924-010-0260-6
44. Hashim R, Nadhari WNAW, Sulaiman O, Kawamura F, Hiziroglu S, Sato M, Sugimoto T, Seng TG, Tanaka R (2011) Characterization of raw materials and manufactured binderless particleboard from oil palm biomass. Mater Des 32(1):246–254. https://doi.org/10.1016/j.matdes.2010.05.059
45. El-Mously HI (2016) Innovating green products as a mean to alleviate poverty in upper Egypt. In: 1st international joint symposium on "product development and innovation," Ain Shams University
46. Ghofrani M, Ashori A, Mehrabi R (2017) Mechanical and acoustical properties of particleboards made with date palm branches and vermiculite. Polym Testing 60:153–159. https://doi.org/10.1016/j.polymertesting.2017.03.028
47. Amirou S, Zerizer A, Pizzi A, Haddadou I, Zhou X (2013) Particleboards production from date palm biomass. European J Wood Wood Prod 71(6):717–723. https://doi.org/10.1007/s00107-013-0730-3
48. Mohamed Ahmed MVO, Bouna ZEO, Mohamed Lemine FM, Djeh TKO, Mokhtar T, Mohamed Salem AO (2011) Use of multivariate analysis to assess phenotypic diversity of date palm (Phoenix dactylifera L.) cultivars. Scientia Horticulturae 127(3):367–371. https://doi.org/10.1016/j.scienta.2010.11.011
49. Khristova P (2005) Alkaline pulping with additives of date palm rachis and leaves from Sudan. Biores Technol 96(1):79–85. https://doi.org/10.1016/j.biortech.2003.05.005
50. Nasser RA, Al-Mefarrej HA (2011) Midribs of date palm as a raw material for wood-cement composite industry in Saudi Arabia. J Appl Sci Res 15(12):1651–1658
51. Saadaoui N, Rouilly A, Fares K, Rigal L (2013) Characterization of date palm lignocellulosic by-products and self-bonded composite materials obtained thereof. Mater Des 50:302–308. https://doi.org/10.1016/j.matdes.2013.03.011
52. Raquez J-M, Deléglise M, Lacrampe M-F, Krawczak P (2010) Thermosetting (bio)materials derived from renewable resources: a critical review. Prog Polym Sci 35(4):487–509. https://doi.org/10.1016/j.progpolymsci.2010.01.001
53. Lehmann WF (n.d.) Resin efficiency in particleboard as influenced by density, atomization and resin content. Forest Prod J 20(11):48–54
54. Widyorini R, Xu J, Umemura K, Kawai S (2005) Manufacture and properties of binderless particleboard from bagasse I: effects of raw material type, storage methods, and manufacturing process. J Wood Sci 51(6):648–654. https://doi.org/10.1007/s10086-005-0713-z
55. Okuda N, Hori K, Sato M (2006) Chemical changes of kenaf core binderless boards during hot pressing (I): influence of the pressing temperature condition. J Wood Sci 52(3):244–248. https://doi.org/10.1007/s10086-005-0761-4
56. Okuda N, Hori K, Sato M (2006) Chemical changes of kenaf core binderless boards during hot pressing (II): effects on the binderless board properties. J Wood Sci 52(3):249–254. https://doi.org/10.1007/s10086-005-0744-5

Chapter 7
Date Palm Byproducts in Construction, Insulation and Building Materials

Abstract This chapter reviews several scientific breakthroughs that have tackled the potential of utilizing date palm byproducts for various uses in contemporary construction field as a sustainable and renewable alternative to conventional materials such as concrete and steel. The chapter demonstrates and classifies the fields of utilizing date palm byproducts in construction into usage in the natural form and usage in the processed form. Using date palm midribs in their natural form for light structures was investigated by several studies that showed clear loyalty to traditional techniques whose continuing thriving to the present day indicates their compatibility with the material while sustaining low-cost production. In addition, date palm midribs and trunk were investigated to be used to produce facades and structural sandwich panels. On the other hand, other researches focused on processing date palm fibers to be used in more sophisticated fields such as reinforcing concrete, high strength concrete and mortar to enhance shrinkage resistance and ductility. In addition, the superior insolating properties of date palm fibers led to their involvement in several researches specialized in producing insulation materials, polyester composites and insolating panels which are highly in demand globally to reduce the cooling loads and heating loads and improve the indoor air quality of residential and office spaces. Moreover, crushed date palm midribs ash, midribs and petioles fibers have been introduced as a recommendable stabilizer for traditional building materials such as raw earth blocks and compressed blocks.

Keywords Date palm · Date palm midribs · Light structures · Date palm reinforced soil blocks · Date palm fiber reinforced concrete · Date palm fibers insulation

7.1 Introduction

Up to the time this chapter was authored, date palm byproducts have been utilized in construction of houses and huts in the rural villages and oases where date palms are abundant [1]. Date palm leaves are piled over the roofs of rural houses as a part of the innate human instinct to reduce the heat gain through the roofs during the summer [2]. Date palm leaf sheaths fibers are still being actively used as a

H. EL-Mously et al., *Date Palm Byproducts: A Springboard for Circular Bio Economy*, Materials Horizons: From Nature to Nanomaterials,
https://doi.org/10.1007/978-981-99-0475-4_7

major component of mud blocks, as a filler, stabilizer and enhancer of the thermal conservation performance in Nubia [1]. Date palm petioles are lined along the exterior boundaries of rural houses as a guide to ensure the straightness and verticality of the walls [3]. Date palm midribs are woven into mats as reinforcement below simple concrete roofing in houses and sanitation wells [4]. Date palm midribs and trunks are used as beams under Kershef roofing in Siwa oasis in Egypt [5]. This thriving technical heritage encouraged multiple researchers to tackle more elaborate uses of date palm byproducts in architecture and insulation, in natural or processed forms. This chapter demonstrates several of these attempts as detailed in the next sections.

7.2 Uses of Date Palm Byproducts in Natural Form

7.2.1 Structural Elements

7.2.1.1 Date Palm Midribs Space Truss

Used Byproduct: Midribs

Hassan [6] conducted the first attempt to use date palm midribs for wide-span construction. He depended on using a modern and light wide-span structural system, a space truss. An ideal truss involves only axial tension or compression on its members, as the joints are considered to be pins, or hinge joints, with no resistance to moment [7]. This concept was chosen because the fibers that are responsible for the strength of the date palm midribs are arranged in the longitudinal direction only without any radial cross linking between them [8]. Therefore, the loads that can be carried in date palm midribs have to be axial only, as any transverse or radial loads will cause the splitting of the fibers. This case of loading is available in truss structures. Therefore, trusses were found to be compatible with the microscopic structure of date palm midribs.

The primary unit was assumed to be a square pyramid. The joinery used was inspired from the cruciform gusset plate connector, which is a prefabricated joinery without a node that requires fixing the members right to the plates using welding like in the case of a Mero connector. Hence, the designed joinery depended on custom-made U-shaped high strength bolts that were used to hold down the members to the steel plates, which were welded at different angles onto steel boxes.

The date palm midribs used were from the Siwi species, and they were dried to 8–12% moisture content. The midribs had to be standardized in order to unify the design of the joinery. Firstly, the midribs were cut into 10 * 10 mm cross-section strips, then those strips were bonded together into 1-m-long identical members of 10cm^2 cross-section area using urea formaldehyde (62% conc.) as adhesive and citric acid (4% conc.) as a hardener. Secondly, bolts were used to fasten truss members to steel plates. Then, these plates were welded to a steel box at the specific angles that

Fig. 7.1 3 × 3 m Experimental date palm midribs space truss (Hassan 2001)

allowed the formation of the quadrilateral pyramid units of a 3 × 3 m space truss model, as shown in Fig. 7.1.

In order to test the experimental model, the space truss model was supported on four rigid steel columns on the four corners where they were welded to the connectors of the space truss. The initial own weight of the model was estimated to be 250 kg. Firstly, 25 kg weights were suspended from each upper connection of the model as concentrated loads, and the deflection values were measured, as shown in Fig. 7.2. Secondly, the weights were increased to 50 kg on each upper connection and the deflection values were measured, as shown in Fig. 7.3. Thirdly, the weights were increased to 75 kg on each upper connection and the deflection values were measured, as shown in Fig. 7.4. It was found that the deflection values measured were about 1/3 the predicted values by the Finite Element Method (FEM) model. This indicates that the actual behavior of the connections altered the failure mode due to the unpredictable sliding between the truss members and the U-bolts. This indicated the importance of adding adhesives between truss members and U-bolts.

Fig. 7.2 Model with 25 kg concentrated loads

Fig. 7.3 Model with 50 kg concentrated loads

Fig. 7.4 Model with 75 kg concentrated loads

The results of the validated FEM model showed that the system can cover a 16 m dia. geodesic dome under gypsum boards cladding. However, the validated mechanical properties, obtained from experimentation, exclusively represent the strength of the midribs after the removal of the peripheral layers, glue bonding and using the customized steel connections [9]. Thus, the validated FEM model properties did not represent the actual strength of the midribs.

7.2.1.2 Date Palm Midribs Planar Truss

Used Byproduct: Midribs

The challenges, reported in [6], led to the suggestion that it would be cheaper and simpler to build planar trusses in one direction and depend on purlins to span the trusses in the other direction. El-Sherbiny [10] designed the main concept by using simple and common joinery that can be constructed by the common artisans without specific skills like the customized U-bolt joints. This design introduced various alternatives of simple planar truss joinery. Scaled models of the alternatives were tested while covering the span of 3 m and the depth of 0.5 m according to the testing machine limitations.

The midribs were cut into semi-standard cross-section pieces from the middle portion of air-dried date palm midribs while sustaining the outer layer as much as possible. Five truss models were developed as followed (Figs. 7.5 and 7.6):

- Truss1 (Fig. 7.7) was an N-truss divided into four equal parts, and the members were two date palm midribs joined together by plastic wirings and the joinery by

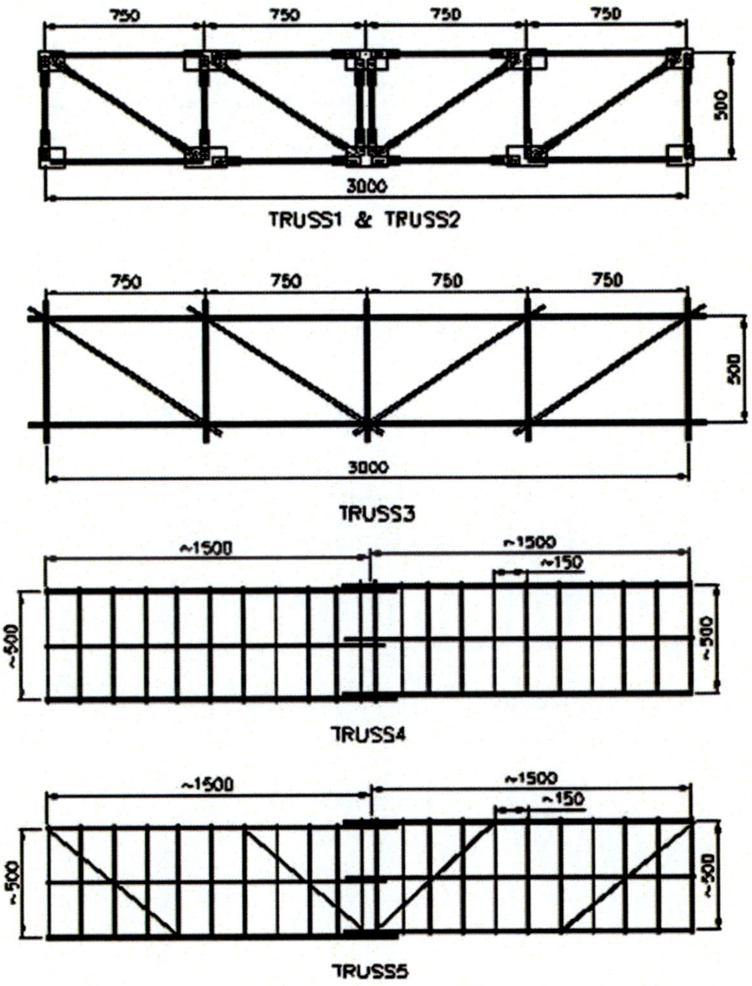

Fig. 7.5 Elevations of the trusses. *Source* [10]

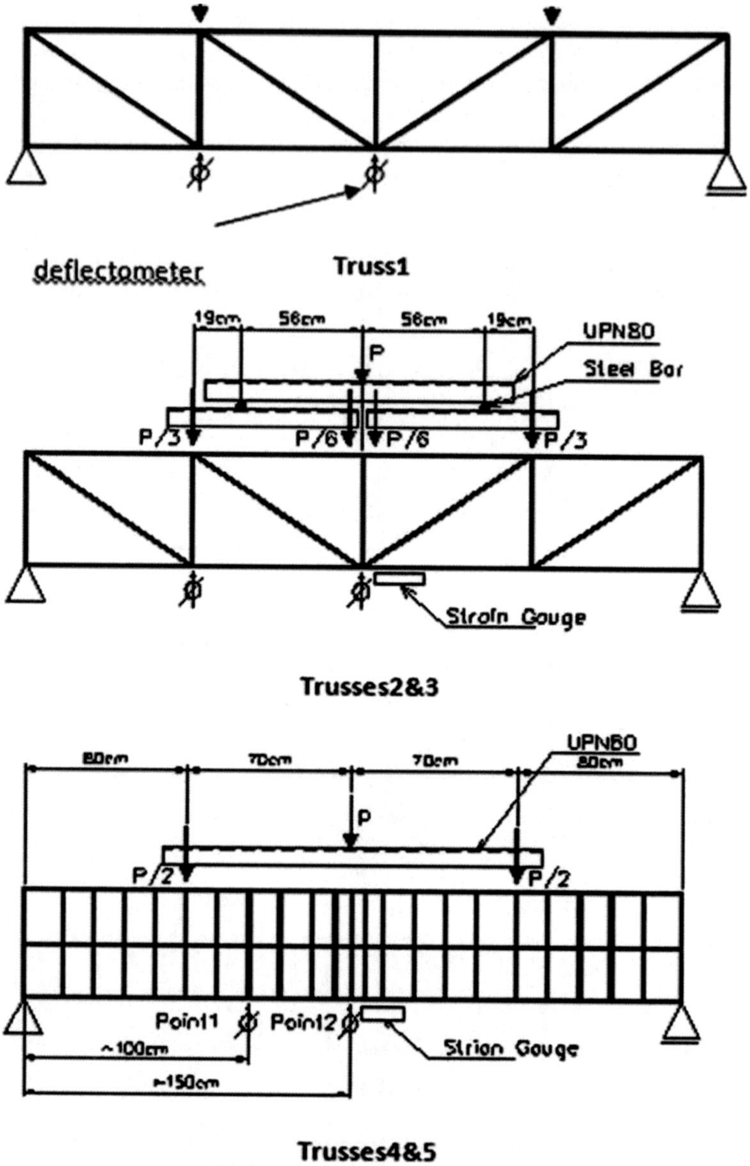

Fig. 7.6 Loading types of the trusses. *Source* [10]

2-mm-thick steel plates and steel bolts (Figs. 7.8, 7.9, 7.10, 7.11, 7.12, 7.13, 7.14, 7.15, 7.16, 7.17).

- Truss2 was similar to Truss1 but with 6-mm-thick steel plates.
- Truss3 (Fig. 7.10) differed from Truss1 and Truss2 in the type of joinery. The N-truss depended on steel bolts at the connections between the bracings and the chords. This means that the chords were made double so the spacing in between would host the bracings.
- Truss4 (Fig. 7.13) was a truss that was made as a traditional crate in three coherent layers.
- Truss5 was similar to Truss4 but with using additional diagonals.

Fig. 7.7 Joinery in Trusses1 and 2. *Source* [10]

Fig. 7.8 Testing frames of Trusses1 and 2. *Source* [10]

Fig. 7.9 Splitting of members in Truss2. *Source* [10]

Fig. 7.10 Joinery in Truss3. *Source* [10]

Fig. 7.11 Testing frames of Truss3. *Source* [10]

Fig. 7.12 Movement of joints in Truss3. *Source* [10]

Fig. 7.13 Three layers of Trusses4 and 5. *Source* [10]

Fig. 7.14 Testing frames of Truss4. *Source* [10]

Fig. 7.15 Severe deflection of Truss4. *Source* [10]

Fig. 7.16 Joinery failure in Truss5. *Source* [10]

Fig. 7.17 Testing frames of Truss5. *Source* [10]

The trusses were tested in a universal testing machine of 1 ton capacity using deflectometers and strain gauges. Steel frames were used to prevent out-of-plane movement, in order to imitate that the experimented truss was an intermediate truss in a continuous truss roofing structure as shown in Fig. 7.6.

- Truss1: Load was installed on two symmetrical points.
- Truss2 and Truss3: Load was distributed evenly on the upper middle three points.
- Truss4 and Truss5: Load was distributed evenly at two points that were 80 cm away from the truss ends.

The testing results showed that the maximum load Truss1 carried until failure was 250 kg. The failure happened due to the bending that happened in the thin steel plates. The bending caused twisting of the truss, yet no buckling was noticed in the compression members. In Truss2, the thicker steel plates gave higher stiffness to the truss. However, the failure happened at the supports due to the out-of-plane bending, although the steel frames were used to cancel this out-of-plane bending, as shown in Fig. 7.9. The maximum load was measured for Truss3. The failure occurred because of the lateral twisting of the joints at the end of the truss at the top chord. This truss showed higher lateral stiffness than Truss1 and Truss2. The failure of Truss3 is shown in Fig. 7.12. Truss4 acquired the least reported load at failure. The location of failure could not be specified, yet it was indicated by the sound of the cracking in the members and severe deflection, as shown in Fig. 7.15. The failure of Truss5 occurred due to the twisting of the truss at the connection between the diagonals and the bottom chord, as shown in Figs. 7.16 and 7.18.

Thus, Truss3 was the most durable, as the bolts used for the connections were simple and highly functional. However, the failure of the developed trusses occurred because of the weak joinery design and the low lateral stiffness of the trusses.

Fig. 7.18 Deflection of Truss5. *Source* [10]

7.2.1.3 Date Palm Midribs Tri-Arched Space Truss

Used Byproduct: Midribs

The significance of the established mechanical properties of date palm midribs, reported by El-Mously [11, Darwish et al. 12], encouraged more studies to investigate the potential of using date palm midribs in light wide-span construction while depending on simpler joinery than the joineries proposed by Hassan [6] and El-Sherbiny [10].

The design of the 12 m span Tri-Arched Space Truss TAST was inspired from the spontaneous genius of the local builders demonstrated in the Iraqi Mudhif [13]. The Iraqi Mudhif is a traditional wide-span public building that is used to cover public meeting halls in the Ahwaz region in Iraq. The main building material used in building the Mudhif is the reeds resourced from the Marsh lands in the Ahwaz region, where the reeds are grouped into 70 cm diameter bundles that are shaped to create consecutive parabolic arches to host the wide-span vault halls. Although the mechanical properties of the reeds differ greatly from those of date palm midribs, the similar long and thin geometrical shape of reeds and date palm midribs elements led to realizing that mimicking the concept of bundling the reeds and shaping them to arches that are directly planted in the soil would minimize the need for standardization or using complicated corner joinery as in the cases of the space trusses and the trusses proposed by Hassan [6] and El-Sherbiny [10]. This assumption was enforced by reviewing the works of Piesik [14].

Piesik [14] demonstrated an on-site trial of building a date palm midribs vault using consecutive 25 cm diameter arches made from date palm midribs. The original span of the developed vault was 13 m. This difference between the diameter of the reeds bundles in [15] and the midribs bundles in [14] could be justified because the midribs are solid members unlike reeds, which would increase the total weight of the bundles making the workability of moving and carrying the 70 cm diameter midribs bundles unfeasible. On the other hand, the hollow geometry of the reeds decreased the total weight of the bundles allowing for high labor workability despite the large diameter of the bundles. The completed vault was found to be simple and built by traditional and local labor without the need for standardization or complicated joinery. However, the completed vault was found to depend on intermediate palm trunk columns planted along the span of the arches with 3.25 m spacing [16]. Those columns were used because the on-site trial showed that the 25 cm diameter arched bundles lacked the structural stability to cover the span without intermediate columns. Thus, the span decreased from 13 m in the original design to 3.25 m only.

Piesik et al. [17] overcame the need for the intermediate columns by using a cross grid of interlaced arched bundles to make a modular space dome with the span of 8 m × 8 m, as shown in Fig. 7.19. Each arched bundle carried a set of other arched bundles while also being carried by another set of arched bundles. This reciprocal design was designed theoretically to cover a food shelter where sailing fabric was tensioned and fixed to the intersection points of the space dome using special steel connectors [17]. This design succeeded to cover the modular 8mx8m space without

Fig. 7.19 Interlacing arched bundles

intermediate columns. However, the strict modular design and the reduced heights at the sides would limit the continuity and future expansion of the space hosted by the dome. In addition, the connectors used to fix the fabric to the arched bundles showed signs of rusting that clearly affected the integrity of ropes fastening the midribs in the bundles [16], as shown in Fig. 7.20.

Furthermore, it was concluded that shaping the date palm midribs to create arches was not only favorable to minimize using complicated corner joinery, but also because the bundling technique, which is actively used to store the midribs, is based on the workable limits of weight and the diameter of the bundles [13]. In addition, the natural curved shape of the midribs would cause high deformations if used as straight building members such as beams and columns under own weight as reported in [13]. Furthermore, there is sufficient similarity between the axial structural action of arches and the axial fibers in midribs [9, 16].

Arches are form-active structural elements. Form-active structural elements are elements where a codependence occurs between their structural performance and their initial shape. Ideally, an arch with coplanar foundation and under distributed loads converts all the generated moments to axial compression that begins at the apex of the arch and increases along the arch down to the foundation [7]. The increase happens as the vertical component of the compression increases with the accumulative resistance of the arch's own weight under gravity toward the foundation. This increase draws thrust curves that make the arch tend to open at the foundation. The shape of the thrust curve generated is parabolic. The more an arch longitudinal axis coincides with the ideal parabolic thrust curve, the more uniform axial compressive stress is generated on the arch with minimal tension and minimal moment [18]. Thus, a parabolic arch is the ideal shape of a form-active arch with uniform axial compressive stress exclusively. This characteristic is highly critical specially while using date palm midribs bundles as the main material of the arch. This is because midribs consist mainly of longitudinal fiber bundles extending from the base of the

Fig. 7.20 Deteriorated ropes at the steel connectors

midrib at the trunk to the far end of the midrib. These fiber bundles are the only reliable channels of stresses in the midribs [8]. Therefore, maximum strength of midribs is found parallel to the fiber bundles, which is significantly higher than the strength perpendicular to the fiber bundles which would cause cross than disintegration of the fibers.

Another form-active structural element with uniform axial stresses is truss. Trusses consist of a triangulated assembly of members connected with pin joints [7]. The pin joints have no moment resistance. Therefore, the stresses generated on the members are exclusively axial without moment stresses. In addition, the triangulation enhances the lateral stability of the relatively light trussed structures when compared to conventional column and beam structures [18]. Therefore, trusses are commonly used in wide structures when utilizing light structural elements. Moreover, space trusses were developed to cover wide spans due to their lightness, flexibility and lateral stability [19]. Hence, Hassan [6] and El-Sherbiny [10] chose trusses to investigate the potentials of the light date palm midribs in structures.

As a result, Mansour et al. [2] merged the axial compression of parabolic arches and the lateral stability of trusses to introduce the TAST design, where the midribs bundles are used to create the parabolic shaped chords of trusses. The midribs, under high compression to create the parabolic chords, will exert friction on the ends of the

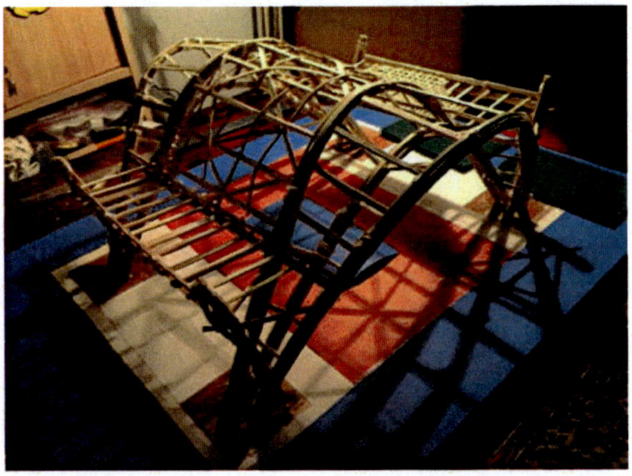

Fig. 7.21 Primary TAST design

trussing bracings inserted through the chords. This friction was assumed theoretically to be enough to fix the bracings into the chords without the need for complicated steel joinery. This assumption was inspired from the friction joinery in the traditional bread crates and bird cages made from date palm midribs in Egypt. Thus, the design was based on traditional techniques to ensure its cost-efficiency, high workability and simplicity as well as acquiring wide-span structural integrity and light lateral stability. These considerations were necessary to minimize the processing costs of the midribs and to facilitate the workability of the construction that it can be simply built and communicated by the common know-how of traditional artisans. The primary TAST design is shown in Fig. 7.21.

The primary design of the TAST was developed to increase the efficiency of the design by Darwish et al. [16]. The development was focused on enhancing three aspects: productivity, self-stability and the aesthetic value. Standardizing the parabolic ratios and designing the bracings to be equilateral reduced the number of the different prototypes. Inverting the space truss inclination to make the bracings at the foundation the widest increases the self-stability of the design. The aesthetic value of the design was reinforced by applying the visual design basics such as unity, emphasis and rhythm [20]. The developed TAST design is shown in Fig. 7.22.

In order to validate the design and determine the structural performance of the date palm midribs in TAST, 1:3 specimens were built and tested in the Concrete Laboratory in the Faculty of Engineering, Ain Shams University [9]. The specimens were subjected to vertical loading upon the apex of the arches. The specimens demonstrated high ductility with slow and gradual failure under the maximum load of 4 KN as shown in the sequence in Fig. 7.23. The failure (Figs. 7.24 and 7.25) was based on the lower parts of the middle section of each chord above the overlapping areas between the three sections that constitute the chords. The bending occurring at the middle section of the chords was resisted by the stiffer lower section of the chords, which led to the failure in the middle chords above the overlapping areas.

Fig. 7.22 Developed TAST design [16]

Nevertheless, the high ductility caused that the specimens bounced back to their original form after removing the load.

The performance was recorded and compared to the predicted performance in the Finite Element Method FEM structural analysis software SAP2000. The mechanical properties were defined into the model settings according to the properties published in [10, 12, 16]. The error between the simulation results and the actual results was minimal in the load value and 4% in the displacement value which was acceptable in the scale of the model. Thus, the model settings were validated to be used to predict the structural performance for the full scale structures. The results of the structural performance of the 12 m TAST showed that the structure was safe under its own weight and extreme wind loads according to the Egyptian code for design and construction of concrete structures [9, 21].

Consequently, the 12 m TAST was implemented in various structural arrangements that were selected according to their structural efficiency and the general demands of the market, to host functions that are allowed by the restrictions of

Fig. 7.23 Gradual vertical displacement sequence of TAST until failure (left to right)

Fig. 7.24 Bending failure above the overlapping area

Fig. 7.25 Closeup view of bending failure above the overlapping area

the Egyptian code of fire protection [22]. More importantly, those designs mainly depend on simple techniques that are commonly employed in such functions in order to introduce a new fusion that merges between date palm midribs and commonly used wide-span building techniques. Tensile fabric structures were chosen to implement the date palm midribs TAST in covering traditional wide-span market spaces *Souqs* that allow for transverse and longitudinal expansion [23]. The saddle tent design was based on fixing a cotton fabric in the form-active shape of the hyperbolic parabola tensioned along the chords of two main TASTs, as shown in Fig. 7.26. Fixing the fabric to the chords can be by using stainless steel tubes around the chords of each TAST where each tube has a hook that passes through the steel rings in the fabric, as shown in Fig. 7.27, or by using ropes through the steel rings to be coiled around the chord of each TAST as a lighter and a cheaper alternative, as shown in Fig. 7.28.

The saddle tent was analyzed structurally using the validated FEM model properties [9] under extreme wind loads by the Egyptian code. The in-plane and out-of-plane wind loads on the system are shown in Fig. 7.29a, b. The shell loads are shown in Fig. 29c. The deformations under the in-plane and out-of-plane wind loads on the system are shown in Fig. 7.29d, e. The design of the cross-sections of the system was safe by a large factor, as shown in Fig. 7.29f. The deformations of the model were small and recoverable. This illustrates the ability of the saddle vault design to be altered to cover wide spans in a different alternative. Thus, the saddle tent was proved to be safe under own weight and wind loads [22].

Fig. 7.26 Perspective of the saddle tent [22]

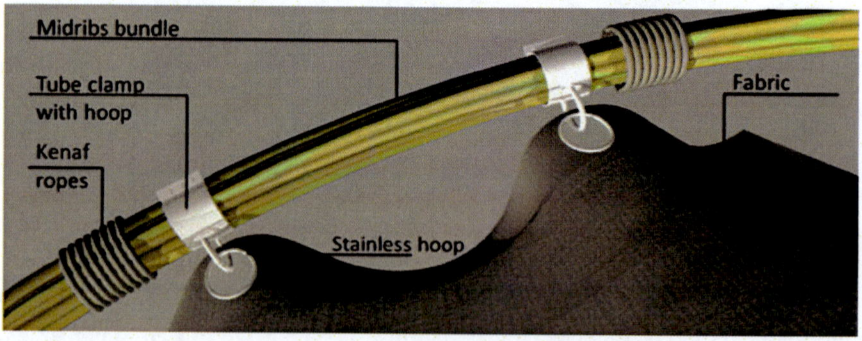

Fig. 7.27 Connection between the tensile fabric and the chords using stainless tubes [22]

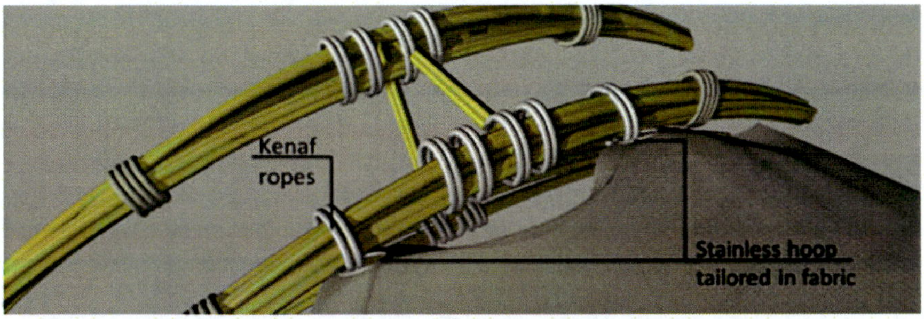

Fig. 7.28 Connection between the tensile fabric and the chords using ropes [22]

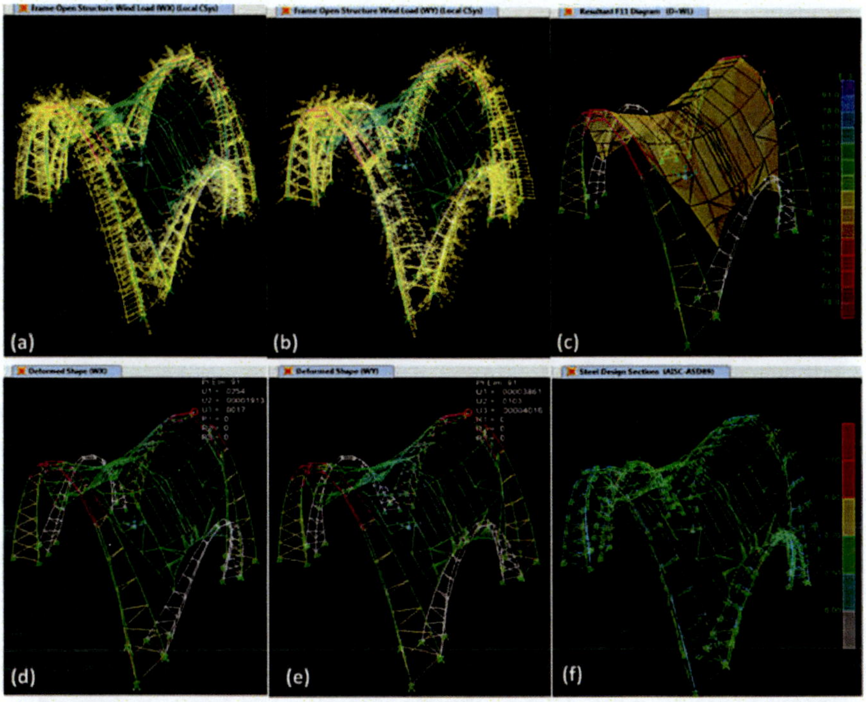

Fig. 7.29 Digital analysis of the saddle vault under own weight, wind loads and shell load of 100 kg/m^2 [22]. **a** The in-plane wind loads. **b** The out-of-plane wind loads. **c** The shell loads. **d** Deformation resultant from the in-plane wind loads. **e** Deformation resultant from the out-of-plane wind loads. **f** Design is safe

In addition, braced barrel vaults structures (Fig. 7.30) were chosen to implement the date palm midribs TAST in covering multi-functional social spaces that allow for transverse and longitudinal expansion [22]. The braced barrel vault design was based on using date palm midribs bundles as diagonal lamella shell vault that is fixed along two TASTs and stiffened by palm trunk purlins. The proposed fixation of the purlins, as the halved trunks shown in Fig. 7.31, can consist of ropes and stud nuts to fasten the grip over the diagonal arched bundles and the purlins.

The braced barrel vault was analyzed structurally using the validated FEM properties [9] under extreme wind loads by the Egyptian code and 100 kg/m^2 shell loads to imitate the weight of simple roofing [24]. The in-plane and out-of-plane wind loads on the system are shown in Fig. 7.32a, b. The shell loads are shown in Fig. 7.32c.

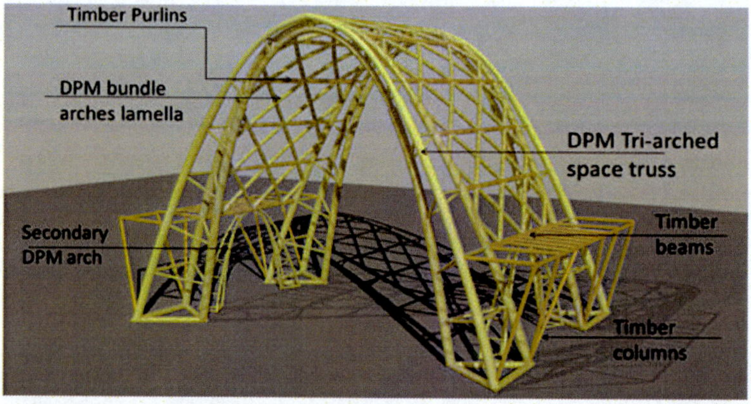

Fig. 7.30 Perspective of braced barrel vault with date palm trunk posts and beams [22]

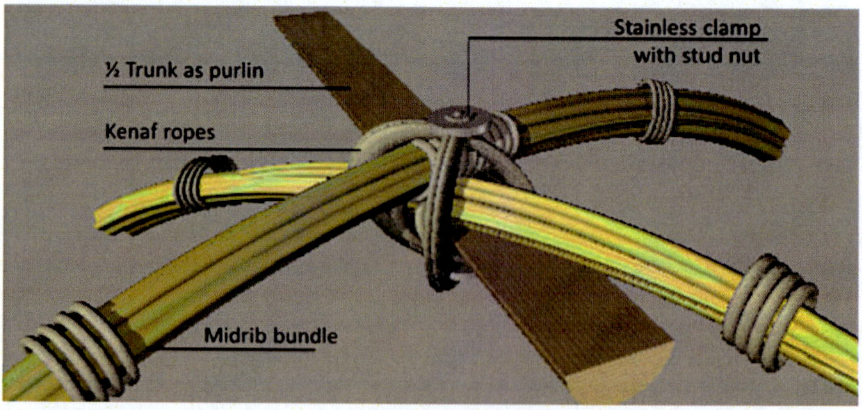

Fig. 7.31 Detail of the connection in the Zollinger lamella bracing between the bundles and the purlins [22]

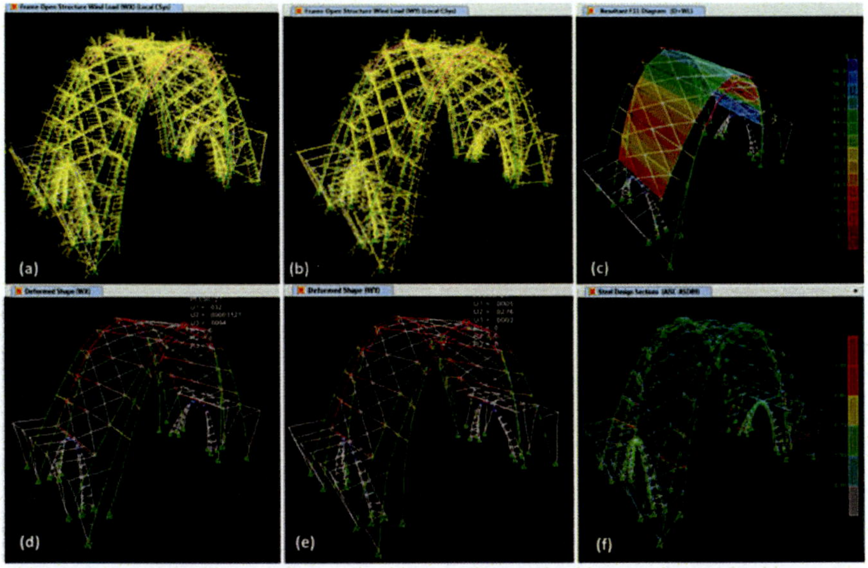

Fig. 7.32 Digital analysis of the braced barrel vault under own weight, wind loads and shell load of 100 kg/m^2 [22]. **a** The in-plane wind loads. **b** The out-of-plane wind loads. **c** The shell loads. **d** Deformation resultant from the in-plane wind loads. **e** Deformation resultant from the out-of-plane wind loads. **f** Design is safe

The deformations under the in-plane and out-of-plane wind loads on the system are shown in Fig. 7.32d, e. The design of the cross-sections of the system was safe by a large factor, as shown in Fig. 7.32f. The deformations of the model were small and recoverable. The system proved to be safe under own weight, extreme wind loads and standard sheathing load [22].

7.2.2 Facades and Panels

7.2.2.1 Date Palm Facades

Used Byproduct: *Midribs*

Darwish [25] worked on developing sustainable date palm midribs lattice to work as a low-cost solar shading facade on the southern elevation of a faculty building in Aswan, Egypt. This facade was required to act as a thermal barrier to decrease the indoor solar heat gain and, consequently, saving energy by providing a better indoor air quality and, thus, reducing the annual cooling loads.

The design of the date palm facade aimed to focus on the following challenges:

- The need of a contemporary design of the facade while depending on the traditional low-cost techniques of using date palm midribs in architecture.

- The wide size of the façade panels which would increase the panels' required resistance for wind loads.
- The fibrous nature of the midribs core which decreases the durability of using steel nailing or steel connections to join the assembled members.

Therefore, Darwish [25] designed the date palm facades to be flexible enough to adapt to the wind loads and to depend on the sliding mechanism where the midribs panels can be slid and fixed along grooves made in a main wood frame.

The sliding mechanism was chosen in order to facilitate maintenance by replacing damaged panels with new panels and to concentrate on using conventional steel connections on the wood frame while the joinery used for the midribs is exclusively traditional joinery that is established to be suitable for the fibrous nature of the midribs core. The shop drawing of the panel is shown in Fig. 7.33.

The facade was divided into wood frame panels where each panel consisted of date palm units that were designed to be slid along the grooves in the members of the wood frame. The designed date palm midribs units varied between basic horizontal midribs and diagonal lattice arrangement (Fig. 7.34) that was inspired from the traditional reeds huts in the Manzala lake region in [26].

The wood frames of the panels were made from beech wood which is common in the Egyptian market. The vertical frame members were grooved to allow for sliding the midribs units. In addition, each panel had longitudinally divided wood members with the ability to be removed to replace and slide the damaged midribs units in the concerned panel only. Thus, the maintenance was facilitated to reduce time consumption and to minimize the negative impact on the aesthetic look of the faculty elevation during maintenance. Two prototypes of the panels were constructed in the carpentry workshop in the faculty of engineering, Ain Shams University. The main difference between both prototypes is the type of the horizontal beams used in the wood frames.

The first prototype (Figs. 7.35 and 7.36) depended on using the midribs as horizontal beams with the addition of small wooden sections to fix the midribs to the vertical wood frame members by steel nails. Although the dependence on using wood in this prototype was minimized, the efficiency of maintenance was also reduced in the case of damaged midribs members or small wooden section which their durability is compromised by the steel nails and their rusting.

On the other hand, the second prototype (Figs. 7.37 and 7.38) depended on using the frame and the horizontal beams entirely from wood in order to facilitate maintenance, which was simply based on replacing the midribs units by sliding along the grooving of the wood members. The second prototype was deemed to be more efficient due to the following reasons: depending on sliding instead of steel nails to fix the midribs units to the frame, the ordinated design enjoyed high aesthetic value that matched the modern design of the faculty building and the simplified maintenance. Furthermore, not using steel nails through the midribs sustains their structural integrity and allows for a wider range of reuse of midribs pieces for the manufacture of *Mashrabiah* products or strips to be used as a core-layer in blockboards, etc.

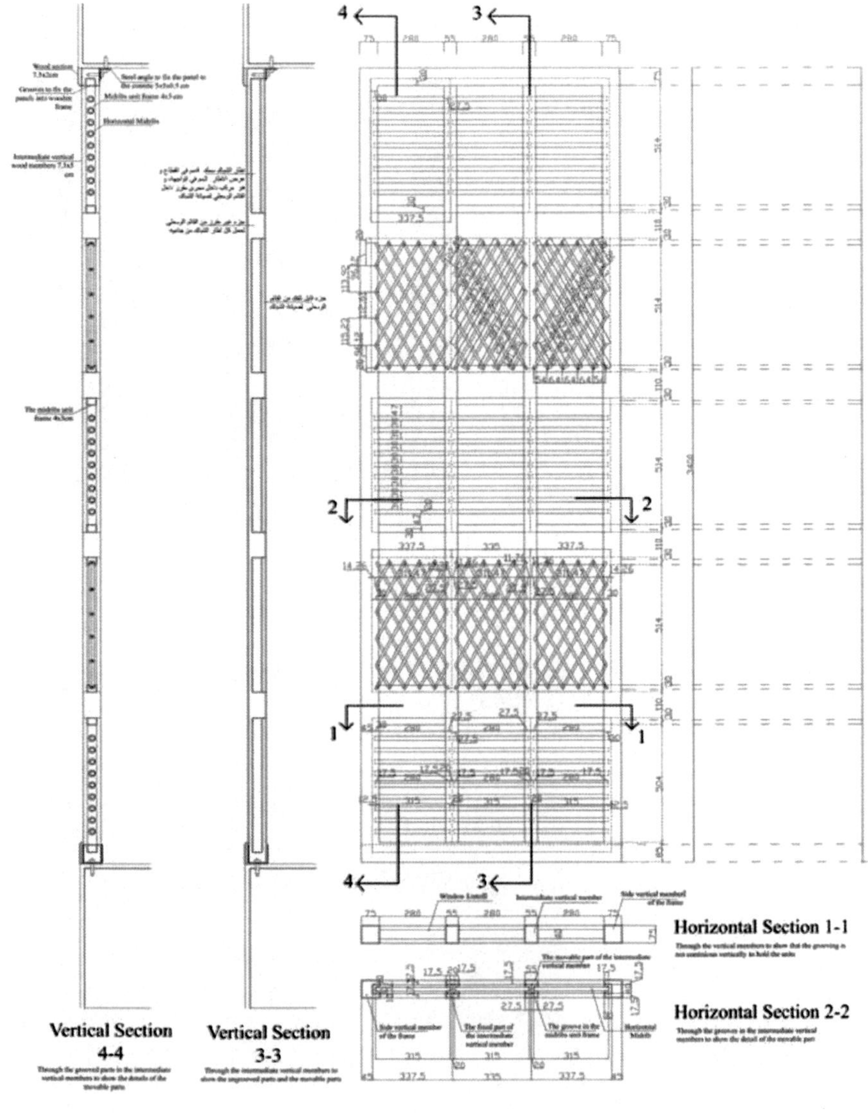

Fig. 7.33 Shop drawings of the panel

Fig. 7.34 Diagonal lattice arrangement of a midribs unit

Fig. 7.35 First prototype of the panel using horizontal midrib beams

7.2.2.2 Date Palm Sandwich Panels

Used Byproduct: *Trunk*

The trend of using residues in developing sandwich panels for structural, construction and furniture studies is becoming highly in demand globally. This type of sandwich panels consists of two stressed faces such as plywood, oriented strand board (OSB) or veneer glued and compressed to unstressed and lightweight core. This core can consist of insulation polymer, such as in the case of the extruded polypropylene core in Structural Insulated Panels (SIP), or wood residues strips in the case of blockboards

Fig. 7.36 Using separate wood sections in the first prototype to fix the midribs to the intermediate vertical wood members by steel nails

Fig. 7.37 Second prototype using horizontal beams made from wood

and batten-boards. The mechanical properties of the composite sandwich panels are enhanced by using the high strength face layers. This allows for more experimenting in selecting the core. Haseli [27, 28] focused on using date palm trunk as the raw material of the core of a batten-board and a blockboard. In addition, the impact of the manufacturing details, such as the adhesive type, the core strips width and orientation, on the modulus of elasticity of the developed panels was investigated.

The date palm trunks were cut into short lumbers, dried and split to strips. All the used strips had the thickness of 20 mm. The widths of the blockboard and batten-boards strips were 25 mm and 60 mm, respectively. The face layers of the date palm trunk blockboard and batten-board were made of thin MDF sheets. The bonding between the strips and the face layers depended on isocyanate-based polyurethane or polyvinyl acetate. The boards were fabricated by pressing the core strips to create

Fig. 7.38 Using grooved wood members in the second prototype to fix the horizontal midribs units and the diagonal lattice units

the core, and then the MDF face layers were placed and cold pressed onto the glued surface of the core.

The developed blockboards were classified into two types, where the core grain orientation in the first type was parallel to the main axis of the sample and the core grain orientation in the second type was perpendicular to the main axis of the sample. The same classification applied to the developed batten-boards. Three samples were produced for each classification. In addition, two different orientation samples were prepared for each type of adhesive.

As the sandwich panels consist of various materials (date palm trunk core, glue, face layers) with different properties which render them heterogeneous, the heterogeneous nature of the sandwich panels makes it more suitable to measure modulus of elasticity using the wave propagation velocity measurement. This way, the modulus of elasticity can be measured by finding the best transmission path through the samples. The best results were obtained for the parallel-core grain batten-boards either when using isocyanate-based polyurethane or polyvinyl acetate as adhesives.

7.3 Uses of Date Palm Byproducts in Processed Form

7.3.1 Date Palm Fibers Reinforced Structural Elements

Several researches investigated the performance of concrete, reinforced by date palm fibers as a natural and local alternative to steel fibers and synthetic fibers such as polypropylene [29, 30]. The most common used type of reinforcing fibers is steel fiber. However, steel fiber is vulnerable to corrosion which limits its uses [31]. Therefore, synthetic fiber is used to solve the problem of corrosion but with increased environmental and economic costs [32]. That trend of research was motivated by the reported improved performance aspects of fibers reinforced high strength concrete (HSC) such as improved flexural strength, enhanced ductility and reduced thermal conductivity [33].

7.3.1.1 Concrete

Used Byproduct: *Surface fibers around trunk*

Hot-dry climates were previously reported to have a negative effect on the mechanical properties of concrete, specially flexural strength, causing high levels of shrinkage and cracking [34]. Therefore, investigating the effect of adding natural fibers to the concrete mix was encouraged in order to enhance the concrete performance in hot-dry climates. Kriker et al. [35] investigated the performance of date palm fiber reinforced concrete. The fibers used were taken from the surface fibers around the palm trunk. Kriker et al. [35] found that male date palm surface leaf sheaths fibers acquired the highest tensile strength. Therefore, Kriker et al. [35] chose to investigate the mechanical properties of male date palm fibers reinforced concrete in hot-dry climates.

The fiber contents of the reinforced samples were 2% and 3%. The samples were cured in three types of environment: in the first environment, the samples were placed in water at 20–25 °C. In the second environment, the samples were cured in uncontrolled hot-dry climate with severe field conditions. In the third environment, the samples were placed in a steam room in the laboratory with relatively high temperature and low humidity. The samples were tested at 7 days, 28 days, 90 days and 180 days of curing for compressive and flexural strength. It was reported that the addition of the fiber reinforcements decreased the voids and cracks in the early age of the sample curing. These voids are further reduced when curing the samples in water environment due to the existence of the hydration products. The results of the compressive strength in curing environment samples showed that the compressive strength decreased with increasing the fiber content with improved ductile performance, while in each sample the compressive strength increased along with the curing age. The same profile applies for the samples from the second environment curing method. However, the compressive strength of the samples that were cured in the uncontrolled hot-dry weather (second environment) was much lower than the samples cured in water (first environment). Regarding flexural strength, although the ductility performance improved, the first crack strength also decreased with the increase of the amount and lengths of fiber content. The strength of each sample continues to increase with the curing age. And again, the flexural strength of the samples that were cured in the steam room (third environment) was much lower than the samples cured in water (first environment) [35].

This reduction between water cured samples and hot-dry climate samples or steam room samples could be justified by the lack of hydration which would lead to faster evaporation of water during the curing causing cracking in the concrete and larger voids around the fibers [35].

In addition, Kriker et al. [36] investigated the durability of date palm fiber concrete in alkaline environment. Samples of the male date palm surface fibers were submerged in two alkaline solutions NaOH and $Ca(OH)_2$ for six months and were tested along the submerging duration to determine the effect of the alkaline solutions on the fibers.

The effect of the alkaline solutions on the fibers was investigated using comparative tensile strength before and after immersion. The results of the tensile strength testing on the immersed fibers showed that the fibers had poor resistance to the alkaline solutions. The fibers became increasingly brittle with the increase of immersion duration, where immersion in $Ca(OH)_2$ was ten times more destructive to the fiber pores than immersion in NaOH. This deterioration of the fibers integrity led to the decrease of the tensile strength of the fibers. This difference could be justified because of the diffusive attack of $Ca(OH)_2$ which caused decomposition of fiber surface leading to the excessive mineralization of the fibers and the deterioration of the filaments. On the other hand, the attack of the NaOH penetrated the cells and the pores of the fibers locally creating local holes on the surface without diffusive crystallization as in the case of the $Ca(OH)_2$ attack [36]. Therefore, it was concluded that the calcium hydroxide, released during the hydration process of the concrete, would affect the durability of the male date palm surface fibers in Portland cement.

Hence, although reinforcing concrete with date palm fibers decreased the strength of the concrete, the main benefits of using date palm fibers in concrete reinforcement were enhancing its ductility, and more importantly, decreasing thermal conductivity [37]. Thus, date palm fiber reinforced concrete can be used for better thermal insulating capacity.

7.3.1.2 Concrete

Used Byproduct: *Fibers extracted from petioles and midribs*

Kareche et al. [38] investigated the durability of mortar reinforced with fibers extracted from crushed date palm petioles and midribs. Samples were prepared with three different weight fractions of fibers (control, 5%, 10% and 15%). Open porosity, resistance to swelling, resistance to drying shrinkage and resistance to acidic attack and alkaline salt solution exposure were tested after 28 days of curing.

The open porosity increased with increasing fibers content due to the increase of the voids that is accompanied with the fibers porosity and the spaces created in between the fibers and the mortar. In addition, the samples were submerged in water for 4 days, and then the samples were re-measured to determine the swelling effect. It was revealed that the water absorption and swelling increased along the increase of the fiber content as a result of the increasing porosity [38].

Regarding the concrete shrinkage, concrete generally shrinks in the drying process which results in generating stress and cracks. Therefore, special additives such as carbon fibers and silica fume are generally added to concrete to reduce the drying shrinkage [39]. Therefore, although the shrinkage increased in the fiber reinforced samples with the water loss along with the curing age, the shrinkage itself decreased with the increase of the fiber content. This revealed that adding more date palm fibers led to better shrinkage resistance and less cracking during the drying process [38].

In addition, to determine the resistance of the samples to acidic attacks, the samples were soaked in sulfuric acid solution (5 wt% H_2SO_4) at room temperature for 28 days, and then the samples were measured to determine the weight loss resultant from the acidic attack. It was noticed that increasing the date palm fiber content lowered the weight loss, while the weight loss was the highest in the case of the pure non-reinforced mortar [38].

The effect of alkalization on the reinforced samples was measured by determining the compressive strength of samples using pure mortar, non-treated samples, 28 days sodium chloride solution (5 wt% NaCl) treated samples and 28 days sodium hydroxide solution (5 wt% NaOH). Generally, it was found that the compressive strength decreased with increasing the fiber content which increases porosity of the matrix. The poor fiber/concrete adhesion also reduces the compressive strength. However, each NaCl treated sample demonstrated higher compressive strength than the non-treated samples with the same fiber content. Moreover, the NaOH treated samples demonstrated even higher compressive strength than the corresponding NaCl treated samples. This increase of the compressive strength after alkali treatment was justified by the surface modification that occurred to the fibers by the alkaline salt which reshapes the surface and improves the interfacial bonding between the fibers and the mortar matrix [38]. Regarding tensile strength, previous studies showed that 5% NaOH solution gives optimum tensile strength to the date palm fibers, while 10% NaOH solution treatment would cause excessive surface alteration which damages the fibers and decreases their tensile strength [40].

7.3.1.3 High Strength Concrete (HSC)

Used Byproduct: *Fibers extracted from leaves*

HSC has a lower strain corresponding to high stress and, consequently, a higher elastic modulus than traditional concrete. As a result, HSC is characterized to be noticeably brittle although it has high compressive strength and enhanced durability [41]. Saad et al. [30] targeted improving the brittle behavior of HSC using banana fibers (BF) and date palm leaf sheath fibers (PLSF). The fibers used were treated using the alkaline solution of NaOH by submerging for four hours in room temperature, and then the fibers were dried under sunlight [42]. The samples designed contained BF and PLSFs reinforcements in 1%, 2% and 3% fiber volume fractions. The samples were tested to determine the compressive, tensile and flexural strengths as opposed to the non-reinforced control sample. The slump test revealed that using fibers in HSC decreases the workability of the concrete. However, PLSFs samples showed significant lower workability than the samples with BF because the PLSFs were thicker and taller that BF which led to the weak workability of the PLSFs samples [30].

The compressive strength was measured for all samples at 7 days, 28 days and 56 days of curing. The compressive strengths measured for PLSFs were higher than those of BF, where the compressive strength decreased with the increase of the added percentage of the fibers. However in both cases, the compressive strength testing showed gradual decrease along the age of curing with ductile failure instead of the brittle failure of the non-reinforced samples [30]. The modulus of elasticity showed a gradual decrease along with increasing the added fibers percentage. Interestingly, the modulus of elasticity of the PLSFs samples were generally higher than the BF samples due to the stronger inter-adhesion between the concrete and PLSFs [30].

In contrary, tensile strength increased with the addition of 1% and 2% PLSFs than in the case of the non-reinforced sample, while the tensile strength decreased with the addition of 1%, 2% and 3% BF. However, the 3% PLSFs samples showed a decrease of the tensile strength. The failure after reaching the maximum tensile strength showed higher ductility without symmetrical split as in the case of the non-reinforced sample [30]. In addition, the flexural strength of the 1% PLSFs sample was slightly higher than the non-reinforced sample, whereas the flexural strength decreased for the 2% and 3% PLSFs samples and the BF sample [30].

7.3.2 *Date Palm Fibers Insulation*

The natural hot habitat of date palms led to their natural adaptation to high temperature. The date palm fibers adapted by acquiring high thermal insulation to reach more suitable internal temperature in order to grow the dates safely in temperatures over 50 °C in shaded areas in tropical and subtropical climates. Agoudjil et al. [43] reported that the low thermal conductivity of date palm fibers is less than hemp and close to sisal, which means that date palm fibers are highly suitable for utilization in thermal insulation.

7.3.2.1 Mortar Composite Insulation

Used Byproduct: *Fibers extracted from petioles and midribs fibers*

Benmansour et al. [44] investigated the thermal and mechanical properties of mortars reinforced with date palm fibers extracted from dried date palm petioles and midribs. The developed mortar mixes were tested to determine their potential as thermal insulators in buildings. The cement, sand and water samples were reinforced by three collections of fibers: fine fibers with a mean diameter of 3 nm, wide fibers with a mean diameter of 6 nm and mixture of both sizes. For every group of samples reinforced with one of the three fiber collections, the fiber contents varied to the following values: 0%, 5%, 10%, 15%, 20%, 25% and 30%.

The fibers were tested to determine the impact of the fibers size on their water absorption. In addition, the samples were tested to determine their thermal conductivity and compressive strength.

Date palm fibers demonstrated a high ability for water absorption that they can absorb water up to three times their dry weight due to their porous structure. In addition, it was found that the 3 nm diameter fiber collection samples absorbed higher levels of water than the 6 nm diameter fiber collection samples and in a faster rate. This was justified by the increased length to diameter ratio of the 3 nm diameter fiber collection which had more surfaces with more pores to absorb water than in the case of the 6 nm diameter fiber collection.

Moreover, it was once again confirmed that the addition of the date palm fibers increases the voids in the mortar matrix which decreases the density of the matrix. This decreased density, along with the low thermal conductivity of the fiber, reduces the overall thermal conductivity of the composite. It was found that for the samples with fiber content up to 15%, the thermal conductivity of the 6 nm diameter fiber collection was higher than the 3 nm diameter fiber collection or the mixed diameter collection, as the shorter fibers generated more voids because they are more difficult to align densely. However, for fiber contents higher than 15%, all fibers collections behaved similarly in the thermal performance, except for the samples with 30% fiber content where the 6 nm diameter fiber collection had lower thermal conductivity than the 3 nm diameter fiber and mix fiber size collections. The slope of the thermal conductivity reduction became almost flat beyond 20% fiber content. Therefore, it was recommended to limit the thermal content to 20% in order to achieve balance between the thermal conductivity benefit and the risk of high water absorption.

In addition, it was reported that the thermal conductivity increased with increasing the moisture to the limit that the wet thermal conductivity is almost two times dry thermal conductivity. This was justified because water, now filling the voids in the mortar matrix, has higher thermal conductivity than air. Regarding of the impact of the fibers size on the wet thermal conductivity, it was found that the impact was negligible as the wet thermal conductivity was more dependent on the density of the composites and the used fiber contents. The compressive strength test showed that the compressive strength decreased with increasing the fiber content, where the fiber content had higher negative impact on the compressive strength than the size of the fiber diameter collection. Finally, it was realized that the obtained compressive strengths at low fiber contents (5%, 10% and 15%) qualify the developed mortar composite to be used as structural and bearing insulators. Therefore, it was recommended that using 5–15% date palm fiber content is optimum to achieve balance between low thermal conductivity, adequate water absorption rate and reliable compressive strength.

7.3.2.2 Gypsum-Based Insulation Composite Panels

Used Byproduct: *Fibers extracted from petioles and midribs*

Chikhi et al. [45] investigated the development of a composite for thermal insulation in construction or in solar thermal panels based on date palm fiber and gypsum material. The gypsum material used was hemihydrate gypsum. The date palm fibers used were extracted from crushed dried petioles and midribs and were classified into two collections of fiber sizes (3 mm diameter and 6 mm diameter). The samples produced were prepared using seven different fiber contents (0%, 1.2%, 3%, 5%, 7%, 8% and 10%) and tested at 14 days and 28 days of curing. The samples were tested to determine their water absorption, compressive and flexural strengths and thermal conductivity.

Regarding the water absorption, it was found that the increase in the fiber content led to a longer time until full saturation. This was justified because increasing the fiber content meant increasing voids in the composite matrix. Under the osmosis pressure, those voids would firstly attract water to fill in rapidly. However, the water movement would eventually slow down until reaching full saturation. On the other hand, lower fiber contents would create narrower voids, leading to a faster flow of the water into the voids under the capillary effect. In addition, increasing the fiber content was associated with higher water absorption, due to the hydrophilic nature of the fibers. Thus, it was realized that increasing the fiber content in the gypsum composites increases the water absorption and the time to reach full saturation. Furthermore, it was realized that the samples containing 3 nm diameter fiber collections had a higher rate of water absorption than the samples containing the 6 nm diameter fiber collection, because the smaller fibers had higher surface to length ratio, with more porous surfaces exposed to water.

The compressive strength of the 0% fiber content samples was found to be higher than the samples with positive fiber content. However, it was interesting to find that the loss of the compressive strength at 28 days of curing was lower than the loss reported at 14 days because of the lower rate of hydration in the samples. In addition, the compressive strengths of the samples with 3 mm diameter fibers were higher than those of the samples with 6 mm diameter fibers were higher. The highest compressive strengths reported for samples with positive fiber contents at 28 days was for 1.2% fiber content of the 6 mm diameter collection followed by 5% fiber content of the 3 mm diameter collection. The results of the flexural strength tests showed that increasing the fibers content decreased the flexural strength. This loss of the flexural strengths was higher at 14 days than at 28 days of curing. In addition, the flexural strength of the samples with the 3 mm diameter fiber collection was higher than the 6 mm diameter fiber collection. The maximum obtained flexural strength value with positive fiber content was reported or 3% fiber content of 3 mm diameter fiber collection. Thus, the compressive and flexural strengths of the date palm fibers reinforced gypsum composites indicated that these composites can replace conventional gypsum boards in construction.

The thermal conductivity was found to decrease significantly with increasing the fiber content due to the high difference between the thermal conductivities of the fibers and the gypsum and the voids created inside the matrix around the fibers. The lowest thermal conductivity was reported for the 10% fiber content samples. These results were also confirmed by Amara et al. [46], who focused on using 10% date palm fibers in gypsum composites to produce high thermal insulation plasters. The impact of the fibers size on the thermal conductivity was found to be negligible.

In conclusion, it was recommended to use 5% date palm fiber content in gypsum matrix to produce gypsum biocomposite panels with efficient thermal performance and adequate mechanical properties.

7.3.2.3 Lime-Based Insulation Composite

Used Byproduct*: Surface fibers extracted from around female palm trunks*

Researchers [47] worked on developing a sustainable low-cost insulation material based on date palm fibers and lime. The fibers used in this study were crushed fibers extracted from the surface fibers around the trunks of female date palms. This fibrous cover creates a thick buffer that naturally protects the palm against severe hot weather. The particle size distribution ranged from 0.063 to 5 mm. The estimated mean size was 2 mm. The binder used to produce the samples was lime. The dry components of the samples were mixed to achieve a homogeneous mixture, and then water was added and mixed gradually according to the estimated water absorption rate of the fibers and the needed water amount to hydrate the lime. The samples were molded and left to dry in open air. The samples were tested to determine the impact of the fiber/lime ratio on the mechanical, acoustic, thermal and hygric properties. The thermal conductivity was found to decrease with increasing the fiber content for dried and undried samples. However, it was realized that the thermal capacity of the dried samples was higher than the undried samples. On the other hand, the thermal diffusivity of the dried samples was lower than the undried samples. Thus, the thermal behavior was improved by increasing the fiber content due to the increase of the porosity of the samples caused by the voids between the fibers and the lime matrix which consequently reduced the samples density.

Regarding the water absorption rate, increasing the fiber content has led to increasing the moisture buffer value. The moisture buffer value indicates the amount of the water lost or gained per open surface area in a specific time. In addition, increasing the fiber contents decreased the compressive strength. The compression did not lead to the fracture of the specimens with the increasing stress, the stress–strain curves indicated that linear elastic region continues until the compressive loading reaches the maximum compressive strength limit. At this point, the lime matrix fails due to voids around the fibers. These voids cause the plastic behavior, where the strain increases while the stress remains almost unchanged and the voids close and the composite becomes more rigid until failure. This extended plastic behavior until the gradual failure increases the modulus of elasticity, indicating higher ductility.

Nevertheless, the lime-date palm fiber composite can support its weight to be used successfully as thermal insulation material. This insulation material can be used for thermal insulation as well as acoustic insulation. Reducing the lime content for the sake of increasing the fiber content was found to enhance the acoustic absorption coefficient in medium and high frequencies.

7.3.2.4 Polyester Composite Insulation

Used Byproduct: *Date Palm Pits*

Researchers [48] conducted a research to determine the efficiency of insulation composites that consisted of unsaturated polyester and date pits. The designed ratios ranged from 0 to 70% polymer/pits. The developed composites showed a promising performance in terms of water retention. The mechanical properties of the composite specimens, including compressive strength, tensile strength, thermal conductivity, water retention and density, were investigated. The results showed that the thermal conductivity increased due the high thermal conductivity of the pits. However, the overall thermal conductivity of the polymer remained sufficiently comparable to the conventional thermal insulators. On the other hand, the compressive strength and the tensile strength were reduced with increasing the pits ratio. However, these values remained superior to those of the conventional thermal insulators. Accordingly, date palm pits were found to be a successful addition to polymers that would result in producing cheap and effective insulators with promising mechanical properties and efficient thermal performance.

7.3.2.5 Wood–Cement Composite Panels

Used Byproduct: *Midribs* and *Spadix Stems*

Nasser and Al-Mefarrej [49] and Nasser [50] investigated the suitability of date palm midribs to be used in wood–cement composite panels. The tolerance of date palm midribs toward the internal curing of the cement in the composite panels was determined by a series of hydration tests. The midribs were air-dried, grinded and passed through 40-mesh screen under untreated condition, hot water extraction and cold water extraction. Several chemical additives combinations (0%, 3% $CaCl_2$, 3% $MgCl_2$) were added as accelerators to the wood–cement matrix in order to conduct the curing process in the shortest time possible to minimize the adverse effect on the date palm midribs particles.

The particles extracted from date palm midribs in untreated conditions were found to be unsuitable to be used with cement and were classified as extremely inhibitory according to the Inhibitory Index (I). However, the best results were obtained in the case of treated particles in hot water extraction or with the addition of 3% $CaCl_2$ in the case of untreated particles.

Another attempt of reinforcing wood–cement panels depended on using cellulose fibers extracted from the spadix stems [51]. The average length of the cellulose fibers was 4 mm. The fibers were mixed with Portland cement mortar and processed in a wooden mold. The specimens produced were classified into two categories, where the first category had 0.05 cellulose fiber/cement weight ratio and the second category had 0.08 cellulose fiber/cement weight ratio. The boards were oven-dried for half an hour, and then were exposed to CO_2 curing with the following CO_2 gas concentration: 0%, 30% or 100% for 2 days.

The specimens underwent flexural tests. The flexural strength, toughness and stiffness were found to be the highest in the case of using the 0.05 fiber/cement ratio and the 100% CO_2 concentration curing. It was concluded that the CO_2 curing led to higher matrix cohesiveness between the cement and the fibers. This has led to a better performance in terms of flexural strength, flexural stiffness and flexural toughness.

7.3.3 *Date Palm Fibers Reinforced Masonry*

7.3.3.1 Date Palm Ash Reinforced Mortar Blocks

Used Byproduct: *Burnt midribs ash*

Acknowledging the fact that reinforcing concrete with date palm fibers (DPF) decreased its density and reduced its thermal conductivity [44], Ashraf et al. [52] focused on evaluating the thermal and energy performance of date palm ash reinforced mortar blocks. The date palm ash used was produced as a secondary product of burning date palm midribs as a fuel during charcoal production from firewood [53]. The tested blocks were divided into four samples with added date palm ash content ranging from 0 to 30% as a percentage of the total Portland cement content.

The equivalent thermal conductivity (k), density and thermal resistance (R) were measured for the samples experimentally using the Transient Hot Bridge (THB) method. The equivalent thermal conductivity was found to be the highest in the control sample (0% DPF), while the equivalent thermal conductivity decreased along increasing the ash content to be the lowest in the control sample (30% DPF). The same pattern applied for the density which decreased along increasing the ash content, which has led to creating larger voids in the blocks which reduced the thermal conductivity. As a result, thermal resistance increased with increasing the ash content. These results were used to produce energy simulations to assess the energy saving performance of the developed blocks.

The blocks were used in the simulation program as a structural component in an external wall in a typical office building in Dhahran, KSA. The annual energy consumption of the simulated building using the control sample was calibrated to be the nearest to the annual energy consumption of the actual building. The simulation results of using the other three samples showed that the energy saving increased slightly with increasing the ash content. This was justified because the heat gain through a wall is lower than other building structural components. Cooling loads and the average monthly indoor temperature were also found to decrease with increasing the ash content. In addition, the unit cost of the masonry blocks was found to decrease significantly with increasing the ash content, where the control sample block was the most expensive and the ash content percentage sample block was the cheapest.

7.3.3.2 Date Palm Fibers Reinforced Raw Earth Bricks

Used Byproduct: *Fibers extracted from leaves*

Abanto et al. [54] reported that using biocomposite materials in raw earth blocks could help reduce cooling loads due to their enhancement of the thermal insulation of raw earth bricks. In addition, such biocomposite materials provide an affordable and eco-friendly alternative to conventional building materials. Raw earth bricks, commonly known as adobe, are manufactured by molding prepared earth to a brick-like shape and drying the bricks in open air. Raw earth bricks are affordable, eco-friendly and thermally efficient. However, their sensitivity to water would reduce their durability [55]. Therefore, raw earth blocks are often stabilized to improve their mechanical behavior under the influence of water. This stabilization can be achieved by adding cement or lime or both to stabilize the brick mixture chemically [56], or by incorporating fibers into the mixture physically [57]. Bouhicha et al. [57] reported that adding straw reduced the shrinkage and curing and increased the flexural strength, shear strength and ductility of the reinforced bricks when compared to straw-free bricks. In addition, the compressive strength was found to increase at using a specific optimal straw content in the brick mixture. Moreover, Vatani Oskouei [58] reported that adding of palm fibers to a mixture of clay soil, sand and gravel, along with straw, wood shavings and rice hulls improved the compressive and tensile strengths as the palm fibers prevented the formation of deep cracks in bricks during shrinkages and curing. Therefore, Khoudja et al. [55] investigated the mechanical, physical and thermal properties of raw earth blocks stabilized with lime and reinforced by date palm fibers.

The tested bricks samples consisted of soil, crushed sand, quicklime, water and date palm waste. The used date palm fibers were aggregates produced from palm grove maintenance. Six samples were produced where the date palm waste content percentage were as 0%, 2%, 4%, 6%, 8% and 10%. The samples were tested to determine their compressive strength, flexural strength, capillary absorption, total absorption, ultrasonic testing, thermal conductivity and specific heat. As predicted, the apparent density, and consequently, the speed of the ultrasonic waves in the bricks decreased along with increasing the date palm waste content. The voids resultant from increasing the waste content led to decreasing the capacity of the bricks which slowed down the propagation of the ultrasonic waves through the bricks. This indicated the high potential of using date palm waste reinforced raw earth bricks for acoustic insulation.

Capillary absorption and total absorption increased along with increasing the date palm waste content due to the increasing voids that were generated by the fibers in the clay matrix and the hydrophilic nature of the fibers. Remarkably, the bricks after immersion in water for 4 days remained stable with their original capacity without degradation. In addition, the compressive strength decreased along with decreasing the sample density resulting from increasing the waste content. The highest compressive strength was reported for the control sample without any date palm waste content. Nevertheless, all the samples were qualified by the Turkish Standard institute (Turkish Standard Institution, Ankara, 1985—TSE: Adobe Blocks and Production Methods TS 2514) for adobe construction. The failure of the fibers reinforced samples was found to be gradual and ductile.

In addition, it was reported that the peaks of the stress–strain curves of the compressive test flattened more with increasing the fibers content which demonstrated increasing ductility of the fibers reinforced samples in comparison with the control sample. This ductility was also accompanied with the significant reduction of the apparent modulus of elasticity values along with increasing fibers content. Similarly, flexural strength and flexural modulus of elasticity dropped significantly with the increase of the fibers content. However, this high ductility, where the strain increases with increasing the fibers content, qualifies date palm waste reinforced samples to be used in seismic construction.

Furthermore, the thermal conductivity decreased with the increase of the fibers content, which led to increasing thermal insulation. This was justified by the decrease of the density of the fiber reinforced bricks and the increase of the fibers which have low thermal conductivity by nature. In addition, the specific heat was found to increase with the addition of the waste content, which would improve the heat conservation of the date palm fibers reinforced bricks.

7.3.3.3 Date Palm Surface Fibers Reinforced Compressed Earth Blocks

Used Byproduct: *Surface fibers around trunk*

Compressed earth blocks are a modern brick type derived from traditional raw earth "adobe" bricks. Compressed earth blocks are based on the ideas of compacting earth to improve the mechanical properties of the molded earth bricks [59]. Previous studies reported that reinforcing compressive earth block with natural fibers reduces shrinkage cracking, improves ductility, increases tensile strength and decreases thermal conductivity of the composite blocks [57, 60–63]. These natural fibers included hibiscus cannabinus, sisal, coconut, jute, straw, bamboo and date palm leaf sheaths fibers. Therefore, Taallah et al. [64] investigated the mechanical properties and hydroscopicity behavior of date palm fiber reinforced compressed earth blocks.

The compressed earth block samples used consisted of soil, ordinary Portland cement, water, crushed sand and date palm fibers. The fibers used were male date palm leaf sheaths fibers as used previously in [36]. The composition of the samples varied according to the cement content (5%, 6.5%, 8%) and the fibers content (0%, 0.05%, 0.1%, 0.15%, 0.2%) creating a matrix of 15 different samples.

The samples were tested to determine their dry and wet compressive strength, dry tensile strength, total absorption performance and swelling by immersion performance. The result of the dry compressive strength test showed that the dry compressive strength increases with increasing the cement content. The addition of fibers to the 8% cement content samples slightly increased the compressive strength, where the highest value was recorded to the 0.05% fibers content addition. However, compressive strength was slightly decreased by the addition of the fiber contents to be around 6.5% and 5% of the cement content samples. This was justified by that the lower cement contents allowed for the impact of the fibers to be more evident on creating voids in the brick matrix which decreases the compressive strength. On the other hand, the 8% cement content released calcium silicate hydrates, calcium aluminate hydrates and the calcium aluminosilicate hydrates, which filled the voids in the 0.05% fibers content matrix and enhanced the rigidity of the samples and increased the compressive strength. On the other hand, it was found that increasing the fibers content in the samples was associated with the decrease of the wet compressive strength.

While it was reported earlier that the addition of natural fibers to adobe reduced shrinkage cracking and enhanced ductility, this was not the case for compressed earth blocks. The tensile strength was found to decrease along with increasing the fiber content. This reduction was justified by the low adhesion between the fibers and the brick matrix due to the decompression after removing the loads used during manufacturing the compressed blocks. This low adhesion causes slipping of the fibers against the matrix and the poor transfer of forces between the fibers and the matrix. However, applying compacting pressure on the samples during mixing was found to decrease the internal voids, increase the density of the samples and enhance dry and wet compressive strengths.

Similarly as in the case of date palm fibers reinforced concrete and raw earth bricks, water absorption and swelling increased with the increase of the fiber content in the samples due to the hydrophilic nature of the fibers and the generated voids around the fibers in the block matrix. These voids are generated essentially during the mixing and drying stages, where the hydrophilic fibers tend to absorb water and swell pushing the brick matrix away. Eventually, the fibers shrink back after losing the absorbed water leaving voids around the fibers which would create a porous matrix highly prone for water absorption and swelling.

7.4 Conclusion

In their natural form, date palm midribs were found to be suitable to be used as façade lattices and truss members after standardizing the cross-section of the midribs or as curved truss chords. Similarly, date palm trunks were used as the core of sandwich panels. On the other hand, fibers extracted from date palm byproducts allowed for extensive fields of utilization. Concrete reinforced by fibers extracted from trunk, midribs and leaves were found to have enhanced tensile strength and ductility. However, the addition of the fibers reduced compressive strength performance of concrete because of the cavities formed around the fibers in the matrix during curing. These cavities reduce the overall density of the concrete leading to a reduced thermal conductivity. This reduction has led to suggesting the use of the fiber reinforced concrete as insulators for its enhanced thermal performance. Other insulators involving the use of date palm fibers; depended on mortar, gypsum, lime, polyester and wood–cement composites; showed promising thermal performances. This enhanced thermal performance was long recognized in traditional construction techniques such as using them in the mud brick mixtures. This performance was achieved by developing fibers reinforced mortar bricks, raw earth bricks and compressed earth bricks. Thus, it can be concluded that date palm byproducts possess great potentials in the field of construction, especially in the fields of light structural members, insulation composites and brick mixtures.

References

1. El-Mously H, Darwish EA (2020) Date palm byproducts: history of utilization and technical heritage. In: Midani M, Saba N, Alothman OY (eds) Date palm fiber compos. Springer Singapore, Singapore, pp 3–71. https://doi.org/10.1007/978-981-15-9339-0_1
2. Mansour Y, Elmously H, Darwish EA (2017) Utilizing palm rachis for eco-friendly and flexible construction in Egypt. In: WSBE17 Hong Kong—transform. Our Built Environ Innov Integr Putt Ideas Action, Construction Industry Council and Hong Kong Green Building Council Limited, Hong Kong. https://www.wsbe17hongkong.hk/download/WSBE17%20Hong%20Kong%20-%20Conference%20Proceedings.pdf
3. Azzam OA (1960) The development of urban and rural housing in Egypt. Swiss Federal Institute Of Technology, Zurich
4. Naguib EE, Hanafy AW (2008) Traditional crafts encyclopedia: vernacular architecture. Asala Organization for Traditional Crafts, Cairo
5. Ahmed RM (2014) Lessons learnt from the vernacular architecture of bedouins in Siwa Oasis, Egypt. In: 31st International symposium automation robotic construction, Sydney. https://doi.org/10.22260/ISARC2014/0123
6. Hassan NI (2001) Use of local raw materials in low cost roofing. Ain Shams University, Cairo, Institute of Environmental Studies & Research
7. Ambrose JE, Tripeny P (2012) Building structures, 3rd edn. Wiley, Hoboken, N.J.
8. El-Shabasy AB, El-Mously HI (1997) A study of the variation of the tensile strength across the cross section of date palm leaves' midrib. In: Sarton LAJL (eds), European conference on advanced materials and processes and applications, federation of European materials societies, Nederlands Society for Materials Science, Materials Funct Des Proc 5th Eur Conf Adv Mater Process Appl vol 4 Characteristics Production, Nederlands Society for Materials Science, Zwijndrecht, The Netherlands, pp 4/597–4/600. https://www.tib.eu/en/search/id/BLCP%3ACN024422797/Study-of-the-Variation-of-Tensile-Strength-across/
9. Darwish EA, Mansour Y, Elmously H, Abdelrahman A, Moustafa A (2020) Structural performance of arched space trusses using date palm midribs for light and cost-effective construction in Egypt. Buildings 10:106. https://doi.org/10.3390/buildings10060106
10. El-sherbiny TR (2010) Use of palm midribs as structural elements. Ain Shams University, Cairo, Faculty of Engineering

11. El-Mously H (2001) The Industrial use of the date palm residues: an eloquent example of sustainable development. In: United Arab Emirates University, El Ain, p 21. http://citeseerx.ist.psu.edu/viewdoc/download?doi=10.1.1.588.1154&rep=rep1&type=pdf
12. Darwish EA, Mansour YM, Elmously H (2018) Date palm rachis as a local and renewable structural material for rural communities in Egypt. Int J Environ Sci Sustain Dev 2:115. https://doi.org/10.21625/essd.v2i1.173
13. Darwish EA, Mansour Y, Elmously HI (2018) Development of the structural uses of date palm rachis for cheap wide span construction for rural communities in the middle east. IOP Conf. Ser. Mater. Sci. Eng. 371:012023. https://doi.org/10.1088/1757-899X/371/1/012023
14. Piesik S (2012) Arish: palm-leaf architecture. Thames & Hudson, London
15. Almusaed A, Almssad A (2015) Building materials in eco-energy houses from Iraq and Iran. Case Stud Constr Mater 2:42–54. https://doi.org/10.1016/j.cscm.2015.02.001
16. Darwish EA, Mansour Y, Elmously H, Abdelrahman A (2019) Development of sustainable building components utilizing date palm midribs for light wide-span multi-purpose structures for rural communities in Egypt. J. Build. Eng. 24:100770. https://doi.org/10.1016/j.jobe.2019.100770
17. Piesik S, Abdelouahhab Z, Coleman J, Sheehan P, Mangelsdorf W, Alkhouri S, Sunday P, A paradigm of agro-ecosystems of date palm oases understood in the wider context of economic and culture ecosystems through transfer of knowledge and technology in the deseret regions. In: Cancun, Mexico. https://www.researchgate.net/publication/281773609_A_Paradigm_of_Agro-ecosystems_of_Date_Palm_Oases_Understood_in_the_Wider_Context_of_Economic_and_Culture_Ecosystems_through_Transfer_of_Knowledge_and_Technology_in_the_Deseret_Regions/stats
18. Macdonald A (2001) Structure and architecture, 2nd edn. Architectural Press, Oxford
19. Lan TT (2005) Handbook of structural engineering, 2nd edn. CRC Press, Boca Raton, Fla
20. Ching FDK (2012) A visual dictionary of architecture, 2nd edn. Wiley, Hoboken, N.J.
21. The Egyptian Code for Design & construction of concrete structures (2006)
22. Darwish EA (2020) Developing wide-span building techniques by using low cost traditional date palm midribs bundles. Ain Shams University, Cairo, Faculty of Engineering
23. Elmessiry N (2012) Al-Souq: traditional Egyptian markets. General Egyptian Book Organization, Cairo
24. Neufert E, Herz R (1970) Architects' data, 1st English, language. Lockwood, London
25. Darwish EA (2017) Design of building elements for fast construction utilizing agro-components in Egypt. Ain Shams University, Cairo, Faculty of Engineering
26. Helal A (1989) Spontaneous architecture in Egypt: analytical study of reed architecture in Manzala lake region. Helwan University, Cairo, Faculty of Fine Arts
27. Haseli M, Layeghi M, Zarea Hosseinabadi H (2018) Characterization of blockboard and battenboard sandwich panels from date palm waste trunks. Measurement 124:329–337. https://doi.org/10.1016/j.measurement.2018.04.040
28. Haseli M, Layeghi M, Zarea Hosseinabadi H, Evaluation of modulus of elasticity of date palm sandwich panels using ultrasonic wave velocity and experimental models. Measurement 149:107016. https://doi.org/10.1016/j.measurement.2019.107016
29. Aljalawi NMF (2019) Effect of sustainable palm fiber on high strength concrete properties. IOP Conf Ser Mater Sci Eng 518:022004. https://doi.org/10.1088/1757-899X/518/2/022004
30. Saad M, Agwa IS, Abdelsalam Abdelsalam B, Amin M (2020) Improving the brittle behavior of high strength concrete using banana and palm leaf sheath fibers. Mech Adv Mater Struct 1–10. https://doi.org/10.1080/15376494.2020.1780352
31. Islam MS, Alam S (2013) Principal component and multiple regression analysis for steel fiber reinforced concrete (SFRC) beams. Int J Concr Struct Mater 7:303–317. https://doi.org/10.1007/s40069-013-0059-7
32. Islam MS, Ahmed SJ (2018) Influence of jute fiber on concrete properties. Constr Build Mater 189:768–776. https://doi.org/10.1016/j.conbuildmat.2018.09.048
33. Thielemans W, Wool RP (2004) Butyrated kraft lignin as compatibilizing agent for natural fiber reinforced thermoset composites. Compos Part Appl Sci Manuf 35:327–338. https://doi.org/10.1016/j.compositesa.2003.09.011

34. Alekrish AA, Alsayed SH (1994) Shrinkage of fibre and reinforced fibre concrete beams in hot-dry climate. Cem Concr Compos 16:299–307. https://doi.org/10.1016/0958-9465(94)90042-6
35. Kriker A, Debicki G, Bali A, Khenfer MM, Chabannet M (2005) Mechanical properties of date palm fibres and concrete reinforced with date palm fibres in hot-dry climate. Cem Concr Compos 27:554–564. https://doi.org/10.1016/j.cemconcomp.2004.09.015
36. Kriker A, Bali A, Debicki G, Bouziane M, Chabannet M (2008) Durability of date palm fibres and their use as reinforcement in hot dry climates. Cem Concr Compos 30:639–648. https://doi.org/10.1016/j.cemconcomp.2007.11.006
37. Mo KH, Alengaram UJ, Jumaat MZ, Yap SP, Lee SC (2016) Green concrete partially comprised of farming waste residues: a review. J Clean Prod 117:122–138. https://doi.org/10.1016/j.jclepro.2016.01.022
38. Kareche A, Agoudjil B, Haba B, Boudenne A (2020) Study on the durability of new construction materials based on mortar reinforced with date palm fibers wastes. Waste Biomass Valorization 11:3801–3809. https://doi.org/10.1007/s12649-019-00669-y
39. Xu Y, Chung DDL (2000) Reducing the drying shrinkage of cement paste by admixture surface treatments. Cem Concr Res 30:241–245. https://doi.org/10.1016/S0008-8846(99)00239-2
40. Oushabi A, Sair S, Oudrhiri Hassani F, Abboud Y, Tanane O, El Bouari A (2017) The effect of alkali treatment on mechanical, morphological and thermal properties of date palm fibers (DPFs): study of the interface of DPF–Polyurethane composite, South Afr. J Chem Eng 23:116–123. https://doi.org/10.1016/j.sajce.2017.04.005
41. Gümüş M, Arslan A (2019) Effect of fiber type and content on the flexural behavior of high strength concrete beams with low reinforcement ratios. Structures 20:1–10. https://doi.org/10.1016/j.istruc.2019.02.018
42. Debnath K, Roy S, Bose T, Adhikari S, Sinha V, Rabha V (2019) A study of chemical treatment of natural fibers and its effect on the mechanical properties of developed composites. In Rakesh PK, Singh I (eds) Process Green Compostion. Springer Singapore, Singapore, pp 65–80. https://doi.org/10.1007/978-981-13-6019-0_5
43. Agoudjil B, Benchabane A, Boudenne A, Ibos L, Fois M (2011) Renewable materials to reduce building heat loss: characterization of date palm wood. Energy Build 43:491–497. https://doi.org/10.1016/j.enbuild.2010.10.014
44. Benmansour N, Agoudjil B, Gherabli A, Kareche A, Boudenne A (2014) Thermal and mechanical performance of natural mortar reinforced with date palm fibers for use as insulating materials in building. Energy Build 81:98–104. https://doi.org/10.1016/j.enbuild.2014.05.032
45. Chikhi M, Agoudjil B, Boudenne A, Gherabli A (2013) Experimental investigation of new biocomposite with low cost for thermal insulation. Energy Build 66:267–273. https://doi.org/10.1016/j.enbuild.2013.07.019
46. Amara I, Mazioud A, Boulaoued I, Mhimid A (2017) Experimental study on thermal properties of bio-composite (gypsum plaster reinforced with palm tree fibers) for building insulation. Int J Heat Technol 35:576–584. https://doi.org/10.18280/ijht.350314
47. Belakroum R, Gherfi A, Kadja M, Maalouf C, Lachi M, El Wakil N, Mai TH (2018) Design and properties of a new sustainable construction material based on date palm fibers and lime. Constr Build Mater 184:330–343. https://doi.org/10.1016/j.conbuildmat.2018.06.196
48. Abu-Jdayil B, Mourad A-H, Hussain A (2015) Date pits and date palm wood-based heat insulator composites. In: Lisboa
49. Nasser RA, Al-Mefarrej HA (2011) Evaluation of using midribs of date palm fronds as a raw material for wood-cement composite panels industry in Saudi Arabia. Agric Eng Res J 1:43–50
50. Nasser RA (2014) Influence of board density and wood/cement ratio on the properties of wood-cement composite panels made from date palm fronds and tree prunings of buttonwood. Alex Sci Exch J 35
51. Al-Anbari R, Hassan MS, Fakri AH (2015) Manufacturing of sustainable cellulose date palm fiber reinforced cementitious boards in Iraq. https://doi.org/10.13140/RG.2.1.3909.6163
52. Ashraf N, Nasir M, Al-Kutti W, Al-Maziad FA (2020) Assessment of thermal and energy performance of masonry blocks prepared with date palm ash. Mater Renew Sustain Energy 9:17. https://doi.org/10.1007/s40243-020-00178-2

53. Jiang Y (2010) Generation of typical meteorological year for different climates of China. Energy 35:1946–1953. https://doi.org/10.1016/j.energy.2010.01.009
54. Abanto GA, Karkri M, Lefebvre G, Horn M, Solis JL, Gómez MM (2017) Thermal properties of adobe employed in Peruvian rural areas: experimental results and numerical simulation of a traditional bio-composite material. Case Stud Constr Mater 6:177–191. https://doi.org/10.1016/j.cscm.2017.02.001
55. Khoudja D, Taallah B, Izemmouren O, Aggoun S, Herihiri O, Guettala A (2021) Mechanical and thermophysical properties of raw earth bricks incorporating date palm waste. Constr Build Mater 270:121824. https://doi.org/10.1016/j.conbuildmat.2020.121824
56. Nagaraj HB, Sravan MV, Arun TG, Jagadish KS (2014) Role of lime with cement in long-term strength of compressed stabilized earth blocks. Int J Sustain Built Environ 3:54–61. https://doi.org/10.1016/j.ijsbe.2014.03.001
57. Bouhicha M, Aouissi F, Kenai S (2005) Performance of composite soil reinforced with barley straw. Cem Concr Compos 27:617–621. https://doi.org/10.1016/j.cemconcomp.2004.09.013
58. Vatani Oskouei A, Afzali M, Madadipour M (2017) Experimental investigation on mud bricks reinforced with natural additives under compressive and tensile tests. Constr Build Mater 142:137–147. https://doi.org/10.1016/j.conbuildmat.2017.03.065
59. Rigassi V (1995) Compressed earth blocks: manual of production. Manuel de production, Germany
60. Ramaswamy HS, Ahuja BM, Krishnamoorthy S (1983) Behaviour of concrete reinforced with jute, coir and bamboo fibres. Int J Cem Compos Lightweight Concr 5:3–13. https://doi.org/10.1016/0262-5075(83)90044-1
61. Coutts RSP, Ni Y (1995) Autoclaved bamboo pulp fibre reinforced cement. Cem Concr Compos 17:99–106. https://doi.org/10.1016/0958-9465(94)00002-G
62. Ghavami K, Toledo Filho RD, Barbosa NP (1999) Behaviour of composite soil reinforced with natural fibres. Cem Concr Compos 21:39–48. https://doi.org/10.1016/S0958-9465(98)00033-X
63. Tolêdo Filho RD, Scrivener K, England GL, Ghavami K (2000) Durability of alkali-sensitive sisal and coconut fibres in cement mortar composites. Cem Concr Compos 22:127–143. https://doi.org/10.1016/S0958-9465(99)00039-6
64. Taallah B, Guettala A, Guettala S, Kriker A (2014) Mechanical properties and hygroscopicity behavior of compressed earth block filled by date palm fibers. Constr Build Mater 59:161–168. https://doi.org/10.1016/j.conbuildmat.2014.02.058

Chapter 8
Date Palm Byproducts in Organic Fertilizers, Compost, Soil Amendment and Coal

Abstract Grinded date palm trees mulch applications have additional benefits of decreasing water evaporation from the soil surface, helping control weed invasion, dust suppression, helping prevent soil erosion loss by wind or water and providing thermal stabilization by keeping soil cooler in hot weather and warmer in cool weather. In an important study, the composting of date palm trees mulch, produced from grinded date palm leaves, trunks and roots and fresh cow manure, was investigated and analyzed in AL Hofuf Stars Recycle Station of AL Hasa City, K.S.A. The composite compost mixtures were built by mixing date palm mulch as a carbon source with fresh cow dung as a nitrogen source with C/N ratio 44.3:1, 50:1 and 44:1 for the mixture 1, 2 and 3, respectively. The turned windrow method has been applied in the composting process. The obtained organic fertilizer had natural soil odor and was brown in color. Thus, it can be concluded that the date palm mulch can be biologically recycled into an organic product having the criteria of organic fertilizers, soil stabilizers and soil plantation. Another research was conducted with the objective of preparing local farm residues, such as date palm leaves (DPL) as a substitute for imported peatmoss. DPL compost has been prepared according to Abu-Alfadhal method with some modifications. Air-dried DPL were cut to pieces of 10 cm length and buried in 2 × 1 m size concrete pit of depth 1.1 m. The compost layers were ~25 cm deep. There were four identical layers of DPL up to 1 m in height. The compost layers were stirred after 6 weeks followed by remixing the compost layers with an interval of 3 weeks. The compost pit was opened after 6 months, and the completely decomposed date palm leaves (compost) was separated from the undecomposed part. The results of the germination experiment in the first growing season showed that there was no significant difference in the number of seeds germinated in the DPL compost and peatmoss. In the second growing season the total number of seeds germinated were significantly higher in the DPL compost than peatmoss. In conclusion, the date palm leaves compost is an excellent substitute for peatmoss. Another study has been conducted to evaluate the feasibility of using composted date palm residues as a growth medium for tomato greenhouse plants production. Compost was prepared by mixing goat manure with crushed date palm residues (1:3 v/v). The date palm residues were sourced from the oasis of Cheneni (Gabes) and the goat manure from a local organic farm in Gabes (south of Tunisia). The

H. EL-Mously et al., *Date Palm Byproducts: A Springboard for Circular Bio Economy*, Materials Horizons: From Nature to Nanomaterials,
https://doi.org/10.1007/978-981-99-0475-4_8

windrow of 7 tons was irrigated by well water having pH 7.08 and electrical conductivity 3.4 mS/cm. The obtained compost was then evaluated for its effects on the growth of Rio Grande tomato seeds under greenhouse conditions at a local nursery in Sfax (south of Tunisia). The ingredients of the windrow were mixed by mechanical rotation to allow for aeration. The mature compost was ready after 6 months of adequate aeration and humidification. The produced compost from date palm leaves and goat manure co-composting had a pH value within the range of 6.0–8.5 and electric conductivity value lower than 3 mS/cm and is thus compatible with most plants growth needs. In addition, the final C/N ratio value indicated the complete biodegradation and the stability of the substrate. The produced compost seems to be the most efficient for tomato seedling, especially leaves numbers when used at 30% in the substrate. Thus, it can be concluded that the date palm residues co-composting could provide a viable ecological and sustainable alternative to conventional fertilizers. Another study has been conducted with the objective of achieving a better understanding of the microbial assemblages involved in the composting of lignocellulosic residues and to estimate the diversity of the microbial community structures responsible for the biodegradation. This has been achieved by analyzing the composition of the lipid fraction and following qualitative and quantitative variations in the levels of different fatty acid methyl esters (FAMEs) identified during composting. The study of compost's organic matter composition has been considered important to understand how biowaste can be processed. Various authors devoted their endeavors to study the humic substances. These lipids, formed during the biotransformation of organic matter by composting, provide useful information about the compost's maturity and stability and play an important role in soil processes. These compounds, mainly of plant and microbial origin, are hydrophobic and are often extracted with humic substances. A study has been conducted with the objective of evaluating the potentiality of use of the date palm midrib in charcoal production. The date palm midribs being a product of annual pruning of date palms are extensively available in most of the governorates in Egypt. Charcoal is needed as a fuel beside its environmental, medical and industrial applications. To conduct this study, samples of date palm midribs were collected from two date palm varieties: Balady and Siwei from EL-Qayat village, Menia governorate, Egypt. Each date palm midrib was divided into end, knee, base, middle and top parts. A pyrolysis reactor has been designed and manufactured. The carbonization cycle to 500 °C was conducted according to FAO standard with a heating rate 5–7 °C/min for Balady and Siwei date palm different parts. The experimental analysis has been conducted according to ASTM standards including the calorific value, fixed carbon, volatile matter%, ash content%, sulfur content% and moisture content% for the date palm midrib parts before and after carbonization. The research results show that for Balady midribs the best samples according to FAO standards are the middle part of the followed by the top, base, knee and end. Regarding Siwei midribs, the best samples are the top part of the midrib followed by the base, middle and end. Taking the whole midrib into consideration, Balady and Siwei midribs realized 86% and 88% of the FAO standard, respectively. Thus, it can be concluded that the Balady and Siwei date palm midribs can be used for the manufacture of charcoal suitable for environmental, medical and industrial applications. A research has been conducted with the objective of investigating the effect of pyrolysis temperature on date palm derived biochar characteristics to use it for agronomic or environmental management. Biochar—as a stable carbon-enriched materials—is produced by thermal conversion using unstable carbon-enriched materials. Date palm products of pruning were sourced from a farm near to Riyadh city in Saudi Arabia including leaves and spadix stems. The samples were pyrolyzed to temperatures 300, 400, 500, 600, 700 and 800 °C at a rate of 5 °C/min. The proximate analysis of the samples has been conducted according to ASTM 1762–84 standard method including moisture, ash and volatile matter in biochar. The research results showed that the highest yield of biochar was achieved at the lowest pyrolysis temperature (300 °C). The fixed C, ash and basic cations of biochar increased while its moisture, volatiles and elemental composition (O, H, N and S) decreased with increasing pyrolysis temperature. Biochars produced at high temperature (> 500 °C) could be more resistant to mineralization through biological processes than biochars pyrolyzed at lower temperature (< 500 °C), thus becoming more effective in mitigating greenhouse gas emission into the environmental. Our results suggest that the date palm biochars produced at low pyrolysis temperature (300–400 °C) that are practically carbonized and have relatively high organic functional groups and lower alkalinity may improve the fertility of arid soil more than those pyrolyzed at high temperature (700 and 800 °C). Thus, the research results may lead to the conclusion that the biochars produced from date palm products pf pruning represent a potential alternative materials for agronomic or environmental management.

Keywords Date palm mulch · Organic fertilizer · Substitute for compost · Tomato seedling · Date palm midribs · Charcoal production · Pyrolysis temperature · Biochar · Agronomic management

8.1 Composting Mulch of Date Palm Trees Through Microbial Activator in Saudi Arabia

A research [1] has been conducted with the objective of producing high-quality organic fertilizer using grinded date palm leaves, trunks and roots. The date palms are widely available amounting to more than 100 million date palms on the world level [2]. Saudi Arabia is considered as one of the main cultivators of date palms possessing ~22.6 million date palms [3]. The annual pruning of date palms results in huge quantities of byproducts (e.g. ~20 kg of dry leaves from each date palm [4]). These byproducts dominantly treated as waste and open-field burnt could be alternatively converted to compost by microbiological processes. The grinded date palm trees mulch applications have additional benefits of decreasing water evaporation from the soil surface, helping control weed invasion, dust suppression, helping prevent soil erosion loss by wind or water and providing thermal stabilization by keeping soil cooler in hot weather and warmer in cool weather [5]. Cattle manure containing limited amounts of nutrients (N, P and K) is not recommended to apply as a single organic fertilizer. Thus, one of the important research issues currently and in the future is the recycling of cow manure with agricultural residues (e.g. date palm byproducts) to produce a high-quality organic fertilizer.

In this study, the composting of date palm trees mulch (2 inch particle size), produced from grinded date palm leaves, trunks and roots and fresh cow manure, was investigated and analyzed in Al Hofuf Stars Recycle Station of Al Hasa City, K.S.A. The date palm trees mulch and fresh cow manure have been analyzed in terms of pH, EC, total dissolved solids, temperature, OC, total nitrogen, total phosphorus and total potassium using standard procedure [6].

In order to conduct this research complete date palm byproducts samples were mechanically grinded to produce mulch of particle size 2 inches using Vermeer grinding machine. Then, a standard compost pile was made with layers of date palm trees mulch and fresh cow manure. Three mixtures have been made: date palm mulch: fresh cow manure mixture 1(1:1), mixture 2(2:1) and mixture 3(3:1). Each pile was with volume 45 m^3 (length 15 × width 2.5 and height 1.2 m). The required calibrated operating conditions of moisture, temperature and air flow were chosen according to standard process [7]. Layers of mulch and manure were alternated and each layer was inoculated by microbial activator for fasting decomposition. Pile moisture was kept constant and turned every 5–7 days. Homogenized samples were collected each interval after turning the pile for chemical measurements. The pile temperature was monitored daily through the center of the compost piles at different locations. Composite moisture was adjusted to 60% by squeeze test in order to keep optimum bacterial growth to decompose hard organic fibrous materials.

The results of chemical analysis show that the date palm mulch serving as a carbon source was highly non-degradable. This may be because of its high content of organic matter (88.9%) and high C/N ratio (76.63) and its being rich in cellulose and hemicellulose [8]. The microbiological analysis show that the date palm mulch is free from pathogens and that fresh cow manure has a high number of total viable bacterial counts and both salmonella and E.coli strains. The fresh cow dung was high in N,P,K and moisture as compared with dry cow manure. However, it had lower organic matter and C/N ratio as compared to date palm mulch [9]. Thus, the composite compost mixtures were built by mixing date palm mulch as a carbon source with fresh cow dung as a nitrogen source with C/N ratio 44.3:1, 50:1 and 44:1 for the mixture 1, 2 and 3, respectively. The turned windrow method has been applied in the composting process. The results indicated that the temperature increased for mixture 1, 2 and 3 up to 66, 69 and 69 °C and then deceased to 44, 45 and 45 °C by the end of the composting period. The obtained organic fertilizer had natural soil odor and was brown in color. The final recorded organic fertilizer C/N ratio was 25:1, 12.5:1 and 23:1 for the mixtures 1, 2 and 3, respectively. The loss in organic matter reached values of 15.6, 29.5, 20.0% for the mixtures 1, 2, and 3, respectively. The results, obtained for mixtures 2 and 3, were found better than mixture 1. Thus, these mixtures are the best for composting date palm mulch. The regulated elements: N,P,K, organic matter, pH, electrical conductivity and TDS were improved for all mixtures. Thus, it can be concluded that the date palm mulch can be biologically recycled into an organic product having the criteria of organic fertilizers, soil stabilizers and soil plantation.

8.2 Use of Date Palm Leaves Compost as a Substitution to Peatmoss

An important research [10] has been conducted with the objective of preparing local farm residues, such as date palm leaves (DPL), as a substitute for imported peatmoss. The date palm is an important element of flora in the whole Arab region including Saudi Arabia possessing around 18.2 million date palms [11]. During annual pruning, each palm produces 10–20 leaves [12]. These leaves have no economic value and are associated with the danger of fire and infestation by dangerous insects. Thus, this study has been conducted to evaluate the potentiality of using DPL in the preparation of compost and to determine its influence on seed germination and growth of several ornamental plants for landscaping development.

Date palm leaves compost has been prepared according to Abu-Alfadhal method (1970) with some modification. Air-dried DPL were cut to pieces of 10 cm length and buried in 2 × 1 m size concrete pit of depth 1.1 m. The compost layers were ~25 cm deep. Firstly, a layer of 96 kg dried DPL was laid and then the desired amount of a mixture of ammonium sulfate, trisuper phosphate, fine (100 μ) calcium carbonate and clay in a ratio of 35:7:35:100 kg, respectively, per ton of dried DPL was distributed homogenously. Each layer was sprayed with 77L of water having a total salinity of 640 mgL^{-1} total dissolved solids (TDS). There were four identical layers of DPL up to 1 m in height. The DPL layers were compacted manually. The compost layers were stirred after 6 weeks followed by remixing the compost layers with an interval of 3 weeks. The compost pit was opened after 6 months, and the completely decomposed date palm leaves (compost) was separated from the undecomposed part mainly due to the hard midribs. The prepared compost (DPLC) was used in the experiments.

The results of the germination experiment in the first growing season showed that there was no significant difference in the number of seeds germinated in the DPLC and peatmoss. The rate of germination was identical for both DPLC and peatmoss except Zinnia where the rate of seed germination was significantly more in DPLC than peatmoss ($LSD_{0.05} = 0.05$). In the second growing season, the total number of seeds germinated were significantly higher in the DPLC than peatmoss for all plants. The rates of germination were significantly higher in DPLC than peatmoss for all the plants except Tagetes where the difference was not significant ($LSD_{0.05} = 1.43$). Concerning the growth experiments the trend for the effect of DPLC and peatmoss on plant height, number of leaves and dry weight of vegetative growth were almost identical to the first growing season except the dry weight of biomass of Tagetes which was significantly more in DPLC than peatmoss during the second growing season ($LSD_{0.05} = 0.63$). In conclusion, the date palm leaves' compost is an excellent substitute for peatmoss. However, further research is needed to improve methods of preparation of compost from date palm leaves, Phragmites australis (reed plant growing on the banks of drainage canals), wheat straw and rice straw on economical grounds.

8.3 Date Palm Wastes Co-Composted Product: An Efficient Substrate for Tomato Seedling Production

The objective of this study [13] was to evaluate the feasibility of using composted date palm residues as a growth medium for tomato (*Solanum Lycopersicum* L.) greenhouse plants production. Growing concerns have been expressed over the deteriorating quality and fertility of agricultural soils due to erosion, nutrient depletion, organic matter exhaustion and increased use of chemical fertilizers [14]. Thus, farmers turned to use peat [15] being a non-renewable resource and causing the destruction of high ecological value areas [16]. But an appropriate growing medium is necessary for providing plants nutrients needs and promoting seedling growth [17]. Within this context, the agricultural residues represent a sustainably available cost-effective resource for the manufacture of composts [16–19]. Many studies have been devoted to the composting of various organic byproducts including olive residues [20, 21], grape marc [22], Acacia residues [23] and green residues [15].

The date palm (*Phoenix dactylifera* L.) is an important element of flora in many countries in the world, including the Middle East, North Africa and Arabian Peninsula. There are more than 100 million date palms in the world with an average age of about 100 years [24]. The annual pruning of these palms generates huge quantities of residues (e.g. 0.2 Mt of leaves [25]). These quantities, most dominantly treated as waste, can be valorized into biostimulants or biofertilizers in organic matter—deficient arid and semi-arid regions as soils in south Tunisia (organic matter < 1%). Nevertheless, few researches were devoted to the use of date palm residues to produce compost under green house for seedling production [26, 27].

To conduct this study, compost was prepared by mixing goat manure with crushed date palm residues (1:3 V/V). The date palm residues were sourced from the oasis of Cheneni (Gabes) and the goat manure from a local organic farm in Gabes (south of Tunisia). The date palm residues were crushed to a size of 5–10 cm, swollen in water for 4 days and then mixed with goat manure. The windrow (1, 3 m height, 10 m length and 1.5 m wide base) of 7 tons was irrigated by well water having pH 7.08 and electrical conductivity 3.4 mS/cm. The obtained compost was then evaluated for its effects on the growth of Riograndi tomato seeds under greenhouse conditions at a local nursery in Sfax (south of Tunisia). The prepared compost was mixed with the greenhouse soil composed of 70% sand, 21%clay and 9%silt and characterized by neutral pH(7.14 $\pm$ 0.26), low electric conductivity (0.025 $\pm$ 0.003 mS/cm) and organic matter content (1.18 $\pm$ 0.02%). The ingredients of the windrow were mixed by mechanical rotation to allow for aeration. The windrow was watered to keep moisture at 50% level and turned once a week during the most active biooxidative phase and once a month for the remaining composting period. The mature compost was ready after 6 months of adequate aeration and humidification.

The produced compost from date palm residues and goat manure co-composting had a pH value within the range of 6.0–8.5 and electric conductivity value lower than 3 mS/cm and is thus compatible with most plants growth needs [20]. In addition, the final C/N ratio value indicated the complete biodegradation and the stability of the substrate [28, 29]. Besides, according to the French organic conditions standard, the investigated compost could be regarded as a vegetal compost with total Kjeldahl nitrogen and organic matter values inferior to 3 and 55%, respectively, [30]. The compost had fine structure and large specific surface, thus providing strong absorbability and nutrient retention [31]. The produced compost seems to be the most efficient for tomato seedling, especially leaves number, when used at 30% in the substrate. A dissimilar result was obtained by Ancuta et al. [32] in a study using different types of composts for tomato seedling production, where the maximum leaves number was obtained when compost was used at a rate of 50% in the substrate. The addition of crushed palm residues increased seed germination percent for both manure and compost. Thus, it can be concluded that the date palm residues co-composting could provide a viable ecological and sustainable alternative to conventional fertilizers.

8.4 Lipid Signature of the Microbial Community Structure During Composting of Date Palm Products of Pruning Alone or Mixed with Couch Grass Clippings

A study [33] has been conducted with the objective of achieving a better understanding of the microbial assemblages involved in the composting of lignocellulosic residues and to estimate the diversity of the microbial community structures responsible for the biodegradation. This has been achieved by analyzing the composition of the lipid fraction and following qualitative and quantitative variations in the levels of different fatty acid methyl esters (FAMEs) identified during composting. The study of compost's organic matter composition has been considered important to understand how biowaste can be processed. Various authors devoted their endeavors to study the humic substances [34, 35]. These lipids, formed during the biotransformation of organic matter by composting, provide useful information about the compost's maturity and stability and play an important role in soil processes [36]. These compounds, mainly of plant and microbial origin [37], are hydrophobic and are often extracted with humic substances [38].

This research has been conducted to valorize the large quantities of green residues resulting from public and private date palm gardens estimated at 38% of Marrakech total city's waste [39] and representing a huge environmental problem. The use of these huge quantities of materials in the manufacture of compost may also reduce the municipal expenditure on imported peat.

To conduct this study, two compost windrows have been prepared. The first windrow (DPW-1) was composed of date palm fronds and trunks, roughly chopped into pieces of length 2–50 cm. The second windrow (DPWC-2) was composed of a mixture of 50% date palm waste and 50% couch grass clippings. These windrows were moistened to 60% humidity and covered with tarpaulins. Aeration was provided by regular manual mixing and the temperature was daily monitored at different levels for a period of 8 months. Follow-up was conducted for 14 months. Sub-samples were taken from ten different points (bottom, surface, side and center) and refrigerated until analysis. Lipids were extracted using an accelerated solvent extractor (Dionex ASE 100) working under pressure with dichloromethane/methanol solvent subsequently evaporated off to yield the lipids. Then a known mass of lipid was dissolved in excess acetic anhydride. A catalytic quantity of pyridine was added (3–4 drops). The acetylation reaction was initiated by heating for about 15 min at 70 °C. After stirring overnight at room temperature, the mixture was hydrolyzed still under stirring for 2 h. Acetylated alcohols were extracted with dichloromethane, and excess acetic acid was neutralized by washing the organic phase with water and saturated sodium bicarbonate. The solution was dried over magnesium sulfate and the solvent evaporated off [40].

The results of study show that the addition of couch grass clippings to date palm residues has led to an increase of DPW biodegradability and favored the development and proliferation of diversified microbial activity. Therefore, there was a higher degree of stabilization and maturation reached in the DPGC mixture than in DPW compost. More diversified microbial activity marked by the pressure of odd-numbered and branched fatty acids prevailed in the DPGC mixture as opposed to the DPW compost. This has led to the deduction that the quantity of bioavailable organic matter influenced the proliferation and evolution of the microbial biomass responsible for degradation during the composting process. The monitoring of different fatty acid methyl esters identified during both composting processes showed a more accelerated degradation in the DPGC mixture than in the DPW. The increase of organic matter decomposition is highly favored by proliferation of G^+ bacteria, which seem sensitive to organic substrate bioavailability. This increase confirms the low level of biodegradation of date palm residues which biodegrade in a slow and gradual way. Indeed DPW compost needs a longer time to reach a similar level of maturity as compared with the DPGC mixture. The calculation of the relationship between the various bacterial groups indicates that bacteria strongly prevail over fungi in both DPW and DPGC composts. The increase of the G^+/G^- ratio toward the end of DPGC composting shows that stabilization has not yet been completed. However, stabilization reached a more mature state in DPGC compost than DPW compost. The rise in G^+/G^- and G^+/fungi ratios was only recorded near the end of process, indicating that composting had just started in DPW. The measurement of various steroidal compounds showed that oxidation and hydrogenation were more extensive in the DPGC mixture than in DPW alone. The cholestanol acetate, methyl cholestadiene and stigmasterol acetate changes show that stimulation of microbes and/or micro-fauna in the DPGC mixture occurred. The level of microbial activity during DPW composting oscillated. The appearance of stigmastadienone near the end of

DPW composting shows that composting has just started. The stigmastadiene variation indicated a correlation between the results obtained during fatty acid methyl ester study and steroidal compound study. Finally, it can be concluded that monitoring steroidal compounds gave a better estimate of the degree of involvement of each of the groups identified as participating in the biological stabilization of the studied substrates.

8.5 A Study of the Potentially of Use of the Date Palm Midrib in Charcoal Production

A study [41] has been conducted with the objective of evaluating the potentiality of use of the date palm midribs in charcoal production. The date palm midribs being a product of annual pruning of date palms are extensively available in most of the governorates of Egypt. Charcoal is needed as a fuel beside its environmental, medical and industrial applications. Charcoal is needed in soil amendment, because it increases carbon concentration in soil and reduces the emissions of green carbon gases.

To conduct this study [41], samples of date palm midribs were collected from two date palm varieties: Balady and Siwei from El-Qayat village, Menia governorate, Egypt. Each date palm midrib was divided into end, knee, base, middle and top parts. A pyrolysis reactor has been designed and manufactured. It is a double-wall chamber, made from steel sheets of thickness 6 mm. Its capacity is 120L. The outer dimensions of the reactor are $70 \times 70 \times 95$ cm^3 and the inner dimensions are $40 \times 40 \times 65$ cm^3. The isolation layers are made from ceramic fiber to isolate the inner kiln. The inner kiln is furnished with three heaters: two in both sides and the third in the bottom. The reactor has a door and a thermocouple to measure the carbonization process temperature. The pyrolysis reactor assembly is connected to a central pipe to a condensing unit. The central pipe in the chamber takes out the evolved gases and vapor directly through the condenser unit. The carbonization cycle to 500 °C was conducted according to FAO standards with a heating rate 5–7 °C/min for Siwei and Balady date palm different parts.

The experimental analysis has been conducted according to ASTM standards including the calorific value, fixed carbon, volatile matter%, ash content%, sulfur content% and moisture content for the date palm midrib parts before and after carbonization.

The research results show that for Balady midribs the best samples according to FAO standards are the middle part of the midrib followed by the top, base, knee and end. Regarding Siwei midribs, the best samples are top part of the midrib followed by the base, middle and end. Taking the whole midrib into consideration, Balady midrib and Siwei midrib realized 86% and 88% of the FAO standard, respectively. Thus, it can be concluded that the Balady and Siwei date palm midribs can be used for the manufacture of charcoal suitable for environmental, medical and industrial applications.

8.6 Biochar Production from Date Palm Waste: Charring Temperature, Induced Changes in Composition and Surface Chemistry

A research [42] has been conducted with the objective of investigating the effect of pyrolysis temperature on date palm derived biochar characteristics to use it for agronomic or environmental management. Biochar—as a stable carbon-enriched material—is produced by thermal conversion using unstable carbon-enriched materials [43]. The biochar is being added to the soil to mitigate greenhouse gas emissions [44]. Many researchers devoted their endeavors to the application of biochar for improving the soil properties and fertility, as well as the remediation of contaminated soils [45–47]. Other researchers [48, 49] suggested the use of biochars as alternative sorbents for the removal of different types of organic and inorganic contaminants from aqueous solutions. But the effects of biochar on long-term carbon sequestration, soil fertility and environmental remediation highly depend on its physicochemical properties [50, 51]. The main factors affecting the quality of biochar and main properties are the feedstock type and pyrolysis temperature [47, 52–54]. The pyrolysis temperature is a key factor in volatilizing some elements such as N and S and in concentrating other elements such as C [55–57].

To conduct this study, date palm products of pruning were sourced from a farm near to Riyadh city including leaves and spadix stems. The material was air-dried and chopped to small pieces. Pyrolysis of this material has been conducted in a steel electric pyrolysis reactor of height 22 cm and diameter 7 cm. The samples were pyrolyzed to temperatures 300, 400, 500, 600, 700 and 800 °C at a rate 5 °C/min, the produced biochars were left to cool inside the furnace overnight. The proximate analysis of the samples has been conducted according to ASTM D 1762-84 standard method [58] including moisture, ash and volatile matter in biochar. The total elemental content of C,H,N and S in biochar samples were measured by CHNS analyzer (seriesII, PerkinElmer, USA). The oxygen percent has been calculated as: 100-(C + H + N + S + ash%) and the atomic ratios of H, C, O/C, O + N/C and O + N + S/C were also calculated as indicative of aromaticity and polarity.

The research results showed that the highest yield of biochar was achieved at the lowest pyrolysis temperature (300 °C). The fixed C, ash and basic cations of biochar increased while its moisture, volatiles and elemental composition (O, H, N and S) deceased with increasing pyrolysis temperature. Based on the research results, the date palm derived biochars at pyrolysis temperature $\geq$ 500 °C with a volatile content less than 10% and O/C of 0.02–0.05 may exhibit a high sequestration potential. Biochars produced at high temperature (> 500 °C) could be more resistant to mineralization through biological processes than biochars pyrolyzed at lower temperature

(< 500 °C), thus becoming more effective in mitigating greenhouse gas emission into the environment. However, applying high pyrolysis biochars to arid soils with high alkalinity may be critical due to its high pH. Our results suggest that the date palm biochars produced at the low pyrolysis temperature (300–400 °C) that are practically carbonized and have relatively high organic functional groups and lower alkalinity may improve the fertility of arid soils more than those pyrolyzed at high temperature (700 and 800 °C), inducing nutrient exchange sites as well as enhancing soil cation exchange capacity. The research results may lead to the conclusion that the biochars produced from date palm products of pruning represent a potential alternative materials for agronomic or environmental management.

References

1. Ashhab AO Al (2012) Composting mulch of date palm trees through microbial activator in Saudi Arabia. https://doi.org/10.13140/RG.2.2.21664.92169
2. Zaid A, Jimeniez EJA (1999) Plant production and Protection-Paper 165. FAO
3. Annual Agriculture Statistical Book (1998). Ministry of agriculture
4. Barreveld WH (1993) Date palm products. FAO Agricultural Services Bulletin (FAO)
5. Ashworth S, Harrison H (1983) Evaluation of mulches for use in the home garden. HortScience 18:180–182
6. APHA American Public Health Association (1992) Compendium of methods for the microbiological examination of foods. Pub. 3rd edition. pp 3–10. APHA American Public Health Association
7. Hassen A, Belguith K, Jedidi N, Cherif A, Cherif M, Boudabous A (2001) Microbial characterization during composting of municipal solid waste. Biores Technol 80(3):217–225. https://doi.org/10.1016/S0960-8524(01)00065-7
8. Al-Khanbashi A, Al-Kaabi K, Hammami A (2005) Date palm fibers as polymeric matrix reinforcement: fiber characterization. Polym Compos 26(4):486–497. https://doi.org/10.1002/pc.20118
9. Goyal S, Dhull S, Kapoor K (2005) Chemical and biological changes during composting of different organic wastes and assessment of compost maturity. Biores Technol 96(14):1584–1591. https://doi.org/10.1016/j.biortech.2004.12.012
10. Ali YSS (2008) Use of date palm leaves compost as a substitution to peatmoss. Am J Plant Physiol 3(4):131–136. https://doi.org/10.3923/ajpp.2008.131.136
11. Annon (1999) Agriculture statistical year book (twelfth issue). Department of economic Studies and Statistics, Ministry of Agriculture, Saudi Arabia
12. Pahsa MAA (1998) Fruit production in the Kingdom of Saudi Arabia. Technical Publication, Dirab Experimental Station, King Saud University
13. Abid W, Magdich S, Mahmoud IB, Medhioub K, Ammar E (2018) Date palm wastes co-composted product: an efficient substrate for tomato (*Solanum lycopercicum L.*) seedling production. Waste Biomass Valorization 9(1):45–55. https://doi.org/10.1007/s12649-016-9767-y
14. Madejón P, Alaejos J, García-Álbala J, Fernández M, Madejón E (2016) Three-year study of fast-growing trees in degraded soils amended with composts: effects on soil fertility and productivity. J Environ Manage 169:18–26. https://doi.org/10.1016/j.jenvman.2015.11.050
15. Aleandri MP, Chilosi G, Muganu M, Vettraino A, Marinari S, Paolocci M, Luccioli E, Vannini A (2015) On farm production of compost from nursery green residues and its use to reduce peat for the production of olive pot plants. Sci Hortic 193:301–307. https://doi.org/10.1016/j.scienta.2015.06.048

16. Carmona E, Moreno MT, Avilés M, Ordovás J (2012) Use of grape marc compost as substrate for vegetable seedlings. Sci Hortic 137:69–74. https://doi.org/10.1016/j.scienta.2012.01.023
17. Van Tam N, Wang C-H (2015) Use of spent mushroom substrate and manure compost for honeydew melon seedlings. J Plant Growth Regul 34(2):417–424. https://doi.org/10.1007/s00344-015-9478-9
18. Kadoglidou K, Chalkos D, Karamanoli K, Eleftherohorinos IG, Constantinidou H-IA, Vokou D (2014) Aromatic plants as soil amendments: effects of spearmint and sage on soil properties, growth and physiology of tomato seedlings. Sci Hortic 179:25–35. https://doi.org/10.1016/j.scienta.2014.09.009
19. Wu G, Kechavarzi C, Li X, Sui H, Pollard SJT, Coulon F (2013) Influence of mature compost amendment on total and bioavailable polycyclic aromatic hydrocarbons in contaminated soils. Chemosphere 90(8):2240–2246. https://doi.org/10.1016/j.chemosphere.2012.10.003
20. Hachicha S, Cegarra J, Sellami F, Hachicha R, Drira N, Medhioub K, Ammar E (2009) Elimination of polyphenols toxicity from olive mill wastewater sludge by its co-composting with sesame bark. J Hazard Mater 161(2–3):1131–1139. https://doi.org/10.1016/j.jhazmat.2008.04.066
21. Chowdhury MB, Md AK, Akratos CS, Vayenas DV, Pavlou S (2013) Olive mill waste composting: a review. Int Biodeterior Biodegradation 85:108–119. https://doi.org/10.1016/j.ibiod.2013.06.019
22. Paradelo R, Moldes AB, González D, Barral MT (2012) Plant tests for determining the suitability of grape marc composts as components of plant growth media. Waste Manage Res J Sustain Circ Econ 30(10):1059–1065. https://doi.org/10.1177/0734242X12451307
23. Brito LM, Mourão I, Coutinho J, Smith SR (2015) Co-composting of invasive *Acacia longifolia* with pine bark for horticultural use. Environ Technol 36(13):1632–1642. https://doi.org/10.1080/09593330.2014.1002863
24. Alawar A, Hamed AM, Al-Kaabi K (2009) Characterization of treated date palm tree fiber as composite reinforcement. Compos B Eng 40(7):601–606. https://doi.org/10.1016/j.compositesb.2009.04.018
25. El may Y, Jeguirim M, Dorge S, Trouvé G, Said R (2012) Study on the thermal behavior of different date palm residues: characterization and devolatilization kinetics under inert and oxidative atmospheres. Energy44(1):702–709. https://doi.org/10.1016/j.energy.2012.05.022
26. El Fels L, Zamama M, El Asli A, Hafidi M (2014) Assessment of biotransformation of organic matter during co-composting of sewage sludge-lignocelullosic waste by chemical, FTIR analyses, and phytotoxicity tests. Int Biodeterior Biodegradation 87:128–137. https://doi.org/10.1016/j.ibiod.2013.09.024
27. Zahra El Ouaqoudi F, El Fels L, Lemée L, Amblès A, Hafidi M (2015) Evaluation of lignocelullose compost stability and maturity using spectroscopic (FTIR) and thermal (TGA/TDA) analysis. Ecol Eng 75:217–222. https://doi.org/10.1016/j.ecoleng.2014.12.004
28. Han W, Clarke W, Pratt S (2014) Composting of waste algae: a review. Waste Manage 34(7):1148–1155. https://doi.org/10.1016/j.wasman.2014.01.019
29. Roca-Pérez L, Martínez C, Marcilla P, Boluda R (2009) Composting rice straw with sewage sludge and compost effects on the soil–plant system. Chemosphere 75(6):781–787. https://doi.org/10.1016/j.chemosphere.2008.12.058
30. Association Franc¸aise de Normalisation Amendements Organiques (2006) Association Franc¸aise de normalisation amendements organiques—denominations specifications et marquage, NF U 44–051. Denominations specifications et marquage
31. Zaller JG (2007) Vermicompost as a substitute for peat in potting media: effects on germination, biomass allocation, yields and fruit quality of three tomato varieties. Sci Hortic 112(2):191–199. https://doi.org/10.1016/j.scienta.2006.12.023
32. Ancut AD, Renata S, Sumalan R (2013) Influence of different types of composts on growth and chlorophyll content from tomato seedlings. Hortic For Biotechnol17:43–48
33. Elouaqoudi FZ, El Fels L, Amir S, Merlina G, Meddich A, Lemee L, Ambles A, Hafidi M (2015) Lipid signature of the microbial community structure during composting of date palm waste alone or mixed with couch grass clippings. Int Biodeterior Biodegradation 97:75–84. https://doi.org/10.1016/j.ibiod.2014.08.016

34. Amir S, Hafidi M, Lemee L, Merlina G, Guiresse M, Pinelli E, Revel J-C, Bailly J-R, Ambles A (2006) Structural characterization of humic acids, extracted from sewage sludge during composting, by thermochemolysis–gas chromatography–mass spectrometry. Process Biochem 41(2):410–422. https://doi.org/10.1016/j.procbio.2005.07.005
35. Sánchez-Monedero MA, Roig A, Cegarra J, Bernal MP (1999) Relationships between water-soluble carbohydrate and phenol fractions and the humification indices of different organic wastes during composting. Biores Technol 70(2):193–201. https://doi.org/10.1016/S0960-8524(99)00018-8
36. Stevenson FJ (1982) Humus Chemistry, Genesis, Composition, Reaction, 2nd ed. Wiley Interscience Publication
37. Amblès A, Jacquesy JC, Jambu P, Joffre J, Maggi-Churin R (1991) Polar lipid fraction in soil: a kerogen-like matter. Org Geochem 17(3):341–349. https://doi.org/10.1016/0146-6380(91)90097-4
38. Réveillé V, Mansuy L, Jardé É, Garnier-Sillam É (2003) Characterisation of sewage sludge-derived organic matter: LIpids and humic acids. Org Geochem 34(4):615–627. https://doi.org/10.1016/S0146-6380(02)00216-4
39. Wilaya de la Region de Marrakech Tensift El Haouz (2003) Monographie locale de l'environnement de la ville de Marrakech. Wilaya de la Region de Marrakech Tensift El Haouz (WRMTH). Wilaya de la Region de Marrakech Tensift El Haouz, p 116
40. Durand MC (2003) Caractérisation physico-chimique des produits de l'assainissement pluvial. Origine et devenir des metaux traces et des polluants organiques. [Doctoral]. University of Poitiers
41. Abdelsamie MMA (2018) A study of the potentially of use of the date palm midrib in charcoal production [Master]. Ain Shams University, Engineering
42. Usman ARA, Abduljabbar A, Vithanage M, Ok YS, Ahmad M, Ahmad M, Elfaki J, Abdulazeem SS, Al-Wabel MI (2015) Biochar production from date palm waste: charring temperature induced changes in composition and surface chemistry. J Anal Appl Pyrol 115:392–400. https://doi.org/10.1016/j.jaap.2015.08.016
43. Lehmann J, Rillig MC, Thies J, Masiello CA, Hockaday WC, Crowley D (2011) Biochar effects on soil biota—a review. Soil Biol Biochem 43(9):1812–1836. https://doi.org/10.1016/j.soilbio.2011.04.022
44. Lehmann J, Joseph S (2015) Biochar for environmental management: an introduction. In Biochar for environmental management, 2nd ed. Routledge, p 13
45. Al-Wabel MI, Usman ARA, El-Naggar AH, Aly AA, Ibrahim HM, Elmaghraby S, Al-Omran A (2015) Conocarpus biochar as a soil amendment for reducing heavy metal availability and uptake by maize plants. Saudi J Biol Sci 22(4):503–511. https://doi.org/10.1016/j.sjbs.2014.12.003
46. El-Mahrouky M, El-Naggar AH, Usman AR, Al-WABEL M (2015) Dynamics of CO_2 emission and biochemical properties of a sandy calcareous soil amended with conocarpus waste and biochar. Pedosphere 25(1):46–56. https://doi.org/10.1016/S1002-0160(14)60075-8
47. Yuan J-H, Xu R-K, Zhang H (2011) The forms of alkalis in the biochar produced from crop residues at different temperatures. Biores Technol 102(3):3488–3497. https://doi.org/10.1016/j.biortech.2010.11.018
48. Inyang M, Gao B, Yao Y, Xue Y, Zimmerman AR, Pullammanappallil P, Cao X (2012) Removal of heavy metals from aqueous solution by biochars derived from anaerobically digested biomass. Biores Technol 110:50–56. https://doi.org/10.1016/j.biortech.2012.01.072
49. Tan X, Liu Y, Zeng G, Wang X, Hu X, Gu Y, Yang Z (2015) Application of biochar for the removal of pollutants from aqueous solutions. Chemosphere 125:70–85. https://doi.org/10.1016/j.chemosphere.2014.12.058
50. Ahmad M, Moon DH, Vithanage M, Koutsospyros,A. Lee, SS, Yang JE, Lee SE, Jeon C, Ok YS (2014) Production and use of biochar from buffalo-weed (*Ambrosia trifida* L.) for trichloroethylene removal from water: production and use of biochar from *Ambrosia trifida* for TCE removal from water. J Chem Technol Biotechnol 89(1):150–157. https://doi.org/10.1002/jctb.4157

51. Spokas KA, Koskinen WC, Baker JM, Reicosky DC (2009) Impacts of woodchip biochar additions on greenhouse gas production and sorption/degradation of two herbicides in a Minnesota soil. Chemosphere 77(4):574–581. https://doi.org/10.1016/j.chemosphere.2009.06.053
52. Al-Wabel MI, Al-Omran A, El-Naggar AH, Nadeem M, Usman ARA (2013) Pyrolysis temperature induced changes in characteristics and chemical composition of biochar produced from conocarpus wastes. Biores Technol 131:374–379. https://doi.org/10.1016/j.biortech.2012.12.165
53. Gai X, Wang H, Liu J, Zhai L, Liu S, Ren T, Liu H (2014) Effects of feedstock and pyrolysis temperature on biochar adsorption of ammonium and nitrate. PLoS ONE 9(12):e113888. https://doi.org/10.1371/journal.pone.0113888
54. Yuan H, Lu T, Wang Y, Huang H, Chen Y (2014) Influence of pyrolysis temperature and holding time on properties of biochar derived from medicinal herb (radix isatidis) residue and its effect on soil CO_2 emission. J Anal Appl Pyrol 110:277–284. https://doi.org/10.1016/j.jaap.2014.09.016
55. Bridle TR, Pritchard D (2004) Energy and nutrient recovery from sewage sludge via pyrolysis. Water Sci Technol 50(9):169–175. https://doi.org/10.2166/wst.2004.0562
56. Song W, Guo M (2012) Quality variations of poultry litter biochar generated at different pyrolysis temperatures. J Anal Appl Pyrol 94:138–145. https://doi.org/10.1016/j.jaap.2011.11.018
57. DeLuca TH, Gundale MJ, Derek MacKenzie M, Jones DL (2015) Biochar effects on soil nutrient transformations. In: Biochar for environmental management, 2nd ed. Routledge, p 34
58. American Society for Testing and Materials (ASTM) (1989) Standard methods for chemical analysis of wood charcoal, ASTM D1762–84. PA, USA, Philadelphia

Chapter 9
Date Palm Byproducts for Natural Fodder and Silage

Abstract Waste dates (most dominantly treated as waste) representing ~20–30% of total dates production have been used as a replacement of dietary starch in tilapia feed. This has led to the improvement of growth rate, feed conversion, specific growth rate and protein efficiency ratio. In another study, the addition of date palm seed—most dominantly treated as waste—to the diet of the African catfish has led to a considerable increase of the weight gain, specific growth rate, protein efficiency ratio and protein productive value, as well as the improvement of final fish carcass and hematological indices. The improved growth rate and nutrient utilization could be explained by the presence of a range of digestive enzymes in the date palm seeds. A research has been conducted in Kuwait to evaluate the technical feasibility of use of the locally and sustainably available date palm leaflets as a roughage to substitute the imported barley straw in feeding of Friesian and Holstein cows. The results of feeding experiments revealed that milk yields, milk composition and live weight gains of cows fed by either date palm leaflets or barley straw did not significantly differ. Therefore, it can be concluded that the date palm leaflets—a local sustainable resource treated as waste—can be used as an acceptable alternative roughage in feeding of cows to the expensive imported barley straw. Another research was devoted to the evaluation of ensilage, made from cardboxes, annually available with about 100,000 tons in Kuwait and date palm leaves resulting from annual pruning of date palm and treated as waste. The bioconversion and upgrading of nutritive value of cardboxes and date palm leaves to silages were investigated over 90 days through chemical and microbial treatments. The qualitative aspects of the produced silages for ruminants were evaluated. It can be concluded from the research results that cardboxes and date palm leaves can be effectively ensilaged, and the resultant silages are acceptable and palatable to be used as ruminant feed in arid regions. A research has been conducted to evaluate the potentiality of use of date palm leaves for feeding of Omani sheep. Oman and the whole gulf region suffer as an arid region from the shortage of animal feed representing an obstacle in livestock investment projects. This study has been conducted on 32 one-year old male Omani native sheep of starting body weight ~32 kg in a feeding trial for 120 days. The conducted feeding included two types of roughages: Rhodes grass hay (RGH) and urea-treated date palm leaves (UTPF). The experimental animals fed by the date palm leaves were in good health

H. EL-Mously et al., *Date Palm Byproducts: A Springboard for Circular Bio Economy*, Materials Horizons: From Nature to Nanomaterials,
https://doi.org/10.1007/978-981-99-0475-4_9

throughout the trial. But the sheep fed by UTPF had lower feed intake as compared with those fed by RGH, but feed intake/body weight was similar across diet groups. Therefore, date palm leaves can be used as a component of feed for Omani sheep in case of nutritional shortage frequently experienced in arid zones. A research has been conducted in Algeria to evaluate, in vitro, the ruminal fermentation and nutritive value of the date palm leaves, pedicels, date pits and waste dates. To conduct this study, 20–30 specimens of leaves, pedicels and date pits were taken from Deglet-Nour palm and waste dates from three varieties Bourus, Harchaya and Kentichi in Biskra in Saharan Atlas region in Algeria. Vetch-oat hay was taken as a control reference. All the samples including the control were oven-dried at 50 °C and ground to pass 1 mm screen. The results of this research showed that the date palm leaves have the highest NDF, ADF, lignin and crude protein contents (609, 435, 84, 649 kg^{-1} DM), respectively. The cumulative gas production at 144 h of incubation was greatest for Kentichi dates (330 mLg^{-1} DM) and lowest for date pits (69 mL^{-1} DM). Regardless of the variety, waste dates showed the highest DM effective ruminal degradability and organic matter digestibility. The date pits seemed to be a poorly degradable material. These results indicate that waste dates—though of low-protein content—are highly digestible with energy concentration as high as that of vetch-oat hay. The palm leaves and pedicels can be considered as highly fibrous emergency roughages for low-producing animals. Protein supplements should be added when date palm byproducts are used as feedstuff in order to balance the ruminant diets. An important study has been conducted with the objective of learning about farmer's knowledge south of Tunisia on the use of date palm byproducts (DPBP) as forage resources. To conduct this study, samples of wasted dates, date pits, leaves, peduncles and pedicels of two dominant varieties (Deglet Nour and Kentah) were taken and analyzed for dry matter, crude protein, total ash, crude fiber, natural detergent fiber, acid detergent fiber and acid detergent lignin. The in vitro digestibility measurements have been taken on each sample using rumen juices from three species: sheep, goat and dromedary. An extensive socio-economic survey has been conducted in a sub-region of Nefazoua (EL Faoura) concerning the use of DPBP in livestock feeding. It can be concluded from the results of this study that DPBP, taken individually, are highly unbalanced for animal nutrition: high energy content in the cases of wasted dates and date pits and high fiber content for leaves, peduncles and pedicels and for all: low crude protein content (3–6.5% of dry matter). The in vitro dry matter digestibility for wasted dates were found 74.5, 79.7 and 79.2% for sheep, goats and dromedary, respectively. The corresponding figures for leaves, peduncles and pedicels were found much lower: 12.3, 19.6 and 18.3%, respectively. Thus, the main questions to be asked in use of DPBP in ruminant feeding are how to make a well-balanced DPBP-based forage? And how to conserve this feed to overcome the problem of seasonal unavailability? This indicates the significance of research on DPBP-based silage to improve the technical and economic feasibility of use of DPBP as a local and sustainable forage resource. An important study has been conducted to evaluate the effectiveness of a combined treatment of date palm leaves by sodium hydroxide and lime for conversion into a promising ruminant feed. Date palm leaves (DPL), annually available in Iran with big quantities (e.g., 300,000–400,000 t) are being treated as waste! One of the approaches to solve this problem is to use DPL in feeding of ruminants. But DPL in its natural form has a very low ruminal digestibility (16% in vitro dry matter digestibility). The alkaline pretreatment of DPL to improve the ruminal degradability at high temperature is fast and effective, but it is a capital-intensive process. Therefore, the idea of this research was to replace only a part of costly sodium hydroxide with less expensive lime to realize a cost-effective and efficient pretreatment of DPL at a mild temperature with low concentration of sodium hydroxide. To conduct this study, DPLs were collected in Jam country in Busher province in Iran. Raw milled DPLs were pre-washed, and 10 g of dry matter basis was mixed with a solution of sodium hydroxide and lime at a 1:8 solid-to-liquid ratio in a 250 ml bottle, purged with N_2 stream (60 s) to remove probable CO_2 in headspace which may react with lime, tightly closed and stirred at 500 rpm for 10 min. The statistical analysis of results revealed that the impact of sodium hydroxide, lime and residence reaction time is significant ($p < 0.0001$). Under the optimum pretreatment conditions (sodium hydroxide loading 0.06 g NaOH/g dry biomass, lime loading of 0.09 lime/g dry biomass and residence time 71.4 h), the amount of cumulative gas production after 24 h was 101.4 ml gas/g organic matter, versus 104.6 ml gas/g organic matter for the predicted value. Thus, the results of this study proves the applicability of this low-cost efficient on-farm pretreatment to overcome the recalcitrance of date palm leaves and holds a promise to use date palm leaves—this dominantly wasted resource—for feeding of ruminants particularly in arid and semi-arid regions.

Keywords Waste dates · Dietary starch · Date palm seed · Diet of the African catfish · Digestive enzymes · Date palm leaflets · Roughage · Imported barley straw · Ensilage · Cardboxes · Date palm leaves · Silages for ruminants · Omani sheep · Rhodes grass hay · Pedicels · Peduncles · Date pits · NDF · ADF · Lignin and crude protein contents · In vitro dry matter digestibility · Sheep · Goat and dromedary · Sodium hydroxide and lime · Ruminant feed

9.1 Use of Wasted Dates as a Replacement of Dietary Starch in Feed

In dates processing industries ~20–30% of total dates production is not marketable and considered as waste. Therefore, it may be expedient to use wasted dates in feed, especially in countries poor in animal (fish) feeds supply. Nile tilapia was chosen, because it is most popular in many countries in the world. A study [1] has been devoted to evaluate the effect of replacing starch with date flesh on the Nile tilapia weight gain, feed conversion, specific growth rate and protein efficiency. Four isocaloric-isotrogenous rations containing various levels (0%, 15%, 30% and 45%) of date flesh as a replacement for starch were fed to three replicate groups of Nile tilapia fingerlings with a mean initial weight of 2.5 g.

The results show that the inclusion of date in tilapia feed as a replacement of starch has improved growth rate, feed conversion, specific growth rate and protein efficiency ratio. The tilapia specimens fed with the 30% date test diet were superior to all other groups. This may be probably explained by mixing simple (dextrin) and complex (starch) carbohydrates at an optimum ratio (3:1).

9.2 Feed Additive in the Diets of Juvenile African Catfish from Date Palm Seeds

The African catfish is very popular in many countries of the world (e.g. Egypt, Ethiopia, Ghana, Mali and Nigeria in Africa and China, Indonesia, Malaysia, Philippines, Thailand in Asia). However, it is disease resistant and tolerant to adverse environmental conditions (e.g. temperature, dissolved oxygen and high salinity). A study [2] has been conducted to evaluate the effect of addition of date palm seed to the diet of African catfish on the weight gain, specific growth rate, feed conversion efficiency, protein efficiency ratio, protein productive value and energy retention. A number of 150 juvenile African fish of weight ranging from 85.5 to 91.4 g has been divided into 15 experimental groups: 5 treatments (0% control, 0.5%, 1%, 1.5% and 2% date palm seeds) and triplicate groups. The experimental diets have been prepared by blending the ingredients into homogenous mixture and processing the mixture using a laboratory pellet mill. The feeding trial lasted for 10 weeks.

The results of experiments show that the addition of date palm seed to diet of the African catfish is associated with a considerable increase of the weight gain, specific growth rate, protein efficiency ratio and protein productive value. The positive trend in the nutrient utilization efficiency with the addition of date palm seed indicates that the date palm seed may have the ability to aid the digestion of feed for effective utilization. The improved growth rate and nutrient utilization in the present study could be explained by the presence of a range of digestive enzymes in date palm seed [3]. The effect of the date palm seed supplementation on the final fish carcass and hematology indices followed the same positive trend as revealed in the study. Therefore, it could be concluded that the supplementation of date palm seed as a feed additive to African catfish could lead to improvement of the fish growth performance and nutrient utilization, as well as ensuring production of healthy fish. It is recommended to add date palm seeds at 1.5% level in African catfish feeding to improve the fish performance.

9.3 Use of Date Palm Leaflets as a Roughage for Dairy Cows

A research has been conducted [4] to evaluate the suitability of the use of the date palm leaflets as a roughage for dairy cows. Kuwait has a small dairy industry including Friesian and Holstein cows being imported from Europe or USA. The concentrate used in fodder contains barley (locally grown), maize, wheat bran, soybean and a mineral vitamin mixture. A roughage is needed to buffer the effects of a large amount of concentrate by stimulating rumen activity, the production of saliva and the proportion of acetate in the mixture of acids produced in the rumen. The cereal straw being imported as a roughage from neighbor countries is expensive as compared with the concentrates. Meanwhile Kuwait has palm plantations expanding with time. These palms are being pruned yearly providing increasing amounts of leaves, which are being treated as waste! Thus, the idea of this research emerged: why not evaluate the technical feasibility to use the date palm leaflets as sustainable roughage to substitute the imported barley straw?

The feeding experiments were conducted on 56 Friesian and Holstein cows of mean initial live weight 482 kg for 12 weeks beginning from the 5th week of lactation. In the first experiment, 48 cows were divided into two groups and fed as a group. In experiment 2, eight cows were randomly chosen, individually treated and fed: either by date palm leaflets or barley straw in addition of course to the concentrate and freshly cut alfalfa. Milk yields were recorded 4 days a week and the cows were weighed monthly in experiment 1 and weekly in experiment 2. Samples of feeds were analyzed for dry matter, crude fiber, ash, neutral detergent fiber, acid detergent fiber and acid detergent lignin according to AOAC [5]. In vitro digestibility was determined for concentrate mixture, the alfalfa and date palm leaflets by Tilly & Terry method (1963) as modification by Alexander [6]. The milk samples were analyzed for fat, protein and total solids according to Milko-Scan 203 (Foss electric, Denmark). The date palm leaflets were obtained from 20-year-old date palms at the PAAF station by shredding and stored in jute sacks. Barley straw was imported in a shredded form from Iraq, and alfalfa was freshly cut on a daily basis from a nearby irrigation scheme.

The results of both experiments 1 and 2 revealed that the milk yields, milk composition and live weight gains of cows fed by either date palm leaflets or barley straw did not significantly differ. Therefore, it can be concluded that the date palm leaflets—a resource treated as waste—can be used as an acceptable alternative roughage in feeding of cows to the expensive imported barley straw.

9.4 Ensilage of Cardboard and Date Palm Leaves

A study [7] has been devoted to the evaluation of an ensilage made from cardboards and date palm leaves. The livestock production in hot arid regions suffers from the problem of deficiency of supply of feed [8]. Kuwait, for example is scarcely able to satisfy 15–20% of its needs of forage for growing of sheep and camels [9] and is thus obliged to import feed (concentrates and roughage). The local resources that may be used in the manufacture of feed include corrugated cardboxes and date palm leaves [10, 11]. About 100,000 tons of cardboard boxes is being annually generated in Kuwait. In addition, the palm plantations in Kuwait are expanding with time. The annual pruning of these palms results into big quantities of leaves, which are of no economic value and are dominantly treated as waste. Both the cardboards and date palm leaves are of low nutritive value. But there are big potentials of biotechnological methods of treatment of lignocellulosic materials via ensiling. Ensiling is characterized as the development of anaerobic conditions, the production of organic acid and lowering of pH value, reduction of the number of potential pathogens, enhancement of the nutritional value and improvement of palatability [12]. Ensilage and utilization of lignocellulosic residues as ruminant feed has been extensively reviewed by researchers [9, 13–19]. Al-Awadhi et al. [20] and EL-Nawawy et al. [21] have investigated the potentiality of using cardboxes and poultry wastes for sheep feeding. Oshiot et al. [22, 23] indicated the utilization of NaOH-treated oil palm trunks and leaves with low lignin content and high soluble sugars content as a promising roughage source for ruminants. Improvements in silage fermentation, diet digestibility, food intake and animal performance have been proved following the use of a lactobacillus inoculant [16, 24–32]. In this study, the bioconversion and upgrading of nutritive value of cardboxes and date palm leaves to silage were investigated over 90 days through chemical and microbial treatments. The effects of NaOH alkali-treatment, inclusion of dates and molasses as fermentation carbohydrate additives, and the addition of Lactobacillus plantarum and Saccharomyces cerevisiae inoculants during days of anaerobic ensiling were determined. The qualitative aspects of cardboxes and date palm leaves silages for ruminants were evaluated.

The results of this study indicate relatively minor differences in silage composition among alkali-treated and untreated, washed and unwashed Lactobacillus plantarum and Saccharomyces cerevisiae after 30, 60 and 90 days of anaerobic ensiling. The concentrations of reducing sugars, ether extract, and crude protein tended to be acceptable among treatments. Silages showed a significantly high lactic acid concentration (up to 4.3%) and only traces of butyric acid (below 1%). Thus, it can be concluded that cardboxes and date palm leaves can be effectively ensilaged, and the resultant silages are acceptable and palatable to be used as ruminant feed in arid regions.

9.5 Effects of Feeding Ensilaged Date Palm Leaves and Byproduct Concentrate on Performance and Meat Quality of Omani Sheep

A research [33] has been conducted to evaluate the potentiality of use of date palm leaves for feeding of Omani sheep. Oman and the whole gulf region suffer as an arid region from the shortage of animal feed representing an abstacle in livestock investment projects. But Oman, as well as other gulf countries, has extensive date palm plantations. The annual pruning of these palms results in the accumulation of huge quantities of date palm byproducts including palm leaves. This resource is generally treated as waste and thus has no economic value. Nevertheless, there were serious trials to use this resource for feed. EL-Hag and EL-Khanjari [34] have incorporated date palm byproducts in ruminant diets in considerable proportions.

Mahagoub et al. [35, 36] have successfully used ground date palm leaves as an ingredient of concentrate for feeding local sheep. In Oman, the local vegetation includes Prosopis cineraria and Prosopis juliflora (Meskit, Mesquite), the pods of which contain up to 120 g/kg crude protein, but the digestibility of these pods by livestock was low probably due to large amounts of tannins and other phenolic compounds [37]. Prosopis pods were incorporated at levels up to 200 g/kg without any deleterous influence on animal feedlot performance [35, 38]. The present study aims to investigate the effects of feeding a pelleted concentrate formulated from ensiled date palm leaves on the performance and meat quality of Omani sheep.

This study has been conducted on 32 1-year-old male Omani native sheep of starting body weight ~32 kg in a feeding trial for 120 days. The conducted feeding included two types of roughages: (Rhodes grass hay (RGH) and urea-treated date palm leaves (UTPF)) and two types of concentrates: (commercial concentrate (CC), purchased from Omani feed Mills and a byproduct-based concentrate (BC) including ground date palm leaves, barley, Meskit pods, dried fish sardines, wheat bran, ground limestone and vitamin/mineral mix and salt). The first roughage (Rhodes grass hay) was obtained from grass grown on the Agricultural Experiment Station and chopped using a tubgrinder. The second roughage was prepared by ensiling a mixture of shredded date palm leaves with 30 g/kg commercial grade urea solution in a plastic tank for 5 weeks, air-dried before being fed to animals. On dry matter basis, BC and CC concentrates contained 179 and 180 g/kg crude protein; 241 and 56 g/kg acid detergent fiber; 37 g and 182 g/kg neutral detergent fiber; 118 and 73 g/kg ash; 17.9 and 18.3 kJ/g gross energy, respectively.

The experimental animals fed by the date palm leaves were in good health throughout the trial. But the sheep fed by UTPF had lower feed intake as compared with those fed by RGH, but feed intake/body weight were similar across diet groups. Byproducts fed sheep gained less weight ($p \leq 0.05$) than those fed by RGH and CC. The average daily gain for CC + RGH, BC + RGH, CC + UTPF and BC + UTPF were, respectively, equal to 80, 56, 32 and 10 g/d. Feed conversion (kg feed/kg body weight) was lower for UTPF and BC feed. Feeding UTPF and BC reduced carcass weight, dressing, fat content and carcass measurement. But feeding CC with UTPF improved sheep performance as compared to BC + UTPF. UTPF or BC did not have influence on meet quality. Therefore, it can be concluded that date palm leaves can be used as a component of feed for Omani sheep in case of nutritional shortage frequently experienced in arid zones.

9.6 In Vitro Assessment of Nutritive Value of Date Palm Byproducts as Feed for Ruminants

A research [39] has been conducted on the use of date palm byproducts (e.g., leaves, pedicels, date pits and waste dates) as feed for ruminants. The feeding costs represent the highest economic component of the livestock farming. The arid regions (e.g.,

Algeria) suffer from shortage of forages during the dry season extending for 6–7 months. Thus, natural vegetation should be supplemented with additional feed. In these regions the date palm plantations are the backbone of flora providing—in addition to dates—a huge wealth of byproducts including leaves, pedicels, date pits and waste dates. These byproducts have been historically used for feeding of animals by local farmers [40, 41]. Waste dates, rich in carbohydrates, dietary fiber and minerals, were traditionally used in South Algeria as supplementary feed for livestock [42, 43]. Waste dates are highly palatable and digestible feedstuffs [44]. But the available literature on the potential value of date palm byproducts as a whole is rather limited [45]. The objective of this study is to evaluate, in vitro, the ruminal fermentation and nutritive value of the dry palm leaves, pedicels, date pits and waste dates.

To conduct this study, 20–30 specimens of leaves, pedicels and date pits were taken from Deglet-Nour palm and waste dates from three varieties—Bouarus, Harchaya and Kentichi—in Biskra, in the Saharan Atlas region in Algeria. Vetch-oat hay was taken as a control reference from the ITELV (Technical Institute of Breeding, Ain Mlila, Algeria). All the samples including the control were oven-dried at 50 °C [46] and ground to pass 1 mm screen.

The dry matter was determined by ID method 934.01, ash—ID method 942.05, crude protein—ID954.01 following the methods of AOAC [47]. The neutral and acid detergent fiber and sulfuric acid detergent lignin were determined with ANKOM220 fiber analyzer as described by Mertens [48], AOAC [47]; official method 938.18 and Robertson and Van Soest [3], respectively. All fiber fractions were expressed including residual ash. The non-fiber carbohydrate content of feeds was calculated by subtracting CP, NDF and ash from total DM [49]. The fat content of date palm byproducts is usually very low ($< 1\%$) except for date pits, in which fat may reach up to 12% [43].

The results of this research showed that the palm leaves have the highest NDF, ADF, lignin and crude protein contents (609, 435, 84, 64 gkg^{-1} DM), respectively. The cumulative gas production at 144 h of incubation was greatest for Kentichi dates (330 mLg^{-1} DM) and lowest for dates pits (69 mLg^{-1} DM). Regardless of the variety, waste dates showed the highest DM effective ruminal degradability and organic matter digestibility. The date pits seemed to be a poorly degradable material. These results indicate that waste dates—though of low protein content—are highly digestible with energy concentration as high as that of vetch-oat hay. The palm leaves and pedicels can be considered as highly fibrous emergency roughages for low-producing animals. Protein supplements should be added when date palm byproducts are used as feedstuff in order to balance the ruminant diets.

9.7 Valorization of Date Palm Byproducts for Livestock Feeding in Southern Tunisia: Potentialities and Traditional Utilization

An important study [50] has been conducted with the objective of learning about farmer's knowledge South of Tunisia on the use of date palm byproducts as forage resources in order to be able to target innovations better adapted to real farming conditions. Livestock production is an important activity in oases [51]. But livestock feeding is problematic in arid regions [52]. The date palm byproducts (DPBP) are available in big quantities (e.g., in Saudi Arabia, about 20% of the total date production is unsuitable for human consumption and wasted [53]) and can thus be used as feed sources. There are available studies [54] on the chemical composition and nutritive value of wasted dates and date pits. But there are few literature sources on how DPBP are integrated in the traditional feeding systems in the arid zones.

To conduct this study, samples of wasted dates, date pits, leaves, peduncles and pedicels representing two dominant varieties—Deglet Nour from the Nefzaoua region and Kentah from the EL Hamma region—were taken. These samples have been analyzed for dry matter (DM), crude protein (CP), total ash and crude fiber (CF) by standard procedures [55]. Natural detergent fiber (NDF), acid detergent fiber (ADF) and acid detergent lignin (ADL) were determined according to Goering and Soest [56] using the automated ANKOM fiber analyser. The in vitro digestibility [57] measurements have been taken on each sample, using rumen juices from three species: sheep, goat and dromedary. The microbial inoculum was obtained from mixed rumen juice of three adult animals recently slaughtered, maintained at 38 °C and quickly transported to laboratory. Measurement of DM disappearance after 48 h incubation of each DPBP sample were repeated in triplicate.

An extensive socio-economic survey has been conducted in a sub-region of Nefazoua (El Faoura) concerning the use of DPBP in livestock feeding [58] including the modalities of distribution (types of animals, period of distribution, quantities, feed treatments, commercialization and perceptions and the structure of production unit). Then an explorative survey has been conducted on 84 production units of the Nefazoua region owning livestock in order to characterize the place of DPBP in the alimentary systems of the region and to understand the interests and limits of these forage resources perceived by local farmers.

The herd structure varied widely, but the great majority of livestock owners possessed small-to-medium sized herd of mixed small ruminants with a mean of 28 sheep and 18 goats and dromedaries were raised marginally. The prevailing pattern of ownership of herds is subsistence economy (provision of milk and meat for residents). Only 18% of farmers raised livestock as the first source of income. The use of DPBP in the Nefzaoua region for feeding livestock are as follows: 37% of households used indistinctly the whole DPBP; 33% used only wasted dates and 26% used wasted dates and leaves. Only 4% used pits for dromedaries. The use of DPBP without any treatment was found prevalent (40%), soaking of wasted dates and date pits in water (12 to 14 h) to increase palatability (26%), grinding of dates and date-pits (9% of farms) and ground date pits were also sold. DPBP were also mixed with other feeding sources: bran (67%), straw (17%), green grass and alfalfa (12%) or olive cake (5%) in 25% of households. It can be estimated that 5–30% of the dates production can be considered as waste dates.

It can be concluded that DPBP, taken individually, are highly unbalanced for animal nutrition: high energy content in the cases of wasted dates and date pits and high fiber content for leaves, peduncles and pedicels and for all: low crude protein content (3–6.5% of dry matter). The in vitro dry matter digestibility for wasted dates were found 74.5, 79.7 and 79.2% for sheep, goats and dromedary, respectively. The corresponding figures for leaves, peduncles and pedicels were found much lower 12.3, 19.6 and 18.3%, respectively. Thus, the main questions to be asked in use of DPBP in ruminant feeding are how to make a well-balanced DPBP–based forage? And how to conserve this feed to overcome the problem of seasonal unavailability. This indicates the significance of research on DPBP-based silages to improve the technical and economic feasibility of use of DPBP as a local and sustainable forage resource [59].

9.8 Combination of Sodium Hydroxide and Lime as a Pretreatment for Conversion of Date Palm Leaves into a Promising Ruminant Feed: An Optimization Approach

An important research [60] has been devoted to the pretreatment of date palm leaves to be used in ruminant feed. The date palm leaf (DPL) contains mainly hemicellulose (16–18%), cellulose (27–41%) and lignin (10–19%) [4, 61, 62]. It is annually available from pruning activities with big quantities (300,000–400,000 t) in Iran [52]. These quantities are dominantly treated as waste causing a big problem of disposal. One of the approaches to solve this problem is to use DPL in feeding of ruminants, since ruminal microbial populations are able to digest plant cell walls [63]. But DPL in its natural form has a very low ruminal digestibility (16% in vitro dry matter digestibility) [64], and thus pretreatment is needed before use in feed. It is well known that alkaline pretreatment improves the ruminal degradability of lignocellulosic materials [65, 66]. During alkaline treatment delignification and saponification (hydrolysis of ester bonds) occur making lignocellulosic materials swollen and porous [67]. Many hydrolytic enzymes produced by ruminal microorganisms are similar to commercial enzymes used to convert lignocellulosic materials to fuel ethanol [20].

The alkaline pretreatment at high temperature is fast and effective [65, 68], but it is a capital-intensive process [69]. Therefore, by replacing only a part of costly sodium hydroxide with less expensive lime, it is possible to realize a cost-effective and efficient pretreatment at a mild temperature with a low concentration of sodium hydroxide [70]. In the present study, the overall pretreatment performance was evaluated based on the volume of cumulative gas produced during an incubation period of 24 h, since it is the only in vitro gas production time with an in vivo correlation, to predict metabolizable energy of feedstuff based on the original research of Menke and Steingass [71].

To conduct this study, DPLs were collected in Jam county in Bushehr Province in Iran. DPLs were washed and dried at 45 °C for 72 h and then hammer-milled to a 2-mm screen size. Ash-corrected hemicellulose and cellulose contents were determined according to detergent extraction methods [72], followed by lignin measurement via cellulose solubilization using 72% (w/w) sulfuric acid for 3 h at room temperature.

Raw milled DPLs were pre-washed and 10 g, dry matter basis, was mixed with a solution of sodium hydroxide and lime at a 1:8 solid-to-liquid ratio in a bottle (250 ml), purged with N_2 stream (60 s) to remove probable CO_2 in headspace which may react with lime, tightly closed and stirred at 500 rpm for 10 min. The pretreatment temperature (average = 40 °C) was programmed according to a day-to-night cycle of 12 h: 12 h (50 °C for first 12 h and 30 °C for another 12 h) to mimic around-the—clock temperature of a typical summer day in tropical regions in Iran. It took approximately 26 min to decrease temperature from 50 to 30 °C, while heating from 30 °C to 50 °C took 5 min. The pH of pretreatment solution at the beginning and end of the pretreatment was measured by a portable pH meter.

The statistical analysis of results revealed that the impact of sodium hydroxide, lime and residence reaction time is significant ($p < 0.0001$), under the optimum pretreatment conditions (sodium hydroxide loading of 0.06 g NaOH/g dry biomass, lime loading of 0.09 g lime/g dry biomass and residence time 71.4 h; the amount of cumulative gas production after 24 h was 101.4 ml gas/g organic matter, versus 104.6 ml gas/g organic matter for the predicted value). SEM studies proved the formation of overtures and cracks on the surface of the substrate pretreated under the optimal conditions. Fourier transform infrared spectroscopic analysis confirmed the saponification and the delignification effect of the combined alkaline and lime pretreatment on DPLs. Thus, the results of this study proves the applicability of this low-cost efficient on-farm pretreatment to overcome the recalcitrance of date palm leaves and holds a promise to use this dominantly wasted resource for feeding of ruminants particularly in arid and semi-arid regions.

References

1. Belal IEH, Al-Jasser MS (1997) Replacing dietary starch with pitted date fruit in Nile tilapia Oreochromis niloticus (L.) feed. Aquaculture Res 28(6):385–389. https://doi.org/10.1046/j.1365-2109.1997.00872.x
2. Sotolu O, Kigbu AA, Oshinow AJ (2014) Supplementation of date palm (phoenix dactylifera) seed as feed additive in the diets of juvenile African catfish (Burchell, 1822). J Fish Aquat Sci 9(5):359–365. https://doi.org/10.3923/jfas.2014.359.365
3. Robertson JB, Van Soest PJ (1981) The detergent system of analysis. In: James WPT, Theander O (eds) The analysis of dietary fibre in food. Marcel Dekker, New York, pp 123–158

4. Bahman AM, Topps JH, Rooke JA (1997) Use of date palm leaves in high concentrate diets for lactating Friesian and Holstein cows. J Arid Environ 35(1):141–146. https://doi.org/10.1006/jare.1995.0145
5. Association of Official Analytical Chemists (1980) Official methods of analysis of the Association of Official Analytical Chemists. Washington DC: AOAC. pp 1018
6. Alexander RH (1969) The establishment of a laboratory procedure for the in vitro determination of digestibility. Research Bulletin, No. 42. Ayr: West of Scotland Agricultural College. pp 17
7. Baroon Z, El-Nawawy AS, Al-Othman A (2004) Ensilage of cardboard and date palm leaves. J Environ Sci Health, Part A Toxic/Hazard Subst Environ Eng 39(2):515–533. https://doi.org/10.1081/ESE-120027542
8. Qureshi AW (1987) Strategies for livestock development in Arab Arid zones and FAO's relevant programmes 56–81
9. Razzaque MA, Al-Nasser A, Salman AJ, Aderibigbe AO, Church DC (1986) Corrugated cardboard and dried poultry manure as dietary ingredients for sheep. Anim Feed Sci Technol 14(3–4):265–278. https://doi.org/10.1016/0377-8401(86)90099-4
10. El-Nawawy AS, Allam M, ElRayes E, Al-Daher R, Hmmoud A, Mussallam M, Dashti N (1987) Agroindustrial uses of lignocellulosic wastes bioconversion of lignocellulosic wastes in Kiwait to protein enriched fermented fodder (No. KISR2223). Kuwait Institute for Scientific Research
11. Razzaque MA, Salman AJ (1982) Utilization of cardboard and poultry wastes from municipal solid waste and poultry farm as animal feed (No. KISR738). Kuwait Institute for Scientific Research
12. Kim CW, Lee NH, Song KS, Kim HM, Park BS, Yoon CS, Han CK, Sohail MA (1993) The utilization of date palm leaves and poultry manure as feedstuffs for famm livestock in Saudi Arabia. VI. ETfect of feeding date paim Ieaf silage on growth performance and digestibility in Awassi and Najdi lambs Korcan. J Anim Nutr Feed 299–304
13. Ilian MA, Razzaque MA, Salman AJ (1988) Use of fat in diets of sheep in hot environments. II. Effect on rumen metabolism. Animal Feed Sci Technol 19(4):343–350. https://doi.org/10.1016/0377-8401(88)90024-7
14. Kannan K, Oblisami G, Loganathan BG (1990) Enzymology of ligno-cellulose degradation by Pleurotus sajor-caju during growth on paper-mill sludge. Biol Wastes 33(1):1–8. https://doi.org/10.1016/0269-7483(90)90116-A
15. Martin AM (1996) Role of lactic acid fermenlation in bioconversion of wustes. In Lactic Acid Bacteria: Current Advances in Metabolisn, Genetics and Applications. Springer
16. Morrison IM (1988) Influence of some chemical and biological additives on the fibre fraction of lucerne on ensilage in laboratory silos. J Agric Sci 111(1):35–39. https://doi.org/10.1017/S0021859600082782
17. Neelakantan S, Deodhar AD (n.d.) Biotechnological approaches of stran uriliation by microbial systens for feed and indhustrial purposes. In: Proceedings international workshop, feeding of ruminants on fibrous crop residucs: aspects of treatment, feeding, nutrient evaluation, research and extension. International workshop
18. Negi SS (1985) A review on the utilization of agricultural wastes as feeds for livestock and experimental safeguards against erroneous conclusions. Agricultural Wastes 13(2):93–103. https://doi.org/10.1016/0141-4607(85)90016-2
19. Venkatesh KV (1997) Simultaneous saccharification and fermentation of cellulose to lactic acid. Biores Technol 62(3):91–98. https://doi.org/10.1016/S0960-8524(97)00122-3
20. Anderson WF, Dien BS, Jung H-JG, Vogel KP, Weimer PJ (2010) Effects of forage quality and cell wall constituents of Bermuda grass on biochemical conversion to ethanol. BioEnergy Res 3(3):225–237. https://doi.org/10.1007/s12155-009-9063-9
21. El-Nawawy AS, Puskas K, Baroon, Yateem A, Razaquc MA, Malik RC, Al-Mutawa T (1993) Unconventional Itens as Aninal Fecd. Studies on Protein Enriched Fermeated Fodder: Preparation in Vitro Studics (No. KISR4344). Kuwait Institute for Scientific Rescarch
22. Oshio S, Hassan OA, Jaafar DM (1990) Uriliation of oil palm trees as ruminant feed rescarch highlights. Tropical Agriculture Research Center 16–18

23. Oshio S, Jaafar DM, Hassan OA (1992) The use of oil palm Truiks as runinant feed 125–133
24. Gordon FJ (1989) An evaluation through lactating cattle of a bacterial inoculant as an additive for grass silage. Grass Forage Sci 44(2):169–179. https://doi.org/10.1111/j.1365-2494.1989.tb01924.x
25. Henderson AR, Anderson DH, Neilson DR, Hunter EA (1987) The esfect of a commercial inoculant and add-f applied at Tro Lerels on the chemical characteristics and utilisation of ryegrass silage over two seasons 13–14
26. Joachim V, Frank I, Christine A (1992) New ways of selecting lactic acid bacteria for biotechnological processes. Appl Microbiol Biotechnol 240–243
27. Kung L, Chen JH, Kreck EM, Knutsen K (1993) Effect of microbial inoculants on the nutritive value of corn silage for lactating dairy cows. J Dairy Sci 76(12):3763–3770. https://doi.org/10.3168/jds.S0022-0302(93)77719-X
28. McAllister TA, Selinger LB, McMahon LR, Bae HD, Lysyk TJ, Oosting SJ, Cheng K-J (1995) Intake, digestibility and aerobic stability of barley silage inoculated with mixtures of *Lactobacillus plantarum* and *Enterococcus faecium*. Can J Anim Sci 75(3):425–432. https://doi.org/10.4141/cjas95-062
29. Meeske R, Basson HM (1998) The effect of a lactic acid bacterial inoculant on maize silage. Anim Feed Sci Technol 70(3):239–247. https://doi.org/10.1016/S0377-8401(97)00066-7
30. Rooke JA, Maya FM, Arnold JA, Armstrong DG (1988) The chemical composition and nutritive value of grass silages prepared with no additive or with the application of additives containing either Lactobacillus plantarum or formic acid. Grass Forage Sci 43(1):87–95. https://doi.org/10.1111/j.1365-2494.1988.tb02144.x
31. Venus J, Idler F, Idler C (1995) Evaluation of the performance of lactic acid bacteria for ehsiling green forage based on the results of physiological studies Bioscope 28–32
32. Woolford MK (1984) The microbiology of silage. In: The silage fermentation, pp 35–42. Marcel Dekker, Inc
33. Mahgoub O, Kadim IT, Al-Busaidi MH, Annamalai K, Al-Saqri NM (2007) Effects of feeding ensiled date palm fronds and a by-product concentrate on performance and meat quality of Omani sheep. Anim Feed Sci Technol 135(3–4):210–221. https://doi.org/10.1016/j.anifeedsci.2006.07.011
34. El Hag MG, El Khanjari HH (1992) Dates and sardines as potential animal feed resources. World Anim Rev 15–23
35. Mahgoub O, Kadim IT, Al-Ajmi DS, Al-Saqry NM, Al-Abri AS, Richie AR, Annamalai K, Forsberg NE (2004) The effects of replacing rhodes grass (Chloris gayana) Hay with Ghaf (Prosopis cineraria) pods on the performance of omani native sheep. Trop Anim Health Prod 36(3):281–294. https://doi.org/10.1023/B:TROP.0000016829.63985.2c
36. Mahgoub O, Kadim IT, Johnson EH, Srikandakumar A, Al-Saqri NM, Al-Abri AS, Ritchie A (2005) The use of a concentrate containing Meskit (Prosopis juliflora) pods and date palm by-products to replace commercial concentrate in diets of Omani sheep. Anim Feed Sci Technol 120(1–2):33–41. https://doi.org/10.1016/j.anifeedsci.2005.01.011
37. Horton GMJ, Chesworth JM, Srikandakumar A, Gumaa K, Wohlt JE (1993) Nutritional value of Acacia tortilis and Prosopis cineraria pods for sheep (abstract). J Anim Sci
38. Mahgoub O, Kadim IT, Forsberg NE, Al-Ajmi DS, Al-Saqry NM, Al-Abri AS, Annamalai K (2005) Evaluation of Meskit (Prosopis juliflora) pods as a feed for goats. Anim Feed Sci Technol 121(3–4):319–327. https://doi.org/10.1016/j.anifeedsci.2005.01.016
39. Souhil B, Lyas B, Antonio D, Manuel F, Zahia A, Secundino L (2016) In vitro assessment of nutritive value of date palm by-products as feed for ruminants. Emirates J Food Agri 28(10):695. https://doi.org/10.9755/ejfa.2016-01-104
40. Chehma A, Longuo HF (2004) Bilan azoté et gain de poids chez le dromadaire et le mouton, alimentés à base de sousproduits du palmier dattier, de la paille d'orge du Drin. Cahiers Agric 221–226
41. El Hag MG (1995) Production of animal feeds from dates and date by-products in the Sultanate. a review. J Anim Sci 195–201

42. Mebirouk-Boudechiche L, Araba A, Ouzrout R (2008) Influence du type de complément énergétique (rebuts de dattes vs orge) sur les performances d'engraissement et caractéristiques des carcasses d'agneaux Berbères à l'engraissement. Revue d'élevage et de Médecine Vétérinaire Des Pays Tropicaux 61(3–4):209. https://doi.org/10.19182/remvt.9991
43. Sidhu JS (2006) Date fruits production and processing. In: Fruits and fruit processing, pp 391–419. Blackwell Publishing
44. Al-Dobaib SN, Mehaia MA, Khalil MH (2009) Effect of feeding discarded dates on milk yield and composition of Aradi goats. Small Rumin Res 81(2–3):167–170. https://doi.org/10.1016/j.smallrumres.2008.11.007
45. Dayang JF, Reuben CR, Raji F (2014) Nutritional, socioeconomic and health benefits of dates. J Food Nutr Sci 63–73
46. Makkar HPS (2003) Quantification of tannins in tree and shrub foliage. Kluwer Academic, A Laboratory Manual
47. AOAC, Association of Official Analytical Chemists International (2006) Official methods of analysis, 18th ed. Current through Revision 1. AOAC, Association of Official Analytical Chemists International
48. Mertens DR (2002) Gravimetric determination of amylase-treated neutral detergent fiber in feeds with refluxing in beakers or crucibles: collaborative study. AOAC INTERNATIONA 85(6):1217–1240
49. Sniffen CJ, O'Connor JD, Van Soest PJ, Fox DG, Russell JB (1992) A net carbohydrate and protein system for evaluating cattle diets: II. Carbohydrate and protein availability. J Animal Sci 70(11):3562–3577. https://doi.org/10.2527/1992.70113562x
50. Genin D, Kadri A, Khorchani T, Sakka K, Belgacem F, Hamadi M (2004) Valorisation of date-palm by-products (DPBP) for livestock feeding in Southern Tunisia. I—Potentialities and traditional utilisation. CIHEAM Options Méditerranéennes 221–226
51. Dollé V (1990) Elevage intensif en oasis, une composante intégrante du système de production. Options Méditerranéennes 195–204
52. Tisserand JL (1990) Les ressources alimentaires pour le bétail. Options Méditerranéennes 237–248
53. Belal IEH, Al-Jasser MS, Mustafa IA, Al-Dosari MN (1999) Evaluation of date-feed ingredients mixes. Anim Feed Sci Technol 81(3–4):291–298. https://doi.org/10.1016/S0377-8401(99)00091-7
54. Hadjipanayiotou M, Rihani N (1998) The potential of using non conventional feed resources during dry years in the Mediterranean region. Livestock production and climatic uncertainty in the Mediterranean. Proceedings of the joint ANPA-EAAP-CIHEAM-FAO symposium, Agadir, Morocco
55. AOAC (1975) Official methods of analysis, 12th edn, p 1094
56. Goering HK, van Soest PJ (1970) Forage fibre analysis. ARS/USDA agricultural handbook. USDA, Washington DC
57. Tilley JMA, Terry RA (1963) A two-stage technique for the in vitro digestion of forage crops. Grass Forage Sci 18(2):104–111. https://doi.org/10.1111/j.1365-2494.1963.tb00335.x
58. DYPEN (2000) Observatoires des relations population-environnement en milieu rural tunisien: Pour une gestion durable des ressources, p 449. Rapport Scientifique. IRA-IRD Tunis, Tunisia
59. Kayouli C, Jouany JP, Ben Amor J (1991) Comparison of microbial activity in the forestomachs of the dromedary and the sheep measured in vitro and in sacco on mediterranean roughages. Anim Feed Sci Technol 33(3–4):237–245. https://doi.org/10.1016/0377-8401(91)90063-X
60. Rajaee Rad A, Ahmadi F, Mohammadabadi T, Ziaee E, Polikarpov I (2015) Combination of sodium hydroxide and lime as a pretreatment for conversion of date palm leaves into a promising ruminant feed: an optimization approach. Waste Biomass Valorization 6(2):243–252. https://doi.org/10.1007/s12649-014-9340-5
61. Arhab R, Macheboeuf D, Aggoun M, Bousseboua H, Viala D, Besle JM (2009) effect of polyethylene glycol on in vitro gas production and digestibility of tannin containing feedstuffs from North African arid zone. Trop Subtrop Agroecosyst 10:475–486

62. Pascual JJ, Fernández C, Díaz JR, Garcés C, Rubert-Alemán J (2000) Voluntary intake and in vivo digestibility of different date-palm fractions by Murciano-Granadina (Capra hircus). J Arid Environ**45**(2):183–189. https://doi.org/10.1006/jare.1999.0622
63. Russell JB (2002) Rumen microbiology and its role in animal nutrition. Russell Publishing Co, Ithaca, NY
64. Ziaei N, Hosseini SMMS (2009) Feeding value and in vitro digestibility of date-palm leaves supplemented with different supplementary energy. Pak J Biol Sci 12(10):817–820. https://doi.org/10.3923/pjbs.2009.817.820
65. Ahmadi F, Rad AR, Holtzapple MT, Zamiri MJ (2013) Short-term oxidative lime pretreatment of palm pruning waste for use as animal feedstuff: Short-term oxidative lime pretreatment of palm pruning waste. J Sci Food Agric 93(8):2061–2070. https://doi.org/10.1002/jsfa.5963
66. Ghasemi E, Khorvash M, Ghorbani GR, Elmamouz F (2014) Effects of straw treatment and nitrogen supplementation on digestibility, intake and physiological responses of water intake as well as urine and faecal characteristics. J Anim Physiol Anim Nutr 98(1):100–106. https://doi.org/10.1111/jpn.12052
67. Hendriks ATWM, Zeeman G (2009) Pretreatments to enhance the digestibility of lignocellulosic biomass. Biores Technol 100(1):10–18. https://doi.org/10.1016/j.biortech.2008.05.027
68. Rezende CA, de Lima MA, Maziero P, Ribeiro deAzevedo E, Garcia W, Polikarpov I (2011) Chemical and morphological characterization of sugarcane bagasse submitted to a delignification process for enhanced enzymatic digestibility. Biotechnol Biofuel 1–19
69. Wu L, Arakane M, Ike M, Wada M, Takai T, Gau M, Tokuyasu K (2011) Low temperature alkali pretreatment for improving enzymatic digestibility of sweet sorghum bagasse for ethanol production. Biores Technol 102(7):4793–4799. https://doi.org/10.1016/j.biortech.2011.01.023
70. Xu J, Cheng JJ (2011) Pretreatment of switchgrass for sugar production with the combination of sodium hydroxide and lime. Biores Technol 102(4):3861–3868. https://doi.org/10.1016/j.biortech.2010.12.038
71. Menke KH, Steingass H (1988) Estimation of the energetic feed value obtained from chemical analysis and in vitro gas production using rumen fluid 103–116
72. Van Soest PJ, Robertson JB, Lewis BA (1991) Methods for dietary fiber, neutral detergent fiber, and nonstarch polysaccharides in relation to animal nutrition. J Dairy Sci 74(10):3583–3597. https://doi.org/10.3168/jds.S0022-0302(91)78551-2

Chapter 10
Date Palm Byproducts for WasteWater Treatment

Abstract The pollution caused by heavy metals, phenols and phenolic compounds, as well as synthetic dyes and pesticides is an environmental problem of world concern. A research has been devoted to the use of a sorbent made from date palm leaflets for the removal of Cd^{2+} and Ag^{+} from waste water. The date palm leaflets are annually available from pruning of date palms with big quantities: 180,000 ton and 3 million tons in Oman and the gulf states, respectively. Palm leaflets were sourced from a farm in Muscat (Oman), washed with distilled water, left to be air-dried and then oven-dried to constant weight. The sorption of Cd^{2+} and Ag^{+} from aqueous solutions was investigated, taking the following parameters into consideration: pH value, contact time, metal concentration and temperature. The sorption behavior was different for both metals. The sorption of Cd^{2+} was fast reaching equilibrium within ~ 2 h, whereas that for Ag^{+} was slow and required ~ 60 h. The research results has led to the conclusion that the date palm leaflets, generally of no value in the field and treated as waste, can be successfully used to make a chemically-carbonized sorbent for the removal of copper and silver metals from the waste water. Another research was conducted to investigate the potentiality of use of raw date palm trunks for the removal of cadmium from waste water. The effects of the process variables, such as fiber size, mixing rate, temperature, pH of the solution and absorbent dose on the absorption capacity of trunk fibers were studied. It was found that the adsorption capacity of Cd^{2+} increased from 29.06 to 51.1 mg/g (~2 times) as the particle size decreased from 875 to 100 μm. Concerning the effect of pH value, it was found that the adsorption capacity of Cd^{2+} decreased in the strong acidic medium and rapidly increased as the pH of the solution increased from 1.69 to 3.71. It was found that the equilibrium time of adsorption process is very small: the maximum adsorption capacity was attained only after 10 min. Thus, it can be concluded that the date palm trunk fibers are a potential absorbent for the removal of cadmium from waste water. This opens a great economic and developmental potentiality of use of the trunks of the old and unproductive date palms as a low cost material for the removal of Cd^{2+} and other toxic metals from the waste water. A research has been conducted on the use of date palm seeds for the production of activated carbon via sequential hydrothermal carbonization (HTC) and sodium hydroxide activation. HTC is a distinguished low temperature and environmentally favorable process. To conduct this study, the date

H. EL-Mously et al., *Date Palm Byproducts: A Springboard for Circular Bio Economy*, Materials Horizons: From Nature to Nanomaterials,
https://doi.org/10.1007/978-981-99-0475-4_10

palm seeds (PDS) were locally obtained in Malaysia. An automated stainless-steel hydrothermal reactor of capacity 200 ml was used for HTC, and a sample of 5 g was put in it. The reactor was sealed and heated to 200 °C for 5 h at 5 °C/min heating rate. After cooling, the obtained hydrochar was well washed with distilled water and put in an oven at 105 °C for 24 h. The PDS-HTC hydrochar was impregnated with NaOH at ratios of PDS-HTC: NaOH (w/w): 1:1, 1:2 and 1:3, then oven-dried at 105 °C for 24 h. Then an automatic electric vertical furnace was used to activate the NaOH pretreated hydrochar at 600 °C under a continuous nitrogen (N_2 99.995%) flow at 150 cm^3/min. A 10 °C/min rate of heating was set for 1 h. The produced AC was collected after cooling, repetitively rinsed with hot distilled water to decrease the pH of the washing solution from 6to 7. Then the ACs were dried in an oven at 105 °C for 24 h and stored in tightly closed container. The methylene blue (MB) was chosen in this study for the evaluation of the adsorption performance of the PDS hydrochar AC in aqueous solution. The textural, morphological and chemical properties of the produced hydrochar AC were investigated. NaOH activation enhanced the porosity and surface functionality of the hydrochar. Thus, the prepared AC exhibited a relatively high specific surface area of1282.49 m^2/g, a total pore volume of 0.66 cm^3/g and an average pore width of 20.73 Å. Thus, it can be concluded that it is feasible by a combination of HTC and NaOH activation to produce activated carbon with remarkable adsorptive properties from date palm seeds, a very cheap resource, dominantly treated as a waste. A research has been conducted on the use of date palm seeds for the removal of phenol. The date palm seeds were received from a local farm in Abu Dhabi, washed with deionized water, dried at 105 °C for 24 h, ground to size > 500 μm and stored in a desiccator. Ten grams of the material was mixed with KOH (40% by volume) solution at an impregnation ratio (KOH/biomass) of 2. The produced porous carbon had a Brunauer–Emmett–Teller area of 892 m^2/g, pore volume of 0.45 cm^3/g and average pore diameter of 1.97 nm. This porous carbon was used for the adsorption of phenol at different concentrations (100–400 mg/L) and temperatures (30–50 °C). The adsorption behavior of phenol on porous carbon was found to be well described by Langmuir isotherm model. The monolayer adsorption capacity was found to be 333 mg/g, the highest as compared with date palm seed biomass-based porous carbons. Thus, it can be concluded that it is feasible to use date palm seeds for the preparation of KOH-based activated carbon for the removal of phenol from aqueous solutions as a low-cost process with an extremely high performance. An important paper has been devoted to a review on the preparation of activated carbons from date palm stones and application for wastewater treatments. The agricultural residues have withdrawn the attention of researchers as promising precursors for activated carbon, because of the low cost, great abundance, renewability and high lignocellulosic content. Fruit stones are of particular interest being byproducts of food industries and thus easy to collect. Examples of these fruit stones are peach, apricot, olive, cherry, grape and date stones. Among these fruit stones, the date stones are especially distinguished for their high carbon content, low price and high availability in the Arab countries. The date stone represents ~ 10% of the fruit weight making it the largest agricultural byproduct in the date-producing countries (~700 thousand tons annually). Chemical activation is preferred over physical due to higher yield, single-step treatment, lower temperature, shorter time and better porous structure. Among the different methods of wastewater treatment the adsorption is advantageous because of its simplicity, flexibility, suitability for batch and continuous processes, possibility of regeneration and reuse, low expenses and capability to remove a wide array of pollutants in different concentrations. The activated carbon is distinguished with a high adsorption performance due to its high surface area, well-developed porous structure and favorable surface properties. The chemical composition of date stones is as follows: moisture (5–10%), ash (1–2%), protein (5–7%), oil (7–10%), crude fiber (10–20%) and carbohydrates consisting of 23% hemicellulose, 15% lignin, 57% cellulose and 5% ash. The activation variables affecting the pore characteristics and yield of carbon are in the order of: activation temperature, impregnation ratio and activation time. The surface areas of date stone carbons were in the range from 490 to 1282 m^2/g and yields from 17 to 47% with the highest values obtained by chemical activation. The application of date palm stones-carbons for the adsorption of organic and inorganic pollutants demonstrated maximum capacities of 612.1, 359.1, 238.1 and 1594 mg/g for dyes, phenols, pesticides and heavy metals, respectively. Thus, it can be concluded that the date palm stones are a low-cost, renewable, non-toxic and biocompatible source for the manufacture of activated carbon of high efficiency for removal of pollutants from waste water. Proceeding from the insufficiency of literature on the use of date palm seeds as a suitable precursor for the manufacture of activated carbon a research has been conducted to investigate the impact of process conditions on the preparation of porous carbon from date palm seeds using KOH activation. To conduct this study, date palm seeds were sourced from a local farm in Abu Dhabi, UAR, washed with deionized water and dried in oven at 105 °C for 24 h and then ground to a particle size > 500 µm. Ten grams of date palm seeds dust were mixed with KOH (40% by volume) solution at an impregnation ratio 1–5. The mixture was stirred for 5 h using a magnetic paddle. After complete soaking of KOH solution, the mixture was oven-dried at 105 °C and then activated at a horizontal furnace at 500–900 °C for an activation time 30–120 min in a nitrogen flow of 150 mL/min. The carbonized samples were cooled to room temperature in an inert atmosphere, washed with HCl aqueous solution to remove ash, and then by distilled water to reduce pH to neutral. The final porous carbon was then dried at 105 °C for 12 h. The yield of activated carbon was estimated based on grams bone dry per grams bone dry of date palm seeds. The results of the study have come to the following conclusions:

1. The increase of the activation temperature and the impregnation ratio (KBR) contribute to the increase of the BET surface area, while the yield decreases.
2. The activation duration was found not to influence the BET surface area, and its effect on the yield is marginal.
3. The optimum conditions were found as follows: activation temperature 601 °C, KBR ratio 1.97 with the corresponding BET surface area and yield being 910 m^2/g and 22.9%. The activated carbon possessed a pore volume of 0.45 cc/g, an

average pore diameter of 1.54 nm with the micropore volume proportion being 0.40 cc/g.
4. The activated carbon textural characteristics have a microporous nature. It has been also tested for iodine adsorption.
5. It was found that the date palm seeds activated carbon contains a large number of acid functional groups in the form of carboxylic and hydroxyl groups and thus could be effectively utilized for the treatment of waste water.

Keywords Heavy metals · Copper · Cadmium · Silver · Phenol · Phenolic compounds · Synthetic dyes · Sorbent · Adsorption capacity · Date palm trunks · Methylene blue · Fibers · Date palm leaflets · Date palm seeds · Activated carbon · Hydrochar · KOH activation · Physical activation · Chemical activation · Fruit stones · Langmuir isotherm model

10.1 A Chemically-Carbonized Sorbent from Date Palm Leaflets for the Removal of Cu^{2+} and Ag^{+} from Aqueous Solutions

A research [1] has been devoted to the use of a sorbent made from date palm leaflets for the removal of cu^{2+} and Ag^{+} from an aqueous solution. The pollution caused by heavy metals is an environmental problem of world concern. Various chemical industries are associated with the presence of copper and silver in the waste water. Copper is of high toxicity for marine life and humans [2, 3], and Wilson disease can be caused by the deposition of copper in the human brain, liver, pancreas and myocardium [4]. Silver is also associated with deleterious effects on the human health including stomach pain, respiratory problems and irritation of throat and lungs [5] as well as the deposition of silver-protein complexes in body tissues causing argyria, a permanent discoloring of skin [5].

The prevailing physical and chemical technologies for the removal of heavy metals from waste water (e.g. electro-coagulation, chemical precipitation, membrane filtration and reverse osmosis) are either too expensive or not appropriate for treatment of waste water [6]. The adsorption technologies using activated carbon are too expensive due to the cost of carbon preparation [7]. Different absorbents were tested to remove Cu^{2+} from aqueous solutions including lignocellulosic materials (wheat and soybean straws, corn stalks and corn cops) [8], chitosan resin [9], activated carbon [10] and perlite [11]. Silver was also removed from aqueous solutions using Corynebacterium glutamicum [12], amino methylene phosphoric acid resin [13], perlite [11], coal [14], lignite [15], activated carbon fibers [16] and activated carbon [17].

The date palm (Phoenix dactylifera L.) is the main agricultural crop in Oman and the Gulf States [18]. The pruning of these palms produces annually 180,000 and 3 million tons of leaflets in Oman and the Gulf States respectively, which are treated as waste: either open-field burnt or sent to landfills. The objective of the present study is to make use of palm leaflets-a waste material-to prepare a carbonaceous sorbent and test its technical feasibility for the removal of copper (Cu^{2+}) and silver (Ag^{+}) from waste water.

Palm leaflets were sourced from a farm in Muscat (Oman), washed with distilled water, left to be air-dried and then oven-dried to constant weight. The leaflets samples were then treated with sulfuric acid and heated to 170 °C for 20 min together with stirring for 25 min. The resulting black mixture was cooled and filtered using a Bucher funnel under vacuum. The used sulfuric acid was filtered, and the carbonized material was washed with deionized water until it didn't make a change in methyl orange color indicating the absence of sulfuric acid. Then the carbonized product was oven-dried at 120 °C until constant weight and ground to the size of sieves between 16 and 60 mesh.

The sorption of Cu^{2+} and Ag^{+} from aqueous solutions was investigated, taking the following parameters into considering: pH value, contact time, metal concentration and temperature. The sorption behavior was different for both metals. The sorption of Cu^{2+} was fast reaching equilibrium within ~ 2 h, whereas that for Ag^{+} was slow and required ~ 60 h. The activation energy E_a for Cu^{2+} sorption was ~ 16.1 kJ/mol indicating a diffusion-controlled ion exchange process, whereas E_a for Ag^{+} sorption was ~ 44.3 kJ/mol indicating a chemically controlled process. The sorption of Cu^{2+} and Ag^{+} fitted the Langmuir equation with an increase in metal uptake with an increase of temperature. This can be explained by the expected increase in swelling of the sorbent, which allows more active sites to be available for metal ions. Ag^{+} was reduced to elemental silver on the absorbent surface as shown using scanning electron microscopy and X-ray powder diffraction. As far as Cu^{2+} sorption is concerned, no reduction processes were involved. As a conclusion, the date palm leaflets, generally of no value in the field and treated as waste, can be successfully used to make a chemically-carbonized sorbent for the removal of copper and silver metals from the waste water.

10.2 Use of Date Palm Trunk Fibers as Adsorbents for the Removal of Cd^{+2} Ions from Waste Water

A study [19] has been devoted to the use of date palm trunk fibers as absorbents to remove Cd^{+2} ions from waste water. Cadmium is one of the heavy metals being toxic and harmful for human health [20]. It is not biodegradable and accumulates in the human body causing different diseases including renal damage, excess salvation, muscular cramps, nausea, destruction of the body cells, diarrhea, hypertension and anemia [21]. Cadmium is present in the waste water of many industries, such as batteries, metal coating/plating, fertilizers, pigments and photography [22]. These industries are thus associated with danger of contamination of groundwater and water reused for irrigation. The concentration of cadmium in the drinking water should be less than 0.005 mg/L according to the WHO [23]. It is necessary to reduce cadmium in aqueous industrial waste water to an acceptable level [24]. There are different technologies for the removal of cadmium from waste water such as evaporation, chemical precipitation, ion exchange, chromatography and reverse osmosis [25]. But, these methods have different drawbacks, such as the high capital and operational costs, as well as the necessity of disposal of the residual sludge with toxic cadmium [26]. Therefore, it is necessary to find cheap absorbents that are found in nature with big quantities. There is in literature many applications of use of economic natural adsorbents for the removal of heavy metals from waste water such as zeolite [27] sediments of rivers [28], peanut hull [29], rice husk [30], bentonite [31], waste tea leaves [32], almond shells and olive stones [33]. The date palm is one of the essential elements of flora in the Arab region producing great amounts of byproducts, such as date stones and trunks of old and unproductive palms. The objective of the present study is the investigation of the potentiality of use of raw date palm trunks for the removal of cadmium from waste water.

In order to conduct this study, cadmium nitrate (from BDH) was used to make the standard solution by dissolving 2.74 g of reagent grade Cd to a 1L volumetric flask half full of distilled water. After swirling of the flask, the concentration was 1 mg/L. Sodium nitrate (from BDH was used as an ionic strength adjuster 1SA). 425 reagent-grade sodium nitrate were added to a 1L volumetric flask about half full of distilled water. The flask was swirled to dissolve the solid, filled with distilled water to the mark, and upended several times to mix the solution.

The effects of the process variables, such as fiber size, mixing rate, mixing time, temperature, pH of the solution and absorbent dose on the adsorption capacity of trunk fibers were studied. The obtained results indicate that the adsorption capacity of Cd^{2+} increased from 29.06 to 51.1 mg/g (~ 2 times) as the particle size decreased from 875 to 100 µm. Concerning the effect of pH value, it was found that the adsorption capacity of Cd^{2+} decreased in the strong acidic medium and rapidly increased as the pH of the solution increased from 1.69 to 3.71. The increase of temperature from 12 to 34 °C and the mixing speed from 100 to 500 rpm were found to have insignificant effect on the adsorption capacity. It was found that the equilibrium time of the adsorption process is very small: the maximum adsorption capacity was attained only after 10 min. Thus, it can be concluded that the date palm trunk fibers are potential absorbent for the removal of cadmium from waste water (highly better than activated carbon, olive stones, tree fern and coconut copra). This opens a great economic and developmental potentiality of use of the trunks of the old and unproductive palms as a low-cost material. These findings are especially attractive because of the small absorbent dosage required and the short time needed. This also supports further research on the use of date palm trunk fibers for the removal of other toxic metals from the waste water.

10.3 Mesoporous and Adsorptive Properties of Date Palm Seed Activated Carbon Prepared Via Sequential Hydrothermal Carbonization and Sodium Hydroxide Activation

A research [34] was conducted on the use of date palm seeds for the production of activated carbon via sequential hydrothermal carbonization and sodium hydroxide activation. Activated carbon is usually synthesized using char from the pyrolysis of organic materials and then subjected to a process of physical or chemical activation at a high temperature. The hydrothermal carbonization (HTC) is a more energy-efficient process to produce carbon from pure carbohydrate or lignocellulosic materials with accepted chemical structures [35]. HTC is a distinguished low temperature and environmentally favorable process to convert various precursors to value-added products [36–39]. HTC is an exothermic process, and wet materials can be used directly without any drying process [40, 41]. Three products are available from HTC: solid hydrochar, an aqueous soluble liquid and gas. Most of studies were mainly concerned with bio-oil, whereas few studies were associated with hydrothermal carbon or hydrochar. Falco et al. [35] studied the influences of biomass precursor (i.e., D-glucose, cellulose and rye straw) and HTC temperature on porosity formation in the AC, made by KOH activation. High-performance ACs were generated in gas storage applications regardless of the precursor used. Seatea and Tippayawong [40] conducted physical activation of HTC- sewage sludge by using hot steam. The produced AC was found to have high potential for adsorption or soil conditioning. Pari et al. [42] were able to obtain spherical AC from cassava and tipioca flour via HTC and KOH activations. The obtained AC was dominantly spherical and of excellent textural and morphological properties. AC was also prepared from beer waste via H_3PO_4 activation with improved surface area and micropores [43]. Zhu et al. [44] studied the characteristics and tetracycline adsorption behavior of a novel porous carbon prepared from hydrochar. Unur [45] repeated the use of hazelnut shells for the preparation of functional nonporous carbons by hydrothermal treatment. Regmi et al. [46] produced hydrochar from switch grass and AC via KOH activation of the hydrochar and investigated their absorption of copper and cadmium in aqueous solutions. In this study, the date palm seeds—dominantly treated as waste in the palm growing countries—were subjected to pretreatment by HTC under mild conditions and subsequently by chemical activation using sodium hydroxide (NaOH) to produce mesoporous AC.

To conduct this study, the date palm seeds (PDS), the precursor for hydrochar, were locally obtained in Malaysia. The seeds were well washed with normal and then distilled water. The material was then air-dried at 28 °C, crushed in a grinder to a particle size 1–2 mm. An automated stainless-steel hydrothermal reactor of capacity 200 ml was used for HTC. PDS sample of 5 g was put in the reactor. The reactor was sealed and heated to 200 °C for 5 h at 5 °C/min heating rate. After cooling, the obtained hyrochar was well washed with distilled water and put in an oven at 105 °C for 24 h. The PDS-HTC hydrochar was impregnated with NaOH at ratios of PDS-HTC: NaoH (w/w): 1:1, 1:2 and 1:3, then oven-dried at 105 °C for 24 h. Then, an automatic electric vertical furnace was used to activate the NaoH pretreated hydrochar at 600 °C under a continuous nitrogen (N_2 99.995%) flow at 150 cm^3/min. A 10 °C /min rate of heating was set for 1 h. The produced AC was collected after cooling, repetitively rinsed with hot-distilled water to decrease the pH of the washing solution from 6to 7. Then the ACs were dried in an oven at 105 °C for 24 h and stored in tightly closed container.

The methylene blue (MB) was chosen in this study for the evaluation of the adsorption performance of the PDS hydrochar AC in aqueous solution. Thus, a 1000 mg/L stock solution was prepared by dissolving an exact amount of MB dye (Sigma-Aldrich) in distilled water. The different concentrations of MB for the entire study were prepared from this stock solution. MB had a molecular weight of 373.9 g/mol, molecular formula is $C_{16}H_{18}N_3SCI$ and its solubility in water is 40 g/L. All other chemicals, such as NaOH and HCI were laboratory grade and purchased from different renowned supplies.

The textural, morphological and chemical properties of the produced hydrochar AC were investigated. NaoH activation enhanced the porosity and surface functionality of the hydrochar. Thus, the prepared AC exhibited a relatively high specific surface area of 1282.49 m^2/g, a total pore volume of 0.66 cm^3/g, and an average pore width of 20.73 Å. The results of FTIR analysis show that the functional groups of the AC either shifted to different frequencies or disappeared in some cases when the hydrochar was activated with NaoH. It was found that the MB adsorption is affected by pH: lowest at pH3 (104.88 mg/g), increased to 106.4 mg/g at pH7 and remained constant from pH7 to 11. It was also found that the adsorption capacity of the AC decreased with the increase of temperature attaining a maximum value of 612.1 mg/g. Thus, it can be concluded that it is feasible by a combination of HTC and NaOH activation to produce activated carbon with remarkable adsorptive properties from date palm seeds, a very cheap resource, dominantly treated as waste.

10.4 KOH-Based Porous Carbon from Date Palm Seed: Preparation, Characterization and Application to Phenol Adsorption

A research [47] has been conducted on the use of date palm seeds for the manufacture of porous carbon for the removal of phenol. It is well known that the phenol and phenolic compounds are present in the waste water of heavy chemical, petrochemical, pharmaceutical, paint, paper and pulp and oil refining industries. These compounds may have negative influences on the human life, soil, ground water, as well as microorganisms. That is why the removal of phenolic compounds from industrial waste water is a matter of world significance. There are available phenol removal technologies (e.g., solvent extraction, distillation, adsorption, membrane separation and electrochemical treatments) [48]. But there are growing evidences that the removal of phenols by porous carbon is the best option over the aforementioned methods. The advantages of use of porous carbon include the zero-sludge production, flexibility of the operation, reduction of operating expenses, chemical and mechanical flexibility, high degree of surface reactivity and high adsorption capacity [49]. But, the commercially available porous carbons are expensive. Thus, there is a necessity to search for alternative cheap materials to manufacture activated carbon, such as rice husk [50], sawdust [51], coconut shell [52] and sugar cane bagasse [53]. The present interest in porous carbon can be evidenced from the fact that the researches on porous carbon in open literature account for 30% of the overall research in the field of carbon. The characteristics of porous carbon depend on the precursor characteristics, whether it is either coal or lignocellulosic material, as well as the method of activation [54, 55]. The metal-based activating agents such as KOH are reported to be more suitable for the production of highly porous activated carbons, as compared with other methods. Proceeding from the lack of sufficient literature on utilization of palm seeds as a suitable precursor for the manufacture of activated carbon and the wide availability of this resource in UAE, the present study was devoted to investigate the potentiality of use of date palm seeds- treated as waste in the Gulf region for the production of activated carbon using KOH as the activating agent.

For the conduction of this research the date palm seeds, received from a local farm in Abu Dhabi, were washed with deionized water, dried at 105 °C for 24 h, ground to size $> 500\ \mu m$ and stored in a desiccator. Ten grams of the material was mixed with KOH (40% by volume) solution at an impregnation ratio (KOH/biomass) of 2. The mixture was kept under continuous stirring for 5 h using a magnetic paddle to realize complete soaking of the KOH solution, and then oven-dried at 105 °C until it was completely dry and crisp. The dried precursor was activated in a horizontal furnace at 600 °C for 60 min in nitrogen flow of 150 ml/min [56]. Then the carbonized sample was cooled to room temperature in an inert atmosphere and repetitively washed by hydrochloric acid aqueous solution to remove ash and then with distilled water to reduce the pH value to neutral. The produced activated carbon was then oven-dried at 105 °C for 12 h. The yield of porous carbon was estimated in terms of bone dry porous carbon: bone dry date palm seeds.

The phenol was sourced from Sigma Aldrich for the preparation of stock solution using distilled water and phenol solutions with concentrations 400, 300, 200 and 100 mg/L were prepared. The ultraviolet light spectrophotometer was used at a wave length of 269 nm to estimate the phenol solution concentration before and after adsorption. The equilibrium batch adsorption experiments were conducted using Erlenmeyer flasks of 250 ml capacity. A fixed amount (0.1 g) of adsorbent was placed in each flask and a known concentration and volume (100 ml) of phenol (100, 200, 300, 400 mg/L) was added to each flask. These bottles were kept in a shaker water bath at temperatures 30, 40, 50 °C at 200 rpm. These experiments were conducted for 40 h to ensure equilibrium between the absorbent and the adsorbate.

The produced porous carbon has a Brunauer–Emmett–Teller area of 892 m^2/g, pore volume of 0.45 cm^3/g and average pore diameter of 1.97 nm. This porous carbon was used for the adsorption of phenol at different concentrations (100–400 mg/L) and temperatures (30–50 °C). The adsorption behavior of phenol on porous carbon was found to be well described by Langmuir isotherm model. The mono layer adsorption capacity was found to be 333 mg/g, the highest as compared with date palm seed biomass-based porous carbons. Thus, it can be concluded that it is feasible to use date palm seeds for the preparation of KOH-based activated carbon for the removal of phenol from aqueous solutions as a low-cost process with an extremely high performance.

10.5 Preparation of Activated Carbons from Date Palm Stones and Application for Waste-Water Treatments: Review

An important paper has been devoted to a review on the preparation of activated carbons from date palm stones and application for wastewater treatments [57]. The agricultural biomass has withdrawn the attention of researchers as promising precursors for activated carbon, because of the low-cost, great abundance, renewability and high lignocellulosic content [58, 59]. Fruit stones are of particular interest being byproducts of food industries and thus easy to collect [60, 61]. Examples of these fruit stones are peach [62–64], apricot [65–67], olive [7, 68, 69], cherry [70–72], grape [58, 73, 74] and date [54, 75–78] stones. Among these fruit stones, the date stones are especially distinguished for their high carbon content, low price and high availability in the Arab countries [79, 80]. The date stone represents ~ 10% of the fruit weigh making it the largest agricultural byproduct in the date producing countries (~700 thousand tons annually [81, 82]). Physical, chemical and physiochemical techniques have been developed for the conversion of date palm stones to activated carbons [71, 83]. In physical activation, carbonization of the materials is followed by activating the produced char in the presence of an activator such as CO_2 or steam [84]. In chemical activation, the material is impregnated with activators such as $ZnCL_2$, H_3PO_4, KOH, and heated under inert atmosphere [85]. Chemical activation is preferred over physical due to higher yield, single step treatment, lower temperature, shorter time and better porous structure [86, 87]. Different methods have been utilized for the treatment of waste water including coagulation, ion exchange, biodegradation, oxidation, solvent extraction, adsorption and electrolysis [20]. As compared with these methods the adsorption is advantageous because of its simplicity, flexibility, suitability for batch and continuous processes, possibility of regeneration and reuse, low expenses and capability to remove a wide array of pollutants in different concentrations [88, 89, 90, 91, 92, 93, 94, 95, 96, 97, 98, 99]. The activated carbon is distinguished with a high adsorption performance due to its high surface area, well-developed porous structure and favorable surface properties [59, 62, 100, 101]. For example, activated carbons from fruit stones were effectively used for the removal of metals [102–104], synthetic dyes [105–107], pesticides [108, 109] and phenols [110–112]. The present review paper is devoted to the preparation of activated carbon from date stones with detailed explanation of the effects of preparation variables on pore structures and yields of carbons, as well as the adsorption behavior of various pollutants on prepared carbons.

The chemical composition of date stones [64] is: moisture (5–10%), ash (1–2%), protein (5–7%), oil (7–10%), crude fiber (10–20%) and carbohydrates (55–65%). The crude fiber and carbohydrates consist of 23% hemicellulose, 15% lignin, 57% cellulose and 5% ash [113]. The activation variables affecting the pore characteristics and yield of carbon are in the order of: activation temperature, impregnation ratio and activation time. The surface areas of date stone-carbons were in the range from 490 to 1282 m^2/g and the yields from 17 to 47% with the highest values obtained by chemical activation. The application of date palm stones-carbons for the adsorption of organic and inorganic pollutants demonstrated maximum capacities of 612.1, 359.1, 238.1 and 1594 mg/g for dyes, phenols, pesticides and heavy metals respectively. Thus, it can be concluded that the date palm stones are a low-cost, renewable, non-toxic and biocompatible source for the manufacture of activated carbons of high efficiency for the removal of pollutants from waste water. However, further research is needed for pilot experimentation to evaluate the technical and economic feasibility of manufacture of activated carbons from date palm seeds and their use for wastewater treatment.

10.6 Impact of Process Conditions on Preparation of Porous Carbon from Date Palm Seeds by KOH Activation

A research [114] has been conducted to investigate the impact of process conditions on the preparation of porous carbon from date palm seeds using KOH activation. The porous carbons have attracted the interest of researchers as absorbents for the removal of industrial pollutants from waste water. This can be asserted by the fact that researches on the activated carbons account for nearly 30% of overall researches on carbon. The activated carbons are manufactured using physical or chemical activation. The physical activation is conducted using steam or CO_2 or both. The chemical activation is conducted using dehydrating agents such as phosphoric acid, sodium hydroxide, potassium hydroxide and zinc chloride. Metal-based activating agents such as KOH are known to produce highly porous activating carbons as compared with phosphoric acid or zinc chloride [115]. Few researches were concerned with the use of KOH as an activating agent including apricot stones [116], oil palm fibers [117], and anthracite [115]. In addition, porous carbon was synthesized from date palm seeds utilizing KOH and CO_2 as activating agents possessing a BET surface area of 763 m^2/g and corresponding pore volume of about 0.43 cc/g [118].

Proceeding from the insufficiency of literature on the use of date palm seeds as a suitable precursor for the manufacture of activated carbon, this study has been conducted. The date palm seeds are a byproduct of date processing. This resource is abundantly available in the gulf region and is predominantly treated as waste. The iodine is an important reactant or product in many chemical reactions that are used for analytical applications [119, 120]. Thus, it is extremely important to find reliable methods for the removal of molecular iodine using activated carbon. The objective of the present study is the evaluation of the potentiality of manufacture of activated carbon from the date palm seeds using KOH method of activation, optimization of the process parameters including activation temperature, activation duration and KOH: biomass ratio on Brunauer–Emmett–Teller surface area and yield, as well as assessing the textural characteristics of porous carbon.

To conduct this study, date palm seeds were sourced from a local farm in Abu Dhabi, UAR, washed with deionized water and dried in oven at 105 °C for 24 h and then ground to a particle size > 500 μm. Ten grams of date palm seeds dust were mixed with KOH (40% by volume) solution at an impregnation ratio 1–5. The mixture was stirred for 5 h using a magnetic paddle. After complete soaking of KOH solution the mixture was oven-dried at 105 °C and then activated in a horizontal furnaces at 500–900 °C for an activation time 30–120 min in a nitrogen flow of 150 mL/min. The carbonized samples were cooled to room temperature in an inert atmosphere, washed with HCl aqueous solution to remove ash and then by distilled water to reduce pH to neutral. The final porous carbon was then dried at 105 °C for 12 h. The yield of activated carbon was estimated based on grams bone dry per grams bone dry of date palm seeds.

The results of the study have come to the following conclusions:

1. The increase of the activation temperature and the impregnation ratio (KBR) contribute to the increase of the BET surface area, while the yield decreases.
2. The activation duration was found not to influence the BET surface area, and its effect on the yield is marginal.
3. The optimum conditions were found as follows: activation temperature 601 °C, KBR ratio 1.97 with the corresponding BET surface area and yield being 910 m^2/g and 22.9%. The activated carbon possessed a pore volume of 0.45 cc/g, an average pore diameter of 1.54 nm with the micropore volume proportion being 0.40 cc/g.
4. The activated carbon textural characteristics have a microporous nature. It has been also tested for iodine adsorption.
5. It was found that the date palm seeds activated carbon contains a large number of acid functional groups in the form of carboxylic and hydroxyl groups and thus could be effectively utilized for the treatment of waste water.

References

1. El-Shafey E-SI, Al-Kindy SMZ (2013) Removal of Cu^{2+} and Ag^{+} from aqueous solution on a chemically-carbonized sorbent from date palm leaflets. Environ Technol 34(3):395–406. https://doi.org/10.1080/09593330.2012.698647
2. Moore J, Ramamoorthy S (1984) Heavy metals in natural waters: applied monitoring and impact assessment. Springer.
3. Shawabkeh RA, Rockstraw DA, Bhada RK (2002) Copper and strontium adsorption by a novel carbon material manufactured from pecan shells. Carbon 40(5):781–786. https://doi.org/10.1016/S0008-6223(01)00198-1
4. Paul Chen J, Yoon J-T, Yiacoumi S (2003) Effects of chemical and physical properties of influent on copper sorption onto activated carbon fixed-bed columns. Carbon 41(8):1635–1644. https://doi.org/10.1016/S0008-6223(03)00117-9
5. ASTDR (1999) Agency for toxic substances and disease registry. http://www.atsdr.cdc.gov/toxprofiles/phs146.html
6. Ahluwalia SS, Goyal D (2007) Microbial and plant derived biomass for removal of heavy metals from wastewater. Biores Technol 98(12):2243–2257. https://doi.org/10.1016/j.biortech.2005.12.006
7. Mohan D, Pittman CU (2007) Arsenic removal from water/wastewater using adsorbents—a critical review. J Hazard Mater 142(1–2):1–53. https://doi.org/10.1016/j.jhazmat.2007.01.006
8. Šćiban M, Klašnja M, Škrbić B (2008) Adsorption of copper ions from water by modified agricultural by-products. Desalination 229(1–3):170–180. https://doi.org/10.1016/j.desal.2007.08.017
9. Kondo K, Matsumoto M, Okamoto K (1999) Enhanced adsorption of copper(II) ion on novel amidoxime Chitosan Resin. J Chem Eng Jpn 32(2):217–222. https://doi.org/10.1252/jcej.32.217
10. Youssef AM, Mostafa MR (1992) Removal of copper ions by modified activated carbons. Indian J Technol 30(8):413–416
11. Ghassabzadeh H, Mohadespour A, Torab-Mostaedi M, Zaheri P, Maragheh MG, Taheri H (2010) Adsorption of Ag, Cu and Hg from aqueous solutions using expanded perlite. J Hazard Mater 177(1–3):950–955. https://doi.org/10.1016/j.jhazmat.2010.01.010
12. Sneha K, Sathishkumar M, Mao J, Kwak IS, Yun Y-S (2010) Corynebacterium glutamicum-mediated crystallization of silver ions through sorption and reduction processes. Chem Eng J 162(3):989–996. https://doi.org/10.1016/j.cej.2010.07.006
13. Shu Z-N, Xiong C-H, Wang X (2006) Adsorption behavior and mechanism of amino methylene phosphonic acid resin for Ag(I). Trans Nonferrous Met Soc China 16(3):700–704. https://doi.org/10.1016/S1003-6326(06)60124-6
14. Yu Z, Zhou A, Qu J, Zhang T (2005) Study on behavior and kinetics of sorption of Ag+ by Shenfu3−1 coal. Microporous Mesoporous Mater 85(1–2):104–110. https://doi.org/10.1016/j.micromeso.2005.05.049
15. Baranger P, DisnarJR (1991) Mechanism and kinetics of the reduction of various metallic cations by a lignite. B Soc Geol Fr: 271–276.
16. Ruowen F, Hanmin Z, Yun L (1993) The reduction property of activated carbon fibers. Carbon 31(7):1089–1094. https://doi.org/10.1016/0008-6223(93)90060-N
17. Adams MD (1992) The mechanisms of adsorption of Ag(CN)2− and Ag+ on to activated carbon. Hydrometallurgy 31(1–2):121–138. https://doi.org/10.1016/0304-386X(92)90112-D
18. Al-Yahyai R, Al-Khanjari S (2008) Biodiversity of date palm in the Sultanate of Oman. Afr J Agric Res 3(6):389–395
19. Al-Ghamdi A, Altaher H, Omar W (2013) Application of date palm trunk fibers as adsorbents for removal of Cd+2 ions from aqueous solutions. J Water Reuse Desalin 3(1):47–54. https://doi.org/10.2166/wrd.2013.031
20. Bhatnagar A, Sillanpää M (2010) Utilization of agro-industrial and municipal waste materials as potential adsorbents for water treatment—A review. Chem Eng J 157(2–3):277–296. https://doi.org/10.1016/j.cej.2010.01.007

21. Pinot F, Kreps SE, Bachelet M, Hainaut P, Bakonyi M, Polla BS (2000) Cadmium in the environment: sources, mechanisms of biotoxicity, and biomarkers. Rev Environ Health 15(3): Article 3. https://doi.org/10.1515/REVEH.2000.15.3.299
22. Yu MH (2005) Environmental toxicity- biological and health effects of pollutants (2nd ed.). CRC Press.
23. Namasivayam C, Sangeetha D (2006) Recycling of agricultural solid waste, coir pith: removal of anions, heavy metals, organics and dyes from water by adsorption onto ZnCl2 activated coir pith carbon. J Hazard Mater 135(1–3):449–452. https://doi.org/10.1016/j.jhazmat.2005.11.066
24. Benjamin MM (2000) Water chemistry. McGraw Hill, New York
25. Rashed MN (2006) Fruit stones from industrial waste for the removal of lead ions from polluted water. Environ Monit Assess 119(1–3):31–41. https://doi.org/10.1007/s10661-005-9002-y
26. Tokimoto T, Kawasaki N, Nakamura T, Akutagawa J, Tanada S (2005) Removal of lead ions in drinking water by coffee grounds as vegetable biomass. J Colloid Interface Sci 281(1):56–61. https://doi.org/10.1016/j.jcis.2004.08.083
27. Moattar F, Hayeripour S (2004) Application of Chitin and Zeolite adsorbents for treatment of low level radioactive liquid wastes. Int J Environ Sci Technol 1:45–50
28. Jain CK (2001) Adsorption of zinc onto bed sediments of the River Ganga: adsorption models and kinetics. Hydrol Sci J 46(3):419–434. https://doi.org/10.1080/02626660109492836
29. Ashour EA (2005) An investigaion on the feasibility of peanut-hull as an adsorbent. Alexandria Eng. J 44(1):129–139
30. F. Scarlet Ola Abdelwahab, Ahmed El Nemr, Amany El Sikaily, Azza Khaled (2005) Use of rice husk for adsorption of direct dyes from aqueous solution: a case study of direct. Egypt J Aquat Res, *31*.
31. Faghihian H, Nejati-Yazdinejad M (2009) Sorption performance of cysteine-modified bentonite in heavy metals uptake. J Serb Chem Soc 74(7):833–843. https://doi.org/10.2298/JSC0907833F
32. Ahluwalia SS, Goyal D (2005) Removal of heavy metals by waste tea leaves from aqueous solution. Eng Life Sci 5(2):158–162. https://doi.org/10.1002/elsc.200420066
33. Spahis N, Addoun A, Mahmoudi H, Ghaffour N (2008) Purification of water by activated carbon prepared from olive stones. Desalination 222(1–3):519–527. https://doi.org/10.1016/j.desal.2007.02.065
34. Islam MdA, Tan IAW, Benhouria A, Asif M, Hameed BH (2015) Mesoporous and adsorptive properties of palm date seed activated carbon prepared via sequential hydrothermal carbonization and sodium hydroxide activation. Chem Eng J 270:187–195. https://doi.org/10.1016/j.cej.2015.01.058
35. Falco C, Marco-Lozar JP, Salinas-Torres D, Morallón E, Cazorla-Amorós D, Titirici MM, Lozano-Castelló D (2013) Tailoring the porosity of chemically activated hydrothermal carbons: influence of the precursor and hydrothermal carbonization temperature. Carbon 62:346–355. https://doi.org/10.1016/j.carbon.2013.06.017
36. Li L, Diederick R, Flora JRV, Berge ND (2013) Hydrothermal carbonization of food waste and associated packaging materials for energy source generation. Waste Manage 33(11):2478–2492. https://doi.org/10.1016/j.wasman.2013.05.025
37. Liu Z, Balasubramanian R (2012) Hydrothermal carbonization of waste biomass for energy generation. Procedia Environ Sci 16:159–166. https://doi.org/10.1016/j.proenv.2012.10.022
38. Parshetti GK, Liu Z, Jain A, Srinivasan MP, Balasubramanian R (2013) Hydrothermal carbonization of sewage sludge for energy production with coal. Fuel 111:201–210. https://doi.org/10.1016/j.fuel.2013.04.052
39. Román S, Valente Nabais JM, Ledesma B, González JF, Laginhas C, Titirici MM (2013) Production of low-cost adsorbents with tunable surface chemistry by conjunction of hydrothermal carbonization and activation processes. Microporous Mesoporous Mater 165:127–133. https://doi.org/10.1016/j.micromeso.2012.08.006

40. Saetea P, Tippayawong N (2013) Characterization of adsorbent from bydrothermally carbonized and steam activated sewage sludge. Proceedings of the World Congress on Engineering 2(3):1909–1912
41. Ruiz HA, Rodríguez-Jasso RM, Fernandes BD, Vicente AA, Teixeira JA (2013) Hydrothermal processing, as an alternative for upgrading agriculture residues and marine biomass according to the biorefinery concept: A review. Renew Sustain Energy Rev 21:35–51. https://doi.org/10.1016/j.rser.2012.11.069
42. Pari G, Darmawan S, Prihandoko B (2014) Porous carbon spheres from hydrothermal carbonization and KOH activation on cassava and tapioca flour raw material. Procedia Environ Sci 20:342–351. https://doi.org/10.1016/j.proenv.2014.03.043
43. Hao W, Björkman E, Lilliestråle M, Hedin N (2013) Activated carbons prepared from hydrothermally carbonized waste biomass used as adsorbents for CO2. Appl Energy 112:526–532. https://doi.org/10.1016/j.apenergy.2013.02.028
44. Zhu X, Liu Y, Zhou C, Luo G, Zhang S, Chen J (2014) A novel porous carbon derived from hydrothermal carbon for efficient adsorption of tetracycline. Carbon 77:627–636. https://doi.org/10.1016/j.carbon.2014.05.067
45. Unur E (2013) Functional nanoporous carbons from hydrothermally treated biomass for environmental purification. Microporous Mesoporous Mater 168:92–101. https://doi.org/10.1016/j.micromeso.2012.09.027
46. Regmi P, Garcia Moscoso JL, Kumar S, Cao X, Mao J, Schafran G (2012) Removal of copper and cadmium from aqueous solution using switchgrass biochar produced via hydrothermal carbonization process. J Environ Manage 109:61–69. https://doi.org/10.1016/j.jenvman.2012.04.047
47. Reddy SK, K., Kannan, P., Al Shoaibi, A., & Srinivasakannan, C. (2014) KOH-based porous carbon from date palm seed: preparation, characterization, and application to phenol adsorption. Water Sci Technol 70(10):1633–1640. https://doi.org/10.2166/wst.2014.419
48. Busca G, Berardinelli S, Resini C, Arrighi L (2008) Technologies for the removal of phenol from fluid streams: a short review of recent developments. J Hazard Mater 160(2–3):265–288. https://doi.org/10.1016/j.jhazmat.2008.03.045
49. Azargohar R, Dalai AK (2005) Production of activated carbon from Luscar char: Experimental and modeling studies. Microporous Mesoporous Mater 85(3):219–225. https://doi.org/10.1016/j.micromeso.2005.06.018
50. Yalçın N, Sevinç V (2000) Studies of the surface area and porosity of activated carbons prepared from rice husks. Carbon 38(14):1943–1945. https://doi.org/10.1016/S0008-6223(00)00029-4
51. Malik PK (2004) Dye removal from wastewater using activated carbon developed from sawdust: adsorption equilibrium and kinetics. J Hazard Mater 113(1–3):81–88. https://doi.org/10.1016/j.jhazmat.2004.05.022
52. Yang K, Peng J, Srinivasakannan C, Zhang L, Xia H, Duan X (2010) Preparation of high surface area activated carbon from coconut shells using microwave heating. Biores Technol 101(15):6163–6169. https://doi.org/10.1016/j.biortech.2010.03.001
53. Liou T-H (2010) Development of mesoporous structure and high adsorption capacity of biomass-based activated carbon by phosphoric acid and zinc chloride activation. Chem Eng J 158(2):129–142. https://doi.org/10.1016/j.cej.2009.12.016
54. Reddy KSK, Al Shoaibi A, Srinivasakannan C (2012) Activated carbon from date palm seed: process optimization using response surface methodology. Waste and Biomass Valorization 3(2):149–156. https://doi.org/10.1007/s12649-011-9104-4
55. Reddy SK, K., Al Shoaibi, A., & Srinivasakannan, C. (2013) A comparison of microstructure and adsorption characteristics of activated carbons by CO2 and H3PO4 activation from date palm pits. Carbon 52:623. https://doi.org/10.1016/j.carbon.2012.10.021
56. Cheng F, Liang J, Zhao J, Tao Z, Chen J (2008) Biomass waste-derived microporous carbons with controlled texture and enhanced hydrogen uptake. Chem Mater 20(5):1889–1895. https://doi.org/10.1021/cm702816x

57. Ahmed MJ (2016) Preparation of activated carbons from date (Phoenix dactylifera L.) palm stones and application for wastewater treatments: Review. Process Saf Environ Prot 102:168–182. https://doi.org/10.1016/j.psep.2016.03.010
58. Chayid MA, Ahmed MJ (2015) Amoxicillin adsorption on microwave prepared activated carbon from Arundo donax Linn: isotherms, kinetics, and thermodynamics studies. J Environ Chem Eng 3(3):1592–1601. https://doi.org/10.1016/j.jece.2015.05.021
59. Gupta VK, Nayak A, Bhushan B, Agarwal S (2015) A critical analysis on the efficiency of activated carbons from low-cost precursors for heavy metals remediation. Crit Rev Environ Sci Technol 45(6):613–668. https://doi.org/10.1080/10643389.2013.876526
60. Heschel W, Klose E (1995) On the suitability of agricultural by-products for the manufacture of granular activated carbon. Fuel 74(12):1786–1791. https://doi.org/10.1016/0016-2361(95)80009-7
61. Mechati F, Bouchelta C, Medjram MS, Benrabaa R, Ammouchi N (2015) Effect of hard and soft structure of different biomasses on the porosity development of activated carbon prepared under N2/microwave radiations. J Environ Chem Eng 3(3):1928–1938. https://doi.org/10.1016/j.jece.2015.07.007
62. Gupta VK, Nayak A, Agarwal S (2015) Bioadsorbents for remediation of heavy metals: current status and their future prospects. Environmental Engineering Research 20(1):1–18. https://doi.org/10.4491/eer.2015.018
63. Soares Maia DA, Sapag K, Toso JP, López RH, Azevedo DCS, Cavalcante CL, Zgrablich G (2010) Characterization of activated carbons from peach stones through the mixed geometry model. Microporous Mesoporous Mater 134(1–3):181–188. https://doi.org/10.1016/j.micromeso.2010.05.024
64. Torrellas SÁ, García Lovera R, Escalona N, Sepúlveda C, Sotelo JL, García J (2015) Chemical-activated carbons from peach stones for the adsorption of emerging contaminants in aqueous solutions. Chem Eng J 279:788–798. https://doi.org/10.1016/j.cej.2015.05.104
65. Depci T, Onal Y, Prisbrey KA (2014) Apricot stone activated carbons adsorption of cyanide as revealed from computational chemistry analysis and experimental study. J Taiwan Inst Chem Eng 45(5):2511–2517. https://doi.org/10.1016/j.jtice.2014.05.015
66. Djilani C, Zaghdoudi R, Djazi F, Bouchekima B, Lallam A, Modarressi A, Rogalski M (2015) Adsorption of dyes on activated carbon prepared from apricot stones and commercial activated carbon. J Taiwan Inst Chem Eng 53:112–121. https://doi.org/10.1016/j.jtice.2015.02.025
67. Petrova B, Budinova T, Tsyntsarski B, Kochkodan V, Shkavro Z, Petrov N (2010) Removal of aromatic hydrocarbons from water by activated carbon from apricot stones. Chem Eng J 165(1):258–264. https://doi.org/10.1016/j.cej.2010.09.026
68. Berrios M, Martín MÁ, Martín A (2012) Treatment of pollutants in wastewater: adsorption of methylene blue onto olive-based activated carbon. J Ind Eng Chem 18(2):780–784. https://doi.org/10.1016/j.jiec.2011.11.125
69. Hazzaa R, Hussein M (2015) Adsorption of cationic dye from aqueous solution onto activated carbon prepared from olive stones. Environ Technol Innov 4:36–51. https://doi.org/10.1016/j.eti.2015.04.002
70. Angin D (2014) Utilization of activated carbon produced from fruit juice industry solid waste for the adsorption of Yellow 18 from aqueous solutions. Biores Technol 168:259–266. https://doi.org/10.1016/j.biortech.2014.02.100
71. Nowicki P, Kazmierczak J, Pietrzak R (2015) Comparison of physicochemical and sorption properties of activated carbons prepared by physical and chemical activation of cherry stones. Powder Technol 269:312–319. https://doi.org/10.1016/j.powtec.2014.09.023
72. Olivares-Marín M, Fernández-González C, Macías-García A, Gómez-Serrano V (2012) Preparation of activated carbon from cherry stones by physical activation in air. influence of the chemical carbonisation with H2SO4. J Anal Appl Pyrol 94:131–137. https://doi.org/10.1016/j.jaap.2011.11.019
73. Al Bahri M, Calvo L, Gilarranz MA, Rodriguez JJ (2012) Activated carbon from grape seeds upon chemical activation with phosphoric acid: Application to the adsorption of diuron from water. Chem Eng J 203:348–356. https://doi.org/10.1016/j.cej.2012.07.053

74. Jimenez-Cordero D, Heras F, Alonso-Morales N, Gilarranz MA, Rodríguez JJ (2014) Preparation of granular activated carbons from grape seeds by cycles of liquid phase oxidation and thermal desorption. Fuel Process Technol 118:148–155. https://doi.org/10.1016/j.fuproc.2013.08.019
75. Alhamed YA (2009) Adsorption kinetics and performance of packed bed adsorber for phenol removal using activated carbon from dates' stones. J Hazard Mater 170(2–3):763–770. https://doi.org/10.1016/j.jhazmat.2009.05.002
76. El-Sharkawy EA, Soliman AY, Al-Amer KM (2007) Comparative study for the removal of methylene blue via adsorption and photocatalytic degradation. J Colloid Interface Sci 310(2):498–508. https://doi.org/10.1016/j.jcis.2007.02.013
77. Sekirifa ML, Hadj-Mahammed M, Pallier S, Baameur L, Richard D, Al-Dujaili AH (2013) Preparation and characterization of an activated carbon from a date stones variety by physical activation with carbon dioxide. J Anal Appl Pyrol 99:155–160. https://doi.org/10.1016/j.jaap.2012.10.007
78. Yamina G, Abdeltif A, Youcef T, Mahfoud HM, Fatiha G, lotfi, B. (2013) A comparative study of the addition effect of activated carbon obtained from date stones on the biological filtration efficiency using sand dune bed. Energy Procedia 36:1175–1183. https://doi.org/10.1016/j.egypro.2013.07.133
79. Ahmed MJ, Theydan SK (2015) Adsorptive removal of p-nitrophenol on microporous activated carbon by $FeCl_3$ activation: Equilibrium and kinetics studies. Desalin Water Treat 55(2):522–531. https://doi.org/10.1080/19443994.2014.920731
80. Amor HB, Ismail M (2015) Adsorption of chromium (VI) on activated carbon prepared by acid activation of date stones. J Sci Res: 309–314.
81. Briones R, Serrano L, Younes RB, Mondragon I, Labidi J (2011) Polyol production by chemical modification of date seeds. Ind Crops Prod 34(1):1035–1040. https://doi.org/10.1016/j.indcrop.2011.03.012
82. El-Naas MH, Al-Zuhair S, Alhaija MA (2010) Removal of phenol from petroleum refinery wastewater through adsorption on date-pit activated carbon. Chem Eng J 162(3):997–1005. https://doi.org/10.1016/j.cej.2010.07.007
83. El-Sheikh AH, Newman AP, Al-Daffaee HK, Phull S, Cresswell N (2004) Characterization of activated carbon prepared from a single cultivar of Jordanian olive stones by chemical and physicochemical techniques. J Anal Appl Pyrol 71(1):151–164. https://doi.org/10.1016/S0165-2370(03)00061-5
84. Nowicki P, Pietrzak R, Wachowska H (2010) Sorption properties of active carbons obtained from walnut shells by chemical and physical activation. Catal Today 150(1–2):107–114. https://doi.org/10.1016/j.cattod.2009.11.009
85. Li Y, Du Q, Wang X, Zhang P, Wang D, Wang Z, Xia Y (2010) Removal of lead from aqueous solution by activated carbon prepared from Enteromorpha prolifera by zinc chloride activation. J Hazard Mater 183(1–3):583–589. https://doi.org/10.1016/j.jhazmat.2010.07.063
86. Chen Y-D, Chen W-Q, Huang B, Huang M-J (2013) Process optimization of K2C2O4-activated carbon from kenaf core using Box-Behnken design. Chem Eng Res Des 91(9):1783–1789. https://doi.org/10.1016/j.cherd.2013.02.024
87. Sudaryanto Y, Hartono SB, Irawaty W, Hindarso H, Ismadji S (2006) High surface area activated carbon prepared from cassava peel by chemical activation. Biores Technol 97(5):734–739. https://doi.org/10.1016/j.biortech.2005.04.029
88. Ahmed MJ, Theydan SK (2014) Fluoroquinolones antibiotics adsorption onto microporous activated carbon from lignocellulosic biomass by microwave pyrolysis. J Taiwan Inst Chem Eng 45(1):219–226. https://doi.org/10.1016/j.jtice.2013.05.014
89. Gupta VK, Agarwal S, Saleh TA (2011) Synthesis and characterization of alumina-coated carbon nanotubes and their application for lead removal. J Hazard Mater 185(1):17–23. https://doi.org/10.1016/j.jhazmat.2010.08.053
90. Gupta VK, Ali I, Saleh TA, Nayak A, Agarwal S (2012) Chemical treatment technologies for waste-water recycling—An overview. RSC Adv 2(16):6380. https://doi.org/10.1039/c2ra20340e

91. Gupta VK, Kumar R, Nayak A, Saleh TA, Barakat MA (2013) Adsorptive removal of dyes from aqueous solution onto carbon nanotubes: a review. Adv Coll Interface Sci 193–194:24–34. https://doi.org/10.1016/j.cis.2013.03.003
92. Gupta VK, Nayak A (2012) Cadmium removal and recovery from aqueous solutions by novel adsorbents prepared from orange peel and Fe2O3 nanoparticles. Chem Eng J 180:81–90. https://doi.org/10.1016/j.cej.2011.11.006
93. Gupta VK, Srivastava SK, Mohan D, Sharma S (1998) Design parameters for fixed bed reactors of activated carbon developed from fertilizer waste for the removal of some heavy metal ions. Waste Manage 17(8):517–522. https://doi.org/10.1016/S0956-053X(97)10062-9
94. Jain AK, Gupta VK, Bhatnagar A, Suhas. (2003) A comparative study of adsorbents prepared from industrial wastes for removal of dyes. Sep Sci Technol 38(2):463–481. https://doi.org/10.1081/SS-120016585
95. Mittal A, Kaur D, Malviya A, Mittal J, Gupta VK (2009) Adsorption studies on the removal of coloring agent phenol red from wastewater using waste materials as adsorbents. J Colloid Interface Sci 337(2):345–354. https://doi.org/10.1016/j.jcis.2009.05.016
96. Mittal A, Mittal J, Malviya A, Gupta VK (2009) Adsorptive removal of hazardous anionic dye "Congo red" from wastewater using waste materials and recovery by desorption. J Colloid Interface Sci 340(1):16–26. https://doi.org/10.1016/j.jcis.2009.08.019
97. Mittal A, Mittal J, Malviya A, Gupta VK (2010) Removal and recovery of Chrysoidine Y from aqueous solutions by waste materials. J Colloid Interface Sci 344(2):497–507. https://doi.org/10.1016/j.jcis.2010.01.007
98. Mittal A, Mittal J, Malviya A, Kaur D, Gupta VK (2010) Decoloration treatment of a hazardous triarylmethane dye, Light Green SF (Yellowish) by waste material adsorbents. J Colloid Interface Sci 342(2):518–527. https://doi.org/10.1016/j.jcis.2009.10.046
99. Saleh TA, Gupta VK (2012) Column with CNT/magnesium oxide composite for lead(II) removal from water. Environ Sci Pollut Res 19(4):1224–1228. https://doi.org/10.1007/s11356-011-0670-6
100. Ahmed MJ, Theydan SK (2013) Microwave assisted preparation of microporous activated carbon from Siris seed pods for adsorption of metronidazole antibiotic. Chem Eng J 214:310–318. https://doi.org/10.1016/j.cej.2012.10.101
101. Saleh TA, Gupta VK (2014) Processing methods, characteristics and adsorption behavior of tire derived carbons: a review. Adv Coll Interface Sci 211:93–101. https://doi.org/10.1016/j.cis.2014.06.006
102. Alslaibi TM, Abustan I, Ahmad MA, Foul AA (2013) Cadmium removal from aqueous solution using microwaved olive stone activated carbon. J Environ Chem Eng 1(3):589–599. https://doi.org/10.1016/j.jece.2013.06.028
103. Bouhamed F, Elouear Z, Bouzid J (2012) Adsorptive removal of copper(II) from aqueous solutions on activated carbon prepared from Tunisian date stones: equilibrium, kinetics and thermodynamics. J Taiwan Inst Chem Eng 43(5):741–749. https://doi.org/10.1016/j.jtice.2012.02.011
104. Duranoğlu D, Trochimczuk AW, Beker Ü (2010) A comparison study of peach stone and acrylonitrile-divinylbenzene copolymer based activated carbons as chromium(VI) sorbents. Chem Eng J 165(1):56–63. https://doi.org/10.1016/j.cej.2010.08.054
105. Demirbas E, Kobya M, Sulak MT (2008) Adsorption kinetics of a basic dye from aqueous solutions onto apricot stone activated carbon. Biores Technol 99(13):5368–5373. https://doi.org/10.1016/j.biortech.2007.11.019
106. Ramírez-Montoya LA, Hernández-Montoya V, Montes-Morán MA (2014) Optimizing the preparation of carbonaceous adsorbents for the selective removal of textile dyes by using Taguchi methodology. J Anal Appl Pyrol 109:9–20. https://doi.org/10.1016/j.jaap.2014.07.018
107. Uğurlu M, Gürses A, Açıkyıldız M (2008) Comparison of textile dyeing effluent adsorption on commercial activated carbon and activated carbon prepared from olive stone by ZnCl2 activation. Microporous Mesoporous Mater 111(1–3):228–235. https://doi.org/10.1016/j.micromeso.2007.07.034

108. Bakouri HE, Morillo J, Usero J, Vanderlinden E, Vidal H (2010) Effectiveness of acid-treated agricultural stones used in biopurification systems to avoid pesticide contamination of water resources caused by direct losses: Part I Equilibrium experiments and kinetics. *Bioresource Technology* 101(14):5084–5091. https://doi.org/10.1016/j.biortech.2010.01.131
109. Ioannidou OA, Zabaniotou AA, Stavropoulos GG, Islam MdA, Albanis TA (2010) Preparation of activated carbons from agricultural residues for pesticide adsorption. Chemosphere 80(11):1328–1336. https://doi.org/10.1016/j.chemosphere.2010.06.044
110. Aygün A, Yenisoy-Karakaş S, Duman I (2003) Production of granular activated carbon from fruit stones and nutshells and evaluation of their physical, chemical and adsorption properties. Microporous Mesoporous Mater 66(2–3):189–195. https://doi.org/10.1016/j.micromeso.2003.08.028
111. Beker U, Ganbold B, Dertli H, Gülbayir DD (2010) Adsorption of phenol by activated carbon: Influence of activation methods and solution pH. Energy Convers Manage 51(2):235–240. https://doi.org/10.1016/j.enconman.2009.08.035
112. Rodríguez Arana JMR, Mazzoco RR (2010) Adsorption studies of methylene blue and phenol onto black stone cherries prepared by chemical activation. J Hazard Mater 180(1–3):656–661. https://doi.org/10.1016/j.jhazmat.2010.04.086
113. Barreveld WH (1993) Date palm products. In: FAO Agricultural Services Bulletin (FAO).
114. Reddy SK, K., Shoaibi, A. A., & Srinivasakannan, C. (2015) Impact of process conditions on preparation of porous carbon from date palm seeds by KOH activation. Clean Technol Environ Policy 17(6):1671–1679. https://doi.org/10.1007/s10098-014-0875-8
115. Zhao W, Fierro V, Zlotea C, Aylon E, Izquierdo MT, Latroche M, Celzard A (2011) Optimization of activated carbons for hydrogen storage. Int J Hydrogen Energy 36(18):11746–11751. https://doi.org/10.1016/j.ijhydene.2011.05.181
116. Şentorun-Shalaby Ç, Uçak-Astarlıog˘lu MG, Artok L, Sarıcı Ç (2006) Preparation and characterization of activated carbons by one-step steam pyrolysis/activation from apricot stones. Microporous Mesoporous Mater 88(1–3):126–134. https://doi.org/10.1016/j.micromeso.2005.09.003
117. Hameed BH, Tan IAW, Ahmad AL (2008) Optimization of basic dye removal by oil palm fibre-based activated carbon using response surface methodology. J Hazard Mater 158(2–3):324–332. https://doi.org/10.1016/j.jhazmat.2008.01.088
118. Salman JM, Njoku VO, Hameed BH (2011) Bentazon and carbofuran adsorption onto date seed activated carbon: kinetics and equilibrium. Chem Eng J 173(2):361–368. https://doi.org/10.1016/j.cej.2011.07.066
119. Rashid K, Reddy SK, K., Shoaibi, A. A., & Srinivasakannan, C. (2013) Sulfur-impregnated porous carbon for removal of mercuric chloride: Optimization using RSM. Clean Technol Environ Policy 15(6):1041–1048. https://doi.org/10.1007/s10098-012-0564-4
120. Rashidi NA, Yusup S, Borhan A, Loong LH (2014) Experimental and modelling studies of carbon dioxide adsorption by porous biomass derived activated carbon. Clean Technol Environ Policy 16(7):1353–1361. https://doi.org/10.1007/s10098-014-0788-6
121. Nordberg M, Nordberg GF (2009) Toxicology and biological monitoring of metals. In: Ballantyne B, Marrs TC, Syversen T (eds) General and applied toxicology (p. gat145). Wiley. https://doi.org/10.1002/9780470744307.gat145
122. Obregón-Valencia D, Sun-Kou M del R (2014) Comparative cadmium adsorption study on activated carbon prepared from aguaje (Mauritia flexuosa) and olive fruit stones (Olea europaea L.). J Environ Chem Eng 2(4): 2280–2288. https://doi.org/10.1016/j.jece.2014.10.004
123. Okman I, Karagöz S, Tay T, Erdem M (2014) Activated carbons from grape seeds by chemical activation with potassium carbonate and potassium hydroxide. Appl Surf Sci 293:138–142. https://doi.org/10.1016/j.apsusc.2013.12.117
124. Uysal T, Duman G, Onal Y, Yasa I, Yanik J (2014) Production of activated carbon and fungicidal oil from peach stone by two-stage process. J Anal Appl Pyrol 108:47–55. https://doi.org/10.1016/j.jaap.2014.05.017

Chapter 11
Date Palm Byproducts for Green Fuels and Bioenergy Production

Abstract The annual products of pruning of date palms are predominantly treated as waste: either open-field burnt or sent to landfills. The date palm leaves representing ~ 54.6% by weight of products of pruning have been successfully used for production of bioethanol with a 60% yield using the simulatanous treatment by 3 enzymes: laccase for lignin degradation, xylanase for hemicellulose hydrolysis and cellulase for cellulose hydrolysis. In anthor study the organoslov process has been successfully used to extract lignin and digestible cellulose pulp from date plam leaves for the production of bioethanol. The maxiumam practical yield of bioethanol was 9.92 g ethanol/100 g raw material representing 90.95% of the theoretical value. Aceton, butanol and ethanol are common solvents used in many important industries and thus can replace energy derived from petrochemicals. A study has been conducted on the use of low-quality surplus date palm fruits as a substrate for fermentation in the production of acetone, butanol and bioethanol using the original Egyptian clostridium strains, isolated from agriculatural soil, cultivated with different plants in Assiut Governorte, Egypt. A research has been conducted on the biodesiel production from date palm seeds. The date seed represents 10–15% of the date fruit weight. The date seed is composed of 3.1–7.1% moisture, 2.3–6.4% protein, 5.0–13.2% fat, 0.9–1.8% ash and 22.5–80.2% dietary fiber. According to FAO statistic 2010 the date production in South-Western Asia and North Africa can be estimated at 7.85 million tons including seeds of 785,000 to 1.18 million tons. Biodiesel is generally considered as the most acceptable biofuel for diesel engines, because of its technical, environmental and strategic advantages compared with fossil-based diesel fuels, such as cleaner engine emissions, biodegradability, renewability and superior lubricating properties. In this research the biodiesel has been produced by transesterification with a yield of 98% and the fatty acid esters were analysed by GC/MS. The research results show that the fatty acid composition of biodiesel was similar to biodiesel fuels produced from other vegetable oils. The date seed biodiesel had a considerable amount of low-chain fatty acids, which gives special features to biodiesel, high cetane number (60.3), low viscosity (3.84mm^2/s), flash point (140 °C) and low iodine value (46). The only weak point is its high pouring point (−1 °C), which limits its use in cold weather as compared with other vegetable biodiesel fuels. A study has been devoted to the investigation of production factors to optimize the extraction

H. EL-Mously et al., *Date Palm Byproducts: A Springboard for Circular Bio Economy*, Materials Horizons: From Nature to Nanomaterials,
https://doi.org/10.1007/978-981-99-0475-4_11

of date palm seeds oil, production of biodiesel and testing of biodiesel blends in a compression ignition diesel engine. The date palm seeds were sourced from Sharjah date facility in Sharjah (UAE). The date palm seed oil has been extracted using the Soxhlet extraction method, because it represents the most practical option for analytical scale, where parameters can be appropriately monitored and controlled. The resulting biodiesel was characterized and assessed based on widely used international standards (ASTMD6751 and EN14214). Four biodiesel blends were prepared (B5, B10, B15 and B20) and tested in a compression ignition engine at engine speeds from 1600 to 3600 rpm (200 rpm increments) and three engine loads (50, 75 and 100%). A date seed oil biodiesel yield of 92% has been achieved at the following transesterification conditions: 55 °C, 9:1 AOMR, 1wt% CMF and 90 min. Thus the research results lead to the conclusion of fundamental suitability of date palm seeds as a biodiesel feedstock. An important study has been devoted to the production of ethanol from date palm residues. The fermented ethanol can be produced from three groups of feedstocks: sugary, starchy and ligno-cellulosic compounds. In the first generation (1G) processing, sugary and starchy feedstock is converted to bioethanol. However (1G) processing is in contradiction with needs of hummans and animals. Therefore, it is expected that (2G) production of ethanol will prevail in the near future. Bioethanol production from ligno-cellulosic biomass (G2) process is economical only if the sugar concentration exceeds 40 gL^{-1} and yield is increased. The aim of the present study is to find the optimum ways to convert the date wastes and products of pruning to ethanol and to encourage investors to invest in bioethanol production. Egypt, Iran and Saudi Arabia produce almost ½ the world's dates production with shares of world production ~ 20%, 14%, 14% respectively. The weight of date seeds varies between 0.5 and 4 g (6–30% of fresh fruit weight). Assuming an average of 13% of total fruit weight belongs to the seed, a weight of 120 kg dried seed per ton of date wastes can be estimated. Thus 13 kg of seed oil (10.85% by weight) can be used for biodiesel fuel production (density 0.871 kg/L). The finding of this study show that there is a great potential to produce ethanol from date wastes and date palm products of pruning in the Middle East countries due to their high availability potential. Besides, these date wastes and lignocellulosic residues are dominantly treated as waste and thus representing a source of environmental pollution. Egypt, Iran and Saudi Arabia can produce annually 173.5, 401 and 438 million liters of bioethanol respectively from date wastes and annual products of pruning. Other date producing countries have the potentiality to produce 1248 million liters of bioethanol. Thus 3260 million liters of bioethanol can be globally produced. The global cost of bioethanol production from date waste is 0.68$ per liter of which feedstock accounts for 85.3% of the total cost of production. The corresponding value for date palm products of pruning is 0.34$ per liter due to high availability potential. An important research has been conducted on the use of date pits for the synthesis of green diesel and jet fuel fractions. Date pits were sourced from a local supplier in Muscat, Oman. A mechanical grinder was used to convert the samples to a powder. After sieving, the oil was extracted using hexane as a solvent in a Soxhlet apparatus following the AOCS official method Am 2–3. After extraction of oil a 100 g sample of date pits was dried at 100 °C for 8 h and then carbonized in a furnace at 500 °C for 5 h with a heating rate of 3 °C/min under N_2 gas at a flow rate of 50 ml/min. Active catalysts were thus produced by carbonization and impregnation with P_t and P_d metals. The synthesized catalysts $P_{t/c}$ and $P_{d/c}$ were characterized by XRD, SEM, TEM, EDC, BET and XPS. The activity of the catalyst's performance was evaluated by hydrodeoxygenation of date pits oil. Based on the elemental analysis, the degree of deoxygenation (DOD) of product oil was 97.5% and 89.4% for $P_{d/c}$ and $P_{t/c}$ catalysts respectively. The high DOD was also confirmed by the product analysis that mainly consists of paraffinic hydrocarbons. The results also show that between the two catalysts, Pd/c showed a higher activity towards hydrodeoxygenation. Based on the type of components in the produced oil, the maximum fraction of hydrocarbons formed lay within the range of 72.03% and 72.78% green diesel and 30.39 and 28.25% jet fuel using Pd/c and Pt/c catalysts respectively. Thus it can be concluded that waste date pits can be a promising springboard for the production of catalysts and biofuels (green diesel and jet fuel fractions). A research has been conducted to evaluate the potentiality of use of date palm midribs as an alternative source for energy production. Three healthy 10–15 year-old date palms of cultivars Barhi, Khalas, Khodry, Sukkaria and Sullaj, grown at the experimental station for Research and Agriculture of King Saud University were chosen for the conduction of this research. Three date palm midribs were randomly selected from the pruning residues. Each midrib was divided into three zones (base, middle and top). The content of ash was calculated as a percentage of the residues based on the oven-dried midrib meal weight. The higher heating value (HHV) was determined using a calorimeter based on the oven-dry weight. The statistical ANOVA analysis has shown that the chemical constituents of the date palm midribs differed significantly between the researched 5 date palm cultivars. The date palm midrib contents of cellulose, hemicellulose and lignin ranging from 42 to 46%, 25 to 30% and 26 to 31% respectively were found similar to those found in wood species. The lignin content was found to increase moving along the midrib from the base to the top ranging from 24 to 30%. The inverse trend was recorded for the ash content decreasing from 7.6% at the midrib base to 3.4% at the top. Thus it can be concluded that the date palm midribs include higher total extractives (19.34–21.68%) and ash content (3.31–5,85%) as compared with soft and hard woods. The relatively high heating values found for the date palm midribs (17.3–17.9 MJ/kg) prove that they are promising as an energy source. However, the high ash content of all the parts of the date palm midribs, especially the basal part represents a hinderance in their use as a source of fuel. Besides, the date palm midribs exhibited the lowest fuel value index values (97 to 336) as compared with the corresponding values published in the literature for different wood species. A research has been conducted to characterize the date palm midrib as a potential solid fuel for jet and power generation through various thermal conversion processes. To conduct this study date palm midribs were sourced from Medina (Rothanah & Ajwah varieties) and Jeddah (Jeddah & Sukkaria varieties) in Saudi Arabia and from Atbara (Mishria variety), North Sudan. These samples were individually prepared to evaluate their fuel properties: ultimate analysis, proximate analysis and calorific value analysis. All the midrib samples gave calorific values ranging from 16.2 to 16.9 MJ/kg and thus falling within the range of biomass materials from 15 to 20 MJ/kg and close to that of low rank coal such as peat and lignite. Thus it can be confirmed that the date palm midribs enjoy the potentiality of use as a solid fuel for home and industrial applications. As far as the ultimate analysis results are concerned the range of carbon, hydrogen and oxygen in the date palm midribs is comparable with that for typical biomass materials. But the nitrogen and sulfur contents in all samples were found higher than those in most biomasses. The proximate analysis results indicated a significantly high content of volatile matter in the date palm midribs implying their suitability for pyrolysis and gasification processes. Meanwhile, the ash content in the date palm midribs was found comparable with those values in the featured biomasses and lower than in peat, lignite and bitominous coals making the date palm midribs a highly appropriate fuel for continuous thermal processes, where ash removal and handling is a common technical barrier. The thermal decomposition stages were identified based on TGA trends and DTGA peaks. The high reactivity of midrib samples at low temperatures releasing high amounts of volatiles revealed their potentials for pyrolysis and low temperature gasification puroposes. Thus, it can be concluded that the date palm midribs satisfy the typical requirements as a solid fuel for thermal conversion processes in heat and power generation sector. The date palm midribs enjoy the advantages of high availability, low cost and ease of moisture removal in arid and semi-arid regions and thus have high potentials to be used as a solid fuel for thermal applications in the Arab Peninsula and North African countries. Another research was devoted to the study of date palm midribs as an effective feedstock for energy production. To conduct this research date palm midribs of Sukariah, Ajwah and Jeddah varieties were sourced from Madinah Al Munawarah and Jeddah cities in Saudi Arabia. The moisture content, volatile matter, ash and fixed carbon were determined by proximate analysis using Thermogravimetric Analyser (Pyris 1 TGA). Jeddah had the highest value of volatile matter amounting to 83%, whereas Ajwah had the lowest value of 78.2%. The petiole had the lowest value of volatile matter of 55.3% among the date palm biomass. According to published literature, the date palm midribs values of volatile matter content are comparable with other resources, such as sugar cane bagasse, oil palm midribs and western hemlock. Regarding the heating value Sukariah midribs showed the highest value of 16.8 kJ/kg, whereas Jeddah sample revealed the lowest value of 16.4 kJ/kg compared with the seed value of 18.97 kJ/kg and bituminous coal of 34 kJ/kg. Therefore it can be concluded that the date palm midribs can be used as an alternative feedstock for different energy conversion processes, such as gasification, pyrolysis and torrefaction. The presence of metallic elements can cause problems in the thermo-chemical processing systems and therefore require proper handling and treatment during process. A study has been devoted to the characterization of three date palm residues: leaves, empty fruit bunches and petioles and their fast pyrolysis in a bubbling fluidized bed reactor at 525 °C. The date palm residues were sourced from the American University in Sharjah campus (Sharjah, UAE). The fast pyrolysis experiment was conducted at the European Bioenergy Research Institute in Aston University, Birmingham, UK. The comparison of the proximate and ultimate analysis of the leaves, petioles and empty fruit bunches have shown great similarity, but with the leaves much higher ash was noticed. In comparison with other types of biomass and energy crops, the date palm residues can be classified as of high ash, high oxygen, low volatiles and average heat value. The fast pyrolysis product was found to consist of 38.65% bio-oil (including 10.39% reaction water), 37.23% biochar and 24.02% non-condensable gas. The GC–MS analysis revealed for the first time the detailed composition of the date palm pyrolysis oil consisting of at least 140 detectable compounds with the major ones being D-Allose (monosaccharide), phenols, catechol and apocynin. The latter two compounds are of particular interest due to their antioxidation characteristics. The bio-oil heating value was 20.88 MJ/kg, which falls within the low-intermediate range for most fast pyrolysis bio-oils. However the oxygen content was high, and this may have a negative impact on the oil stability and corrosivity. Future work will be directed to the application side of the pyrolysis products including the potentiality of use of bio-oil as a blend with biodiesel for combustion engines, as well as the long-term aging and stability of the bio-oil and potentiality of use of the biochar for soil amendment, especially in desert conditions. An important study has been devoted to the investigation of the thermal behavior of the date palm residues including: leaflets, midribs, trunk, spadix stems and date seeds. These residues have been obtained from a date palm oasis in Tozeur, Tunisia. The ultimate analysis corresponding to the elemental composition of the samples was performed by Service Central d'Analyses (Vernaison, France). The proximate analysis was carried out using a thermogravitmetric analyzer (CAHN121 thermobalance) with gas following upward through the furance at 12NL/h. The thermal behavior of the biomass has been studied under inert and oxidation conditions using a CAHN 121 thermo-balance. The obtained results show that the tested samples have high content of volatiles, carbon, hydrogen and oxygen, whereas the contents of nitrogen and sulfur were relatively low. The researched materials were found to have typical composition as compared with biomass. The heating values (LHV) were found in the range from 15.2 to 19.0 MJ/kg, which are in the same order of magnitude as those for sawdust, olive solid waste, oil palm fruit bunches, Miscanthus, wood pellets and wood chips. It is interesting to note that among the date palm residues the date palm stones having the highest percentage of volatiles and fixed carbon and the lowest ash content enjoy the highest calorific values in terms of low heating value. The bulk density of date seeds was very high (656 kg/m^3), higher than that of wood chips (550 kg/m^3). The energy density of the date seeds (11.4 GJ/m^3) was found much higher than other date palm residues and approaching that of wood pellets (12 GJ/m^3). It can be thus concluded that the date palm seeds are the most attractive residue for energy production due to its high energy density and thus low cost of transportation. But the heating values of other date palm residues are high enough to overcome the problems associated with low energy density. Thus, the obtained results can be useful for the design of processing systems for the production of energy from date palm leaflets, midribs, trunks, date stones and spadix stems. There are strong expectations that the fossil fuels such as oil, coal and natural gas will be depleted within the next 40–50 years. Therefore, there is a growing interest among researchers to study the potentiality of use of biomass for energy production as a more sustainable substitute for fossil fuels, as well as for the rescue of the environment from CO_2 emissions. Therefore, a study has been conducted for the evaluation of date palm residues combustion in a fixed bed laboratory reactor as compared with sawdust behavior. Samples of Deglet Nour date palm residues were sourced from Djerid region in Tunisia including the leaflets (DPL), rachises (PDR), trunks (DPT), spadix stems (FD) and date stones (DS). The ultimate analyses corresponding to the elemental composition of the date palm residues samples were carried out by Service Central d'Analyses (Vernaison, France) according to the relevant XP CEN/TS15014 standard method. Proximate analysis was conducted using a thermogravimetric analyser (CAHN 121 thermobalance). The high heating values (HHV) were measured following XPCEN/T515103 standard methods using an adiabatic oxygen bomb calorimeter (1KA). The energetic potential for different residues was estimated basing on the calculation of the low heating values, bulk density, energetic density that is the potential of energy per unit of biomass volume. Referring to the results of ultimate and proximate analyses, the weight fractions of the different date palm residues were found of the same order as several biomasses. The comparison between the date palm residues and conventional biomasses has shown higher chlorine content for DPR. FP and DPT samples with values above 1%. Thus future controls of this element in gas and particles in fumes are needed in order to reduce both corrosion impacts and emission factors of persistent organic pollutants as dioxins and furans. The low heating values (LHV) were found within the range from 15.2 to 19.0 MJ/kg, i.e. in the same order of magnitude as for olive solid waste, Miscanthus, wood pellets and wood chips. The date stones (DS) were found to have the highest percentage of volatiles (VM) and fixed carbon (FC) and lowest ash content similar to sawdust. The bulk density is an important characteristic of the biomass materials in relation to transport and storage cost. It was found that the bulk density value for DS is very hight (656 kg/m^3) higher than that for wood chips (550 kg/m^3). In addition the energy density (ED) of DS was found 11.4 near to that for wood pellets (12 GJ/kg). Thus it can be concluded that date seeds (DS) is the most attractive material for energy production because of its high energetic density and therefore low cost of transportation. The date seeds is one of the best biofuels enjoying the advantages of highest bulk density, calorific value and volatile matter content and the lowest ash content close to 1.2%. Its energy density (11.4 GJ/m^3) is much higher than other date palm residues. Although the highest value of LHV was obtained for date palm leaflets, their high ash content (15.2%) represents a hinderance for their development as a biofuel since it may lead to corrosion problems in the combustion chambers. As biofuels the date palm rachis (DPR), date palm trunk (DPT) and spadix stem (FD) have very close energtic density chemical composition. The high amount of chlorine in DPR and DPT may introduce potential risks of corrosion in exchange boiler tubes and the formation of persistent organic pollutants as dioxin during combustion in district or domestic applications. A study has been devoted to the chemical analysis of date palm residues for energy production using ultimate, proximate and thermos-gravimetric techniques: the Sukkari cultivar has been chosen as an important variety cultivated in most areas of Saudi Arabia. The date palm residues were chosen to include palm trunk (PT), palm frond base (petiole)(PFB), palm leaflets (PL), fruit stalk (FS), fruit empty bunch (FEB), date palm stones (DPS) and leaf sheaths fibers (LSF) taking Acacia tortils wood (AT) as a reference. These residues were found to have medium to high cellulose content

(33–48%) and lignin (26–40%) and low to medium hemicellulose content (13–31%). The total extractives were (8–33%) and the ash content (1–15%) (lignin and higher extractive content contribute to a high heating value, whereas ash is considered an undesirable material). The volatile matter content ranged from 74.3% for PL to 87.5% for FEB; fixed carbon ranged from 10.5 for PL to 17.6% for PT and the ash content ranged from 1.4% (DPS) to 15.2 for PL. The heat values of the residues varied from 15.47 MJ/kg for PFB to 19.93 MJ/kg for LSF. However, the heat values based on ash-free dry weight had a wide range from 16.5 MJ/kg for FEB to 22.6 MJ for PL due to the large variation in the ash content (1.3%-11.6%). The fuel value index of DPS was higher (2078) than the value for A.tortilis wood (1170) and other date palm residues. Concerning the ranking of the date palm residues, the date stones (DPS) showed the best value (1.9), followed by leaf sheaths fibers (LSF) (2.5), while the palm frond base (PFB) showed the poorest rating (6.3). Thus it can be concluded that the date palm seeds and leaf sheaths fibers are the most suitable among the date palm residues for energy production. A research has been devoted to the ultrasound assisted oil extraction from date palm kernels for biodiesel production. Three solvent types were used: hexane, isopropyl alcohol and ethanol for oil extraction from date palm kernels and ultrasound was applied for 5 min to 25 min at 5 levels using trans-esterification of oils with methanol and potassium hydroxide. It was found that the ultrasonic-assisted hexane oil extraction provided the highest yield by extracting 85% of the total available oil present in the date palm kernels. The biodiesel samples produced from oils extracted with and without ultrasonification had similar physical and chemical properties. Thus it can be concluded that ultrasonification has a potential to enchance the industrial processes by reducing the oil extraction time and energy. A research has been devoted to the hydrothermal pretreatment of date palm leaflets and midribs to evaluated their potential for bioethanol production. To conduct this research date palm leaves were sourced from Abu Dhabi in 2013. Leaflets were separated from the midribs, dried and stored before use. The dried material was milled using a knife mill to pass through a 1 mm screen, sequential Soxhlet extractions with water and ethanol were performed based on National Renewable Energy Laboratory (NREL) protocol. The hydrothermal pretreatment was performed at 10% w/w dry matter at 4 temperature levels (180, 190, 200 and 210 °C). Processing time was at 10 min. The research results showed that high glucan (> 90% for both leaflets and midribs) and high xylan (>75% for leaflets and > 79% for midribs) recovery were achieved. Under the optimal conditions of hydrothermal pretreatment (210 °C for 10 min) highly digestible (glucan convertibility 100% to leaflets, 78% to midribs) and fermentable (ethanol yield 96% to leaflets, 80% to midribs) solid fractions were obtained. The fermentability test of liquid fractions proved that no considerable inhibators to saccharomyces cerevisisae were produced in hydrothermal pretreatment. Proceeding from the high sugar recovery, enzymatic digestibility, and ethanol yield it can be concluded that the production of bioethanol by hydrothermal pretreatment from date palm residues is feasible. A research has been conducted on the characterization of pyrolytic products of date palm residues. The pyrolysis process is a thermal cracking of the biomass in an inert atmosphere at temperatures ranging from 300 to 700 °C to produce useful liquid biofuel (bio-oil), solid biocombustible fuel (biochar) and renewable syngas. Therefore, the objective of this study was the investigation of the main characteristics of the obtained products from the pyrolysis of the date pam residues to evaluate their potential as a feedstock for renewable energy and chemical industries. To conduct this study, four date palm residues samples were obtained from the National Institute of Arid Zone (IRA-Kebili, Tunisia): midribs, leaflets, empty fruit bunches and spathes. The samples were finely crushed to small pieces with sizes from 2 to 4 mm and air-dried. The pyrolysis experiments were conducted on a laboratory scale pyrolysis plant under the operational conditions: 500 °C as final temperature, 15 °C/min and 300 g mass initial of the used sample. The elemental composition (CHN-O) of date palm residues, bio-oil and biochar were determined using an elemental analyser (LECOCHNSTRu Spec); the O content was determined by difference. The proximate and ultimate analyses have shown that the date palm spathes had the least ash content of 2.4%, whereas the leaflets had the highest ash content (11.58%). But in general the date palm researched residues had high volatile matter content and ash fairly compared with those found in the literature for lignocellulosic materials converted into biofuels using pyrolysis. The calculated high heat values ranged from 17.88 to 19.09 MJ/kg for all studied residues. These values are low for a commercial fuel. The bio-oil yield ranged from 17.03 wt% for leaflets to 25.99 wt% for empty fruit bunches. Concerning the biochar, the highest yield 36.66 wt% was obtained for leaflets, whereas the lowest one (31.66 wt%) was obtained for the spathes, while the syngas production varied from 39.1 wt% for midribs to 46.31 wt% for leaflets. As a conclusion of this study, the bio-oil which represents moderate amounts of carbon and hydrogen compared to petroleum-based fuels, could be used as a biofuel after grading. The biochar could be used as biocoke in industrial applications. The presence of CH_4 and H_2 in significant proportions in the gaseous mixture gives the obtained syngas good combustion properties. A study has been conducted to evaluate the technical feasibility of using seawater instead of freshwater in the pretreatment of date palm leaflets for bioethanol production. The lignocellulosic biorefineries are one of the most promising alternatives for fossil oil. But one of the obstacles of proliferation is the excessive utilization of freshwater (1.9–5.9 m^3 water per m^3 of biofuel), which may be in shortage in arid and semi-arid regions, where the date palms are usually grown. In this study artificial seawater has been used to replace fully the freshwater in the hydrothermal pretreatment of date palm leaflets to produce bioethanol. The results of this study confirm the feasibility of replacing freshwater with seawater in the hydrothermal pretreatment of date palm leaflets to produce bioethanol. However a lower crystallinity of cellulose has been observed after treatment with seawater rather than freshwater. Pretreatment by using seawater produced slightly lower digestibility of solids (glucan-to-glucose conversion) in enzymatic hydrolysis than pretreatment by using freshwater. But there was no significant difference in the bioethanol yield. Moreover, the fermentability test showed no significant difference in the bioethanol yield between liquids from pretreatment by freshwater and seawater. Thus it can be concluded that seawater could be a promising alternative to freshwater in biorefineries processing lignocellulosic materials. Within the expected increase of demand on energy for homes, industries and transportation the expected contribution of biofuels will grow from 50 EJ/year in

2012 to more than 160 EJ/year in 2050. The bioethanol is seen as the main biofuel for the future. Within this framework a study has been devoted for the realization of an experimental solar fermenter for the production of bioethanol from date palm waste (DPW). To conduct this study DPW of Hchef, Kacien and other varieties of date's scraps of the cattle food were sourced from Algerian Sahara. A batch fermenter was designed and installed to operate effectively by using a solar water heater in order to reduce the cost of the bioethanol generation process. The fermenter was realized within the South Society of Metallic Construction (ECOMES), located in Adrar. Experimentation was performed during the cold season of the year, the 1st week of January 2015. The results of this study indicate that DPW constitute a favorable medium for S.cerevisiae growth, due to its sugar content and is thus considered as attractive feedstock. It is thus technically feasible to produce bioethanol using the solar batch fermenter at relatively moderate cost. The DPW distilled juice produced the highest bioethanol concentraction of about 90° with an acceptable productivity of 3.47 ml/kg/h assessing a scale efficiency 33%. These results represent a strong support to continue R&D in the renewable energy field. It is thus necessary to start to build semi-pilot and pilot fermenters and investigate new methods, microorganisms and other byproducts to improve the quantity of bioethanol produced and to reduce the energy consumption during the bioethanol process transformation to improve the economics of bioethanol production. A study has been conducted to explore the anaerobic digestive technology for processing of date palm residues being the most available and sustainable feedstocks for renewable energy. These residues, annually available with huge quantities are most dominantly treated as waste: being open-field burnt or sent to landfills. Therefore the exploitation of these residues for the production of methane gas should be seriously considered. A carbon to nitrogen (C/N) ratio between 20 and 30 is regarded optimum for anaerobic digestion process. To conduct this study midribs, fruit empty bunches and rotten dates were sourced from the southern oases of Libia. In this research two-liters batch digester was used. The results of the study have shown that production of biogas from fresh waste nitrogen source (NH_4CI) gives best biogas production compared to dry waste with nitrogen source (NH_4CI) and fresh waste without nitrogen source (NH_4CI). The biogas production of dry waste nitrogen source (NH_4CI) was better than biogas production from fresh waste without nitrogen source (NH_4CI). Thus it can be concluded that the heat treatment of date palm residues is very useful to improve the production of biogas. The percentage of methane gas in the produced biogas reached 48% for fresh waste with addition of nitrogen source. The maintenance of pH in the digester at range 6.5–8 is more efficient to produce biogas. The retention time for fresh waste inside the digester is less than that for dry waste. The palm tree wastes contain a high ratio of carbon compared to nitrogen specially the dry one and addition of nitrogen source is required. A study has been devoted to investigate the usage of date palm biomass for biogas production. The Zahdi date cultivar has been chosen because of its abundance in Iraq representing 60% of the country production. The Waste Management Lab/Corneal University (USA) methodology for anaerobic digestion was used in this research. The results of the study indicate that the volatile solids of substrate and inoculums was 39.82, 2.37% respectively with a ratio 16.8:1. The nitrogen content of the substrate was found 2.35 indicating the demand for extra amount of nutrients to provide nitrogen for bacteria growth in the fermentation batch. A total gas pressure with 67% Methane was produced from date pulp waste fermentation with a yield of 0.57 Lit for each gram volatile solid of the substrate. The addition of 1% yeast extract solution as a nutrient increased Methane yield in liters by 5.9%. The high volatile solids content in the date palm biomass compared to inoculums indicates a high potential of biogas production from a small amount of biomass. Given the great abundance of date palms in the Middle East and North Africa there are great future potentials for production of biogas and biofuel in a commertial scale. An important study investigated the production of biogas from date palm fruit wastes by measuring the gas volumetric flow rate directly. Samples of Digal date fruit wastes in their final stage of maturity (with hard texture) were sourced from stores of Diyala province in Iraq. A lab-scale digestion system was used in this research. After the samples were pitted, weighted, mixed with a proper amount of water and placed in the digester, eight mixtures of substrates were divided into two groups. Each group contained 4 samples of substrate having ratios of 0.5, 0.2, 0.15 and 0.1 (w/w). They were subjected to anaerobic digestion at 37 °C for mesophilic and 55 °C for thermophilic conditions. The dischange process of biogas was carried out every 3–5 days and the data was saved on a laptop. Concerning the effect of solid mixing ratio the results of the study showed that under mesophilic temperature of 37 °C, the highest biogas yield was achieved in the case of 0.15 w/w (182 L/kg volatile solid mass), whereas the lowest biogas yield was in the case of 0.5 w/w (84 L/kg volatile solids mass). As far as the effect of recycled digestate is concerned, the research results showed that the use of recycled digestate has improved the production of biogas by 12%. Thus it can be concluded that the date palm fruit wastes are a suitable source for biogas production and that a mesophilic system is the best option for producting biogas from date wastes. A maximum biogas production of 203L/kg volatile solids was achieved for a solid concentration of 0.15 (w/w), when the substrate was mixed with recycled digestate at 25% of the substrate content. A study has been conducted to evaluate the potential impact of date seeds on biogas production. To conduct this study seeds of Khalas and Khudri cultivars were selected. Oil was extracted from the date seeds using an automatic Soxhlet extractor. The oil-spent date seeds were recovered from the extractor, dried and then used for biogas production. The primary waste water treatment sludge was sourced from a nearby domestic wastewater treatment plant. The total solids of the primary sludge was adjusted to approximately 2% and stored in a freezer in small containers until use. The volatile solids after total solids adjustment reached 1.64%. The preparation of the date seeds/sludge mixtures was conducted by thawing the frozen sludge from containers, mixing and adding the predetermined date seed quantity and then adjusting the pH if necessary to approximately 7.3. After 14 weeks of incubation the biogas production expressed in terms of the date seed/sludge ratios, were in the following order: 10% > 7.5%≃5%≃2.5%≃0% > 20% > 40%. The size of the date seed particles did not significantly affect biogas production. The specific gas production was in the range of 370–390 mlg^{-1} volatile solids for the 0–10% seed/sludge ratios, 245 mlg^{-1} at 20% and 120 mlg^{-1} for volatile solids at 40%. The relatively low biogas production from the 20 and 40% seed/sludge mixtures indicated inhibation, which was also, shown by the low pH in the mixtures following digestion. Oil extraction from the date seeds reduced, but did not fully overcome, inhibition of biogas production from the 20% and 40% mixtures.

Keywords Annual products of pruning of date palms · Bioethanol · Acetone · Butanol and ethanol · Date waste · Date palm seeds · Date pits · Green diesel · Jet fuel fractions · Date palm midribs · Energy production · Heating values · Ash content · Ulimate analysis · Proximate analysis · Calorific value analysis · Pyrolysis; Gasification · Torrefaction · Bio-oil · Biochar · Non-condensable gas · Syngas · Leaves · Spathes · Empty fruit bunches · Petioles · Leaflets · Midribs · Spadix stems · Leaf sheaths fibers · Trunk- Date seeds · Soil amendment · Heating values · Volatiles · Fixed carbon · Energy density · Bulk density · Chlorine · Thermo-gravimetric technique · Bioethanol · Glucan · Xylan · Hydrothermal pretreatment · Seawater · Biofuels · Bioethanol · Methane gas · Biogas · Anaerobic digestion · Mesophilic temperature · Recycled digestate · Oil-Spent date seeds

11.1 Bioethanol from Date Palm (Fronds)

The date palm belt extends in the Arab world: from Morocco in the far west to Iraq in the far east including more than ~ 111 million palms [1]. Huge quantities of lignocellulosic byproducts result from the yearly process of pruning. These huge quantities (e.g. ~ 810,000 tons in Egypt, air-dry weight) [2] are mostly: either open-field burnt or put in landfills, which represents a great economic loss and a source of environmental pollution. The date palm fronds are annually available by a quantity of ~ 52.4 kg, air-dry weight representing about ~ 54.6% by weight of the annual products of pruning. This means that the date palm fronds are annually available with huge quantities (~442,362 tons, air-dry weight in Egypt) that can be economically utilized.

The date palm fronds contain 58% of cellulose and 22% of hemicellulose. This means that in ideal conditions 80% of palm fronds can be converted to bioethanol, rendering in Egypt a lone ~ 353,900 tons!

What is the main obstacle before using date palm fronds for the production of bioethanol? The traditional method of chemically hydrolyzing cellulose using dilute or concentrated acids leads to the production of toxic byproducts that require detoxification before fermentation.

The replacement of the conventional pretreatment by biological processes using enzymes represents a promising simplification to the overall bioethanol production needing much less energy and being a cost effective alternative to the traditional pretreatment of lignocellulosic materials. In one of the pioneer studies in this area [3], the substrate was prepared by grinding samples of date palm fronds and stirring them well in distilled water. The solids were then vacuum-filtered and dried in an oven. Three enzymes were sequentially added: laccase for lignin degradation, xylasnase for hemicellulose hydrolysis and then cellulase for cellulose hydrolysis. The results indicate that the reducing sugars yield increased from 5.6% using cellulase only to 45.6% in case of treating with laccase and xylanase prior to the use of cellulase for hydrolysis. In the second case the treated products were immediately reacted by the 3 enzymes, yielding a higher conversion of 60%! The higher yield in the last case can be due to the gradual production of accessible substrate, which reduces the effect of substrate inhibition that cellulase is known to encounter. The use of the 3 enzymes together provides a distinguished simplification to the overall bioethanol production making the process more economically feasible.

11.2 Lignin and Bioethanol from Date Palm Fronds

One of the interesting researches [4] has been devoted to the investigation of the potentiality of use of the date palm fronds for the production of lignin and bioethanol. The Organoslov process has been successfully used to extract lignin and digestible cellulose pulp for bioethanol production from such an extensively available resource as date palm fronds: from date palm plantation or as a landscape waste.

The date palm fronds are distinguished with high lignin content (30 g/100 g dry matter). The Organoslov process was positively influenced by temperature (140–200 °C) generating a high yield of Organoslov lignin (12.93 g/100 g raw material). The digestibility of the cellulose pulp was positively influenced by temperature and negatively influenced by ethanol content. The cellulose enriched pulp was highly digestible by enzymes giving a yield of 21.38 g glucose /100 g raw material. The maximum practical yield of bioethanol was 9.92 g ethanol/100 g raw material equal to 90.95% of the theoretical value. The research results provide an opportunity to fully utilize the date palm fronds energy and create a holistic bioenergy via a relatively new "lignin-first" approach.

11.3 Acetone, Butanol and Ethanol Production from Date Waste

The various sources of biofuels that have been explored in the world so far include biomass, biogas, primary alcohols, vegetable oils and biodiesel [5]. Meanwhile, acetone, butanol, and ethanol are common solvents used in many important industries and thus can replace energy derived from petrochemicals [6, 7]. Accordingly, acetone, butanol and ethanol can be produced inexpensively by fermentation in several possible methods such as using a low-cost fermentation substrate (agro-industry waste) [8, 9].

At the same time, the date palm (Phoenix dactylifera L.) is one of the most important fruit trees growing in the Arab world. In Egypt, the production of date palm fruits is increasing, as recorded by the Food and Agriculture Organization, reaching about 1.7 million metric tons in 2017 [10].

When producing butanol and bioethanol from a fermentation process in Egypt, we find that scientists are using non-indigenous Clostridium strains such as Clostridium acetobutylicum ATCC 824 [11, 12], C. acetobutylicum DSM 792, C. acetobutylicum AS 1.224 [13], Clostridium bystrianum DSMZ 525 [14].

However, there is a study using the low-quality and surplus date palm fruits as a substrate for fermentation in the production of acetone, butanol and bioethanol using the original Egyptian Clostridium strains [15].

One hundred and seven isolates of Clostridium fungi were isolated from agricultural soil cultivated with different plants in Assiut Governorate, Egypt. Only eighty isolates demonstrated the ability to produce acetone, butanol, and ethanol on T6 medium ranging from 0.036 to 31.89 g/L. The largest number of isolates were obtained from potato, wheat and onion soil samples 27, 18, and 10, respectively. On the other hand, there were three original isolates that produced more acetone, butanol and ethanol than those produced by the Clostridium acetobutylicum ATCC 824 (11.543 g/L) reference isolate. The three isolates were identified based on the phenotype and gene coding of 16S rRNA as Clostridium beijerinckii ASU10 (KF372577), Clostridium chauvoei ASU55 (KF372580) and Clostridium roseum ASU58 (KF372581) [15].

The sub-standard and surplus date fruits produced the highest level of acetone, butanol and ethanol by C. beijerinckii ASU10 (24.07 g/L) which includes 67.15% (16.16 g/L) butanol, 30.73% acetone (7.4 g/L) and 2.12% ethanol (0.51 g/L), while C. roseum ASU58 and C. chauvoei produced ASU55 ABE at 20.20 and 13.79 g/L, respectively. The yield of acetone, butanol and ethanol by C. acetobutylicum ATCC 824 was 15.01 g/L (4.62, 9.71 and 0.67 g/L respectively). These results show that the local strains C. beijerinckii ASU10 and C. Roseum ASU58 are highly competitive in producing acetone, butanol and ethanol from very cheap substrate as substandard and surplus fruits. Additionally, the use of this substrate without any food ingredients serves as a commercial substrate for the desired acetone, butanol and ethanol production [15].

11.4 Biodiesel Production from Phoenix Dactylifera as a New Feedstock

A research [16] has been conducted on the use of date palm seeds for biodiesel production. Biodiesel is generally considered as the most acceptable fuel for diesel engines, because of its technical, environmental and strategic advantages. Biodiesel is known as a carbon neutral fuel, because the carbon in its exhaust is originally fixed in the atmosphere. Biodiesel can be produced from various vegetable oils including soybean, Jatropha, rapeseed, palm, sunflower, corn, peanut and cotton seed oils, as well as animal fats (beef tallow and lard), waste cooking oil, greases (trap grease and float grease) and microalgae [17, 18]. Biodiesel possesses many advantages, compared with fossil-based diesel fuels, such as cleaner engine emissions, biodegradability, renewability and superior lubricating properties [19–21]. The main hindrance before commercialization of biodiesel is the high cost of the raw material.

The date palm (Phoenix dactylifera L.) is an essential element of the flora in the Arab region. There are historical evidence that the date palm originated somewhere near to the Persian Gulf as early as 4000 BC [22]. The date palm seeds were traditionally used in fodder of sheep, poultry, camel and cattle [23]. The date seed represents 10–15% of the date fruit weight [24]. The date seed is composed of 3.1–7.1% moisture, 2.3–6.4% protein, 5.0–13.2% fat, 0.9–1.8% ash and 22.5–80.2% dietary fiber. It contains a high level of phenolic compounds (3102–4430 mg gallic acid equivalents/100 g of seed powder) and has high amounts of antioxidants (580–929 lm trolox equivalents/g) and dietary fiber (78–80 g/100 g) [25]. It contains oleic acid (between 11.9–58.8%) [26], lauric acid, myristic acid and palmitic acid ranging from 13% and 25.8%; 7% and 13%; 6% and 13% respectively [27, 28]. According to FAO statics 2010[29] the date production in south-western Asia and north Africa can be estimated at 7.85 million tons including seeds of 785,000 to 1.18 million tons. The present research is devoted to the use of date palm seeds-dominantly treated as waste-as a cheap sustainable feedstock for the production of biodiesel.

To conduct this research 10 kg of dates were sourced from farms in Behbaan, south of Iran in summer 2009. Sodium hydroxide 99% (NaOH), n-hexane 99% and methanol 99% were purchased from Merck. Date seeds were separated, washed, dried and milled to 1.2 mm sieve size. The extraction was performed by n-hexane solvent and soxlet device for 4 h. Particles and impurities were removed by passing through a filter and oil was separated from n-hexane by rotary evaporator. Date seed oil 50 g was pre-heated to 60 °C and a mixture of 11 g methanol and 0.375 g NaOH was added. The reaction continued for 2 h at 60–65 °C under reflux and stiring. Thus biodiesel and glycerin were produced. The alcohol was separated by the rotary evaporator. Then the glycerin and the biodiesel were separated by means of a separating funnel according to difference in density. To purify biodiesel (removing catalysts and remaining glycerin) washing was performed by 70 °C hot water. The biodiesel containing water was dehydrated in vacuum through distillation by means of a rotary evaporator. The fatty acid methyl esters of biodiesel were analysed by GC/MS (GC: VARIAN CP-3800, MSVARIAN Saturn 2200) [30]. Saponification value (SV) and iodine value (IV) of oil were calculated from fatty acid methyl ester compositions of oil according to [31]. Cetane number (CN) and higher heating value (HHV) of fatty acid methyl esters were calculated according to [32]. Kinematic viscosity at 40 °C was determined using Cannon–Fenske viscometer in warm water bath. Pour point and cloud point were simultaneously determined using Tanaka Mini Pour/Cloud Point Tester model MPC-101A. For flash point, Tanaka Automatic Cleveland open Cup Flash Point Tester Model ACO-5 was used. Concerning density measurement it was conducted by Metler-Toledo densimeter at 15 °C.

The biodiesel has been produced by transesterification with a yield of 98% and the fatty acid esters were analysed by GC/MS. The research results show that the fatty acid composition of biodiesel was similar to biodiesel fuels produced from other vegetable oils. The properties of the biodiesel were evaluated by fuel standard tests and the results were compared with EN14214 and ASTM D6751 standards. The date seed biodiesel had a considerable amount of low-chain fatty acids, which gives special features to biodiesel, high cetane number (60.3), low viscosity (3.84 mm^2/s), flash point (140 °C) and low iodine value (46). The only weak point is its high pouring point (−1 °C), which limits its use in cold weather as compared with other vegetable biodiesel fuels.

11.5 Desert Palm Date Seeds as a Biodiesel Feedstock: Extraction, Characterization, and Engine Testing

A research [33] has been conducted on the use of date palm seeds as a feedstock for biodiesel production. The high rate of global consumption of oil indicates that the world may suffer oil supply shortages by 2020–2030 [34]. This presents a serious challenge to sustain global economic and technological development without violating the rights of future generations [35]. This leads to the necessity of developing renewable fuels to satisfy the contemporary and future needs of industry and transportation [36]. Biodiesel is one of the most promising renewable energy alternatives having diversified potential feedstocks of an extended array of vegetable oils and animal fats, having a high cetane number, oxidative stability, lubricity characteristics and being biodegradable and compatible with existing transportation infrastructure [37]. But the first- generation biodiesel produced from vegetable oil and animal fats is not sustainable, because of the associated problems of loss of biodiversity, excessive consumption of water resources, high production cost and food shortages [38]. Thus there is a necessity to develop second- generation biodiesel relying on non-food feedstocks such as food waste (30% of food is sent to landfills or incinerators [39]), spent coffee grounds [40, 41] and used cooking oil [37]. In addition date palm seeds [42] have been considered as a new potential feedstock for the production of biodiesel [42]. Nowadays the cultivation of date palms has exceeded the traditional semi-arid environments to other regions including southern Europe, Australia and the Americas [43]with a total of 100 million date palms comprising more than 2000 cultivars [44]. The global production of dates has increased rapidly from 1.88 million t in 1965 to 3.43 million t in 1990 (~1.8 times) to 8.46 million t in 2016 (4.5 times in 51 years!). The seeds are very hard ranging from 5 to 15 mm in length and have an oblong shape and a ventral groove [45]. They weigh 11–18% of the total fruit mass [46]and contain 4–13% oil [25, 47]. Thus in can be concluded that the use of date palm seeds for biodiesel production is a very sustainable alternative with an estimated 1.3 million t of date palm seeds and 127,000 t of date seed oil (a similar amount of biodiesel can be annually produced).

Date palm seeds, in general, are treated as waste [48]. But they were traditionally used in fodder for domestic poultry and other animals [49, 50]. Recently the date palm seeds have been used in the Arabian peninsula for caffeine- free coffee making [51,

52] as well as pharmaceutical and cosmetic products [53]. Most recently few studies proposed the use of date palm seed oil as a potential material for biodiesel production [42, 54, 55]. This study has been devoted to the investigation of production factors to optimize the extraction of date palm seeds oil, production of biodiesel and testing of biodiesel blends in a compression ignition diesel engine.

Date palm seeds were sourced from the Sharjah dates facility in Sharjah (UAE), hand—isolated and soaked in water for 12 h and then washed with tap water to remove any remaining date flesh. These seeds were dried in sun for 7 days, followed by 12 h in an oven (SLN180, POL-EKO) at 60 °C. A heavy- duty crusher was then used to grind the seeds to 0.6 mm mesh size. Scanning Electron Microscopy (VEGA3-SEM) was used to determine the elemental composition of the ground seeds. The microstructure has been examined before and after oil extraction. The date palm seed oil has been extracted using the soxhlet extraction method, because it represents the most practical option for analytical scale, where parameters can be appropriately monitored and controlled.

The resulting biodiesel was characterized and assessed based on widely used international standards (ASTMD6751 and EN14214). Four biodiesel blends were prepared (B5, B10, B15 and B20) and tested in a compression ignition engine at engine speeds from 1600 to 3600 rpm (200 rpm increments) and three engine loads (50, 75 and 100%). The research results lead to the following conclusions:

1. Hexane consistently outperformed petroleum ether in oil yields. The highest oil extraction (10.74 wt%) was observed with hexane at 60 °C, 4 h extraction time and a solvent: seed ratio of 6.
2. A date seed oil biodiesel yield of 92% was recorded at the following transesterification conditions: 55 °C, 9:1 AOMR, 1wt% CMF and 90 min.
3. Phosphoric acid treatment of the date palm seed oil reduced levels of Mg, Ca and P in the biodiesel to the ASTMD6751 allowable level.
4. The cloud point of the date seed oil was relatively high (9.4 °C), so the use of cold flow improvers would be necessary when this fuel is used in cold climates.
5. The BP, BTE and BSFC values of the date seed oil biodiesel blends were comparable with the baseline diesel, though the latter was superior.
6. All biodiesel blends produced lower levels of CO_2, CO and HC emissions.
7. All biodiesel blends produced higher levels of NO_2 emissions.

These conclusions demonstrate the fundamental suitability of date palm seeds as a biodiesel feedstock.

11.6 Ethanol Production from Date Waste: Adapted Technologies, Challenges and Global Potential

An important study [56] has been conducted on ethanol production from date palm wastes. The expected depletion of fossil fuels in addition to the climate change associated with their use has urged many researchers to find alternative biofuels for use in industry and transport [57–61]. The agricultural wastes reaching up to 50% of certain crops may serve as a sustainable feed stock for the production of ethanol [62–65]. The researchers are showing increased interest in the use of organic wastes in biofuel production [66, 67]. The so-called second generation bioethanol produced from lignocellulosic residues shows energetic, economic and environmental advantages as compared with bioethanol from starch or sugar. Thus the world's ethanol production has almost doubled from 2007 to 2014. According to 2003–2005 US average sugar recovery rates one ton of sugar cane and beet is expected to yield 74 and 94L of bioethanol respectively, whereas one ton of molasses (a byproduct of sugar cane and sugar beet processing) may yield about 263L of ethanol. One ton of raw sugar would yield 513.2–534.4L of ethanol [68].

The date palms are traditionally cultivated in arid and semi-arid regions [69]. A date palm fructifies within 7–10 years after planting and produces 68–176 kg of dates every harvest season [69, 70]. Fructose and glucose constitute over 72%of the dry weight of ripe dates [71]. Many processes were used to convert biomass into fuels, such as combustion, pyrolysis, gasification, hydrogenation and liquefaction [72, 73]. Several studies have been devoted to the extraction of sugar and production of ethanol from dates: the important factors to increase productivity being sugar concentration, process time, temperature, PH, yeast and enzymes, shaker rpm, and ethanol concentration [11, 69, 74–79]. Alzain [77] investigated the use of the date wastes (DWs) as a source of bio-ethanol by yeast and enzymes such as cellulase and amylase. The results showed that the bioethanol production is affected by different fermentation parameters such as yeast concentration, PH, temperatures, fermentation periods, moisture content and substrate components. Literature review also indicates that extensive research has been done on production of biofuel from agriculatural wastes and animal product wastes [80], enzymatic saccharifcation and fermentation of cellulosic date palm wastes to glucose and lactic acid [81], etc. The aim of the present study is to find the optimal ways to convert DWs and encourage investors to invest in the energy production from DWs. Egypt, Iran and Saudi Arabia produce almost ½ the world's dates production (47.755% with shares of world production ~ 20, 14, 14% respectively. Their number of date palms is 16 (30 varieties), 27 and 24 millions (320 varieties) respectively. The percentage of waste dates in Iran is about 50%. The weight of date seeds varies between 0.5 and 4 g (6–30% of the fresh fruit weight). Assuming an average of 13% of total fruit weight belongs to the seed weight 120 kg dried seed per t of DWs can be estimated. Thus 13 kg of seed oil (10.85% by weight) can be used for biodiesel fuel production (density 0.871 kg/L). In addition 93.72 kg of carbohydrate (78.1% of date seed) per ton of DWs can be obtained. One ton of carbohydrate may render 350L of ethanol. The annual pruning activity of date palms renders an average of 15–25 leaves per palm of length 4-7 m and weight 1.5–2.5 kg/leaf and thus producing 34 kg from pruning [82–89]. Thus the annual available quantities of leaves from pruning can be estimated by 914,000 t, 544,000 t and 816,000 t respectively in Iran (27 million date palms), Egypt and Saudi Arabia.

The fermented ethanol can be produced by three groups of feedstocks: sugary, starchy and ligno-cellulosic compounds. In the first generation (IG) processing, sugary and starchy feedstock is converted to bioethanol. The sugar juice is extracted from sugarcane, sugar beet or sweet sorghum, treated in order to remove impurities, concentrated by evaporation and finally fermented [90, 91]. However (1G) processing is in contradiction with needs of humans and animals. Therefore, it is expected that (2G) production of ethanol will prevail in the near future [92]. In this case the ligno-cellulosic feedstock is pre-treated to achieve hemicellulose and cellulose compounds, hydrolysis of these compounds, followed by fermentation and ethanol separation [90, 91, 93–95]. A high-preference method combines both the 1G and 2G processes. The lingo-cellulosic feedstock of 1G process is fed into 2G process. Thus the combined heat and power unit, fed with the residues of 2-G process, supplies the steam and electricity for both processes [90].

The conventional or 1-G process includes cooking, liquefaction and saccharification processes for ethanol production. The temperature of the cooking process is higher than 100 °C [96]. The majority of worldwide produced ethanol is extracted by the sugar fermentation method. In the anaerobic fermentation mechanism of ethanol production, glucose primarily enters the glycolysis cycle, goes through a series of reactions, and finally each glucose molecule turns into two molecules of pyruvate. Then pyruvate turns to acetaldehyde in two stages, and acetaldehyde turns into ethanol by alcohol dehydrogenase [97]. Sugary raw materials normally are pretreated and turned into ethanol by microorganisms, while starch and cellulose have to be first pre-heated, hydrolysed and turned into sugar and then converted to ethanol by fermentation [70, 98–100].

Bioethanol production from lingo-cellulosic biomass (G2 process) is economical only if the sugar concentration exceeds $40 gL^{-1}$ and yield is increased [101]. Separate hydrolysis and fermentation, semi-simultaneous saccharification and fermentation and simultaneous combined saccharification and fermentation are the common methods for bioethanol production [102, 103]. Recent developments have reduced the need for cooking temperature above 50 °C, which significantly lowers energy consumption. These possibilities provide simultaneous combined saccharification and fermentation processes, which eliminates the need for separate equipment for saccharification, decreases the probability of bacterial contamination, and reduces the restrainability of high-sugar concentration to microenzyme [96]. Pretreatment plays a significant role in bioethanol production from lingo-cellulosic biomass [101].

One of the most important steps for 2G ethanol production is pre-treatment of biomass, which greatly affects the production yield. The pretreatment is conducted to break down the lignin structure and destroy the crystalline structure of cellulose to enhance enzymes accessibility to the cellulose during hydrolysis [104]. Lignin can be used for generating electricity or heat [105]. There are different pretreatment technologies depending on the characteristics of the biomass including mechanical pretreatment, use of acids (e.g. oxalic acid), metal salts, alkaline pretreatment, extrusion processing, ultrasound for pretreatment, microwave, pulse electric energy pretreatment, organosolv process for biomass pre-treatment, pretreatment of lignoc, hydrotropic pretreatment, hydrothermal/liquid hot water pretreatment, steam explosion, wet explosion pretreatment, carbon dioxide pretreatment, peroxide pretreatment, pre-treatment with ammonia, chemical oxidation with ozone pretreatment, sulphite pre-treatment, enzymatic hydrolysis, biological pretreatment

[106]. Different attempts have been made to find out the best pretreatment method [104, 107, 108]. The major problem of acid pretreatment method is the production of inhibitory compounds such as furfural from degradation of monosaccharides. The steam explosion has been recognized as one of the most efficient pretreatment technologies for date palm ligno-cellulosic residues including petioles and leaflets [104].

In the (2G) ethanol production the polysaccharides (cellulose and hemi-cellulose) are broken down to free sugar molecules such as glucose, mannose, galactose, xylose and arabinose. These products are then used in fermentation. Hydrolysis of cellulosic materials is generally done by the enzymatic, acid, microbial and thermo-chemical processes [109–111]. Enzymatic hydrolysis is advantageous to produce sugars from lingocellulosic residues, because of its mild operating conditions and the avoidance of by-products. Chemical hydrolysis (frequently sulphuric and HCL acid) requires exposure of lingo-cellulosic residues to a chemical for a time interval at a specific temperature and produces sugar-monomers. Acid hydrolysis has lower cost and more effectiveness than enzymatic hydrolysis [109, 112]. Hackimitehrani and Shahbazia used acid hydrolysis in different situations (concentration, times and temperatures) for date palm leaves. The highest percentage of sugars (50%) was attained at a concentration of 3.12% HCL at 70 °C during 24 h [56]. Acid hydrolysis can yield 200–230 L of ethanol per ton of date palm leaves as compared with 350 L of ethanol per ton of wheat straw [113].

The findings of this study show that there is a great potential to produce ethanol from date palm wastes and date palm lignocellulosic residues in the Middle East countries due to their high availability potential. Besides, these date wastes and lignocellulosic residues are dominantly treated as waste and thus representing a source of environmental pollution. The production costs of ethanol from them depend on the type of feedstock, primary energy sources and energy scenarios. The available technologies for converting cellulosic feedstock into ethanol can be grouped under hydrolysis and thermo-chemical conversion. Pre-treatment plays an important role in bioethanol production from lingocellulosic date palm residues. Egypt, Iran and Saudi Arabia can produce annually 173.5, 401 and 438 million liters of ethanol respectively from date palm wastes and products of pruning (basically leaves). Other countries have the potential to produce 1248 million liters of bioethanol. Thus 3260 million liters of bioethanol can be globally produced. The global cost of bioethanol production from date wastes is 0.68$ per liter, of which feedstock accounts for 85.3% of the total cost of production. The corresponding value for date palm products of pruning is $0.34 per liter due to high availability potential.

11.7 Efficient Utilization of Waste Date Palm Pits for the Synthesis of Green Diesel and Jet Fuel Fractions

An important research [114] has been conducted on the use of date pits for the synthesis of green diesel and jet fuel fractions. The limited reserves of oil and the huge environmental consequences of use of petroleum fuels has motivated researchers to find alternative sustainable renewable fuels [55, 115]. The biofuels enjoy the advantages of high availability, carbon neutral nature, economic viability and environmental friendliness [116, 117]. Biofuels can be produced by fermentation, pyrolysis, gasification, transesterification and hydrogenation [116–118]. These processes are mostly used for the synthesis of first generation biofuels such as biodiesel and bioethanol consuming edible feedstock and thus significantly affecting their future economic feasibility. There is a harsh necessity to search for second generation feedstocks for biofuels, such as the date pits of high availability and being predominantly treated as waste.

Date pits were sourced from a local supplier in Muscat, Oman. The date pits samples were cleaned and washed with water and then oven-dried at 100 °C for 12 h. A mechanical grinder was used to convert the samples to a powder. After sieving, the oil was extracted using hexane as a solvent in a soxhlet apparatus following the AOCS official method Am 2–3. After extraction of oil, a 100 g sample of date pits was dried at 100 °C for 8 h, washed with warm deionized water, dried at 100 °C for 3 h and then carbonized in a furnace at 500 °C for 5 h with a heating rate of 3 °C/min under N_2 gas at a flow rate of 50 ml/min. The carbon material was washed with warm deionized water to remove any impurities and then oven-dried at 100 °C for 8 h. A solution of palladium and platinum chloride was prepared to obtain 4 wt% of each on a base carbon material. The produced carbon material was impregnated by wetness incipient impregnation. The resultant materials were allowed to dry overnight at room temperature and subsequently heated to 350 °C at a heating rate 3 °C/min for 4 h.

Active catalysts were thus produced by carbonization and impregnation with P_t and P_d metals. The synthesized catalysts P_t/c and P_d/c were characterized by XRD, SEM, TEM, EDC, BET and XPS. The activity of the catalysts's performance was evaluated by the hydrodeoxygenation of date pits oil. Based on the elemental analysis, the degree of deoxygenation (DOD) of product oil was 97.5% and 89.4% for P_d/c and P_t/c catalysts respectively. The high DOD was also confirmed by the product analysis that mainly consists of paraffinic hydrocarbons. Results also show that between the two catalysts, p_d/c showed a higher activity towards hydrodeoxygenation. Based on the type of components in the produced oil, the maximum fraction of hydrocarbons formed lay within the range of 72.03% and 72.78% green diesel, and 30.39 and 28.25% jet fuel using P_d/c and P_t/c catalysts respectively. Thus it can be concluded that waste date pits can be a promising springboard for the production of catalysts and biofuels (green diesel and jet fuel fractions).

11.8 An Evaluation of the Use of Midribs from Common Date Palm Cultivars Grown in Saudi Arabia for Energy Production

A research [119] has been conducted with the objective of evaluating the potentiality of use of date palm midribs as an alternative source for energy production. The climate change associated with cutting of forests together with the increasing demand of fuel wood on the national and global levels have motivated researchers to search for renewable resources as alternatives to wood harvested from natural forests, such as tobacco stems [120], rice waste [121], common reed [122] and vine prunings [123]. Saudi Arabia is one of the main countries possessing date palm cultivations (23 million date palms including 400 date cultivars [123]). As far as the suitability of any raw material for fuel wood and energy production is concerned, the chemical composition, heating value and density are the most important factors. Wang et al. [124] defined the heat value (MJ/kg oven-dry basis) as a measurement tool for the characterization of the basic thermo-chemical value of bio-based materials. It ranges from 17 and 22 MJ/kg based on oven-dried wood [125]. Lingo-cellulosic materials are composed of three main components of substances: cellulose, hemicellulose and lignin. In addition, they contain secondary components including extractives and ash. Higher extractives and/or lignin contents often contribute to higher heating values [126, 127], which may be attributed to their higher heating value of 18.61 $MJ.kg^{-1}$ [128]. Ash is regarded as an undesirable material in most wood industries, especially for energy production [126, 129]. The specific gravity of a lingo-cellulosic material is one of the most important physical properties. Denser species are preferable for fuel because of their high energy content per unit volume and slow burning rates [126, 129].

The date palm (Phoenix dactylifera L.) is an important element of flora in all the Arab countries [130] playing a pivotal role in the economic, social and cultural life in the Arab region and resisting the harsh climatic and environmental conditions [131]. The products of annual pruning of date palms may reach an average of 35 kg per palm.

These residues have been successfully used for the production of pulp and paper [125], particleboard [132], wood-plastic composites [133], wood-cement composites [134] and lumber-like products [130]. These date palm residues have been considered as an available source for wood industries in Saudi Arabia [135–137]. EL-May et al. [136] measured the gaseous and particulate pollutants during combustion of date palm wastes for energy recovery.

Three healthy 10–15 year-old palms of cultivars Barhi, Khalas, Khodry, Sukkari and Sullaj, grown at the Experimental Station for Research and Agriculture of King Saud University were chosen for the conduction of this research. Three date palm midribs were randomly selected from the pruning residues. Each midrib was divided into three zones (base, middle and top) according to [130]. Samples from the aforementioned locations in addition to the midrib bases were taken, air dried, reduced to small pieces, ground into meal and passed through a 40-mesh screen and retained on a 60-mesh screen and subjected to chemical analysis and determination of heating values. The extractives content was determined according to the ASTM D1105 [138] in 3 steps of 4 h each using a Soxhlet apparatus. The extractives content was calculated as a percentage based on the oven-dry weight basis. The cellulose content was calculated as a percentage of extractives- free meal as described by Nikitin [139]. Approximately one gram of extractives- free midrib meal was treated in a flask with 20 ml of 3% nitric acid and 25 ml of a solution of 3% sodium hydroxide, and boiled for 30 min. The residue was filtered in a Gooch crucible (G3), washed with warm water to obtain a neutral filtrate, oven-dried and weighed. The hemicellicellose content was determined according to [140]. Approximately 1.5 g of extractives- free midrib meal was treated with 100 ml of sulphric acid (2%), then boiled for 1 h under a reflex condenser and filtered in a Gooch crucible (G2). The residue was washed with 500 ml of hot distilled water to remove the acid. The contents were dried in an oven at 105 ± 2 °C, cooled in a desiccator and weighed. To determine the ash content 1 g of an oven-dried sample (40 to 60 mesh) was placed into a crucible and gradually heated, then ignited at 575 ± 25 °C in a muffle furnace for 6 h, or until all the carbon has been eliminated. The content of ash was calculated as a percentage of the residues based on the oven-dried midrib meal weight according to the Chemical Analysis and Testing Task Laboratory Analytical Procedure [141]. The higher heating value (HHV) was determined using a calorimeter based on the oven-dry weight. A Parr model 6300 oxygen bomb calorimeter (Parr Instrument Company, USA) was used to determine the HHV of the midrib on a dry weight basis according to the ASTM standard D2015 [142]. Approximately 1 g of an oven-dried ground sample (20–40 mesh) was pressed into pellets using a hydraulic pellet press and loaded into a Parr model 6300 oxygen bomb calorimeter. The calorimeter was calibrated using standard benzoic acid prior to analysis of the samples. No correction for nitric acid formation was included in the calculations of the heating value. For each material, nine specimens were combusted to estimate the heating value. The fuel value index (FVI) was calculated using the method of Batt and Todria [143] according to the equation: FVI = HHV* density/ash content. Dry ash-free fuel (daf) was calculated using the equation: daf = HHV (1 + ash content/100).

The analysis of variance (ANOVA) showed that the differences between the date palm cultivars for all of the physical properties were highly significant. The average maximum moisture content (MMC) varied from 284% for Sullaj to 378% for Barhi. The fibre saturation point (FSP) representing the point at which the cell wall is completely saturated with water, but no moisture is present within the cell lumen varied between a minimum value of 138% for Sullaj to a maximum value of 378% for Barhi midribs. It is clear from these results that FSP for date palm midribs is equivalent to tenfold that of wood [144, 145]. The basic specific gravity (SG) of the date palm midribs ranged from 0.24 to 0.29 for Sullaj and Khlas respectively, with an average of 0.27. Thus, the date palm midrib can be classified as "light" wood based on Panshin and DeZeeuw [144].

The statistical ANOVA analysis has also shown that the chemical constituents of the date palm midribs differed significantly between the researched 5 date palm cultivars. The total extractives content (TEC) ranging from 19% for Khodry cultivar to 22% for the Sullaj cultivar is much higher than common wood species. This can be explained by the open anatomical structure, which is accessible to the chemicals. Similarly the ash content ranging between 3.3% for Khodry cultivar to 5.9% for Khalas cultivar is much higher than those observed in most soft woods and hard woods, which do not exceed 1% in most species [146]. However the date palm midrib contents of cellulose, hemicellulose and lignin ranging from 42 to 46%, 25 to 30% and 26 to 31%respectively are similar to those found in wood species. These values are closed to the values obtained by other researchers [137, 147].

The lignin content was found to increase moving along the midrib from the base to the top ranging from 24 to 30%. The inverse trend was recorded for the ash content decreasing from 7.6% at the midrib base to 3.4 at the top.

Thus it can be concluded that the date palm midribs include higher total extractives (19.34–21.68%) and ash content (3.31–5.85%) as compared with soft woods and hard woods. Highly positive-significant correlations have been found between HHV and the lignin content and biomass/ash ratio, whereas highly negative-significant correlations were found between the HHV and the ash content and hemicellulose content. The relatively high heating values found for the date palm midribs (17.3–17.9 MJ/kg) prove that they are promising as an energy source. However, the high ash content of all the parts of the date palm midrib; especially the basal part represents a hindrance in their use as a source of fuel. Besides, the date palm midribs exhibited the lowest FVI values (97–336) as compared with the corresponding values published in the literature for different wood species.

11.9 Characterization of Date Palm Frond as a Fuel for Thermal Conversion Processes

A research [148] has been conducted for the characterization of the date palm midrib as a potential solid fuel for jet and power generation through various thermal conversion processes. The date palm is one of the oldest known fruit trees cultivated in North Africa and the Arabian region since 3000 BC [149]. These date palms are annually pruned shedding 14–20 kg of dry leaves being replaced by 12–15 new leaves [150–152]. Thus it can be estimated that in 2014 1.14 million hectares cultivated with date palms with density of 100–150 palm/ha render a crop of dry leaves around 1.6 to 3.4 million tons. The date palm leaf is morphologically similar to coconut and oil palm leaves. Therefore it is significant to evaluate the potentiality of use of the date palm midrib as a solid fuel.

To conduct this study date palm midribs were sourced from Medina (Rothanah & Ajwah varieties) and Jeddah (Jeddah & Sukkari varieties) in Saudi Arabia and from Atbara (Mishriq variety), Northern Sudan. These samples were individually prepared to undergo three essential tests to evaluate their fuel properties: ultimate analysis, proximate analysis and calorific value analysis. Upon delivery, all midrib samples were dried in a convection oven at 105 ± 0.5 °C for 24 h. Around 50 g of each sample was ground and sieved to 250 μm size for calorific value measurement and proximate and ultimate analyses. The calorific value (or heating value) is the amount of energy contained in a fuel and is determined by measuring the amount of heat produced by the complete combustion of a known quantity of it. In this research the calorific value for each sample was determined using an 1KA C-6000 isoperibol calorimeter according to ASTM D-5865–13 standard. Around 2 to 4 mg of sample was placed in the bomb vessel for calorimetric evaluation, which was conducted automatically by the device. The proximate analysis reveals the amount of moisture, volatile matter, fixed carbon and ash content in a fuel. It was conducted on the midrib samples using Pyris 1 TGA thermo-gravimetric analyser by PerkinElmer according to ASTM E1131-08 standard test method. Each ground sample with a known mass was subjected to heating up to 850 °C in an inert gas environment and then in an oxidant gas as the temperature rose to 900 °C. The TGA and derivative TGA (DTGA) trend generated by the test equipment for each sample was then examined for peaks to determine the amount of volatile matter, fixed carbon and ash based on mass loss, as well as the temperature peaks to determine the devolatilization stages during thermal degradation of the date palm midrib. The ultimate analysis is necessary for assessing the elemental components in a biomass in terms of carbon, hydrogen, nitrogen, sulfur and oxygen regarding its suitability as a fuel. The ultimate analysis of the date palm midrib has been conducted using 2400 CNS/O series II system by PerkinElmer where the elementary carbon, hydrogen, nitrogen and sulfur components in the samples were measured by weight percentage. The oxygen content was determined by the difference.

The calorific value is the amount of energy contained in a specific quantity of a fuel. The higher the calorific value of a fuel, the more energy will be obtained during its combustion. That is why, fuel with higher calorific value is more desirable for thermal conversion processes [153]. All the midrib samples gave calorific values ranging from16.2 to 16.9 MJ/kg and thus falling within the range of biomass materials from 15 to 20 MJ/kg and close to that of low rank coal such as peat and lignite. The obtained calorific values are in agreement with previously reported values [154]. Thus it can be confirmed that the date palm midribs enjoy the potentiality of use as a solid fuel for home and industurial applications.

As far as the ultimate analysis results are concerned the range of carbon, hydrogen and oxygen in the date palm midribs is comparable with that for typical biomass materials. The nitrogen content in all samples was found higher than that in peat and lignite coal and as much as in bituminous coal. This nitrogen content in fuel tends to produce nitrogen oxides: a typical toxic and undesirable combustion byproduct [155]. Similarly the sulfur content in all the midrib samples was also found higher than that in most biomasses. No_x emissions have to be regulated below the allowable discharge limit by legal requirement in most countries due to their contributions to climate change and greenhouse gas effects [156]. That is why the utilization of fuels with high nitrogen and/or sulfur may require additional downstream filtration to reduce NO_x and SO_x emissions and may pose a considerable impact to operating expenditures, and may be avoided by thermal power producers [157].

The proximate analysis results indicate a significantly high content of volatile matter in the date palm midribs implying their suitability for pyrolysis and gasification processes, where the yields of solid to liquid and gas products through the thermal conversion processes may be more valuable than the heat it produces from direct incineration due to the combustion problems associated with biomass with high volatile content [155, 158]. The amount of fixed carbon in the midrib samples was found comparatively lower than those in most biomasses. This may be explained by the low amount of lignin content in the midribs as compared with the other date palm products of pruning [142]. Meanwhile, the ash content in the midribs was found comparable with those values in the featured biomasses and lower than in peat, lignite and bituminous coals making the date palm midribs a highly appropiate fuel for continuous thermal processes, where ash removal and handling is a common technical barrier [159].

The thermal decomposition stages were identified based on TGA trends and DTGA peaks. The 1st active pyrolysis stage began at around 141 ± 4.5 °C where light hydrocarbon components and traces of moisture were released. This 1st active pyrolysis stage progressed until the temperature reached around 325 °C for all samples, where the 2nd active pyrolysis stage began resulting into the initial release of heavier hydrocarbon compounds and liquid products such as pyrolysis oil and light tar. The passive pyrolysis began at above 500 °C for all samples except for Rothanah and Jedah varieties, where the stage began at 457 °C and 478 °C respectively, signifying the high reactivity of the samples against rising temperature. Heavier carbon aromatic compounds were released and broken at this stage as the temperature rose. The solid residue at this point in all samples was mainly made of char, where it underwent combustion at temperatures above 600 °C except for Rothanah at 552 °C. All samples were found to have lost half of their initial mass at around 320–340 °C, where the 2nd stage active pyrolysis was ongoing signifying that more than half of the tested midrib samples' mass was composed of hemicellulose and cellulose-the two most reactive natural polymeric fibers that break down at this particular thermal degradation stage. These findings were verified with the reported hemicellulose and cellulose content in Sukkari variety: hemicellulose content was reported to be between 43 and 45% by weight and cellulose content ranged between 28 to 31% by weight [154]. The high reactivity of midrib samples at low temperatures releasing high amount of volatiles revealed their potentials for pyrolysis and low temperature gasification purposes. Thus it can be concluded that the date palm midribs satisfy the typical requirements as a solid fuel for thermal conversion processes in the heat and power generation sector. The date palm midribs enjoy the advantages of high availability, low cost and ease of moisture removal in arid and semi-arid regions and thus have high potentials to be used as a solid fuel for thermal applications in the Arab Peninsula and North African countries. However, the topics of appropriate methods of processing and storage should be addressed to make full use of the date palm midribs as a feedstock for fuel in the Arab region.

11.10 Characterization of Date Palm Fronds as a Fuel for Energy Production

A research [151] has been devoted to the study of the date palm midribs as an effective feedstock for energy production. The use of biomass as a source of energy is attractive because of its renewability nature. The energy in plants are obtained by photo-synthesis during the process of plant growth [160]. The date palm (Phoenix dactylifera L.) is a main element of flora in arid and semi-arid regions of the world [161]. The main dates producing countries are Egypt, Saudi Arabia, Iran, UAR, Pakistan, Algeria, Sudan, Oman, Libya and Tunisia [162]. The annual pruning activity of date palms results in huge quantities of byproducts predominantly treated as waste [163]. The date palm leaves are an important component of date palm byproducts that can be used in many industrial applications [161]. The productive age of a date palm may reach 100 years [162] and an average of 12–15 new leaves are being grown every year. The global production of date has increased from 7.4 million tons in 2009 to 8.3 million tons in 2014. The date palm products of pruning have found different industrial applications such as production of paper and pulp, composite materials and activated carbon [136].

To conduct this research samples of date palm midribs of Sukariah, Ajwah and Jeddah varieties were sourced from Madinah Al Munawwarah and Jeddah cities in Saudi Arabia, air-dried for 2 to 4 days, ground and sieved to a particle size of 250 μm. The moisture content, volatile matter, ash and fixed carbon were determined by proximate analysis [164] using Thermo-gravimetric Analyser (Pyris 1 TGA). During proximate analysis, the mass of substance was heated at a controlled rate and the mass loss was recorded as a function of time and temperature [165]. The ASTME1131-08 standard test method was used to set up for the parameters. During the analysis, the moisture content and volatile matter were removed by heating the samples up to 850 °C in an inert gas environment. Ultimate analysis is important to know the suitability of the feedstock for energy production. Thus the chemical composition in percentage was determined, i.e. carbon, hydrogen, nitrogen and sulfur using Series II CHNS/O Analyzer and a S8 Tiger Brucker X-ray Fluorescence equipment was used to determine the sulfur content. It is preferable for a feedstock to have a lower sulfur content to avoid formation of acidic compounds during its reaction with oxygen, water and oxidants.

A Leco AC-350 Bomb calorimeter was used to determine the heating values. The higher heating value was determined when the product of heat of combustion of a sample was determined relative to liquid water, whereas the lower heating value was determined when the heat of combustion of the sample was determined relative to the gaseous water. A SUPRA 55VP model Field Emission Scanning Electron Microscope (FESEM) analyser was used in combination with Energy Dispersive X-ray Spectroscopy (EDX) in non-conducting specimens for the study of microstructure of each sample [166]. SUPRA 55VP allows to study the surface of the samples examination with nanometer scale in either high vacuum or in Variable Pressure (VP) Model. FESEM was used to visualize very small topographic details of the surface of the specimens. Electrons which are produced by the field emission source, accelerate in a high electrical field gradient. A narrow scan beam is produced within the high vacuum column by primary electrons. Secondary electrons are emitted from each spot on the specimen. The secondary electrons are caught by a detector and produce an electronic signal. Finally, the signal is amplified and transformed to a video scan image that can be seen on the screen.

As far as the proximate analysis results are concerned, the characteristics of a biomass is affected by many factors including the weather, composition of the soil, method of harvesting, season of harvesting and humidity of the environment. Jeddah had the highest value of volatile matter amounting to 83%, whereas Ajwah had the lowest value of 78.2%. The petiole had the lowest value of volatile matter of 55.3% among the date palm biomass. According to published literature, the date palm midribs values of volatile matter content are comparable with other resources, such as sugar cane bagasse, oil palm midribs and western hemlock [64, 167, 168].

As far as the fixed carbon content is concerned Ajwah variety showed the highest value of 14.1%, whereas Jeddah midribs showed the lowest value of 5.2%. Bituminous coal showed the highest value of 57%. The ash content was highest in Jeddah midribs (11.7%) and lowest in Sukariah (3.6%). Among the date palm biomass the petiole exhibited the highest value of ashes (19.2%). Biomass with higher volatile matter has better combustion [169], whereas the biomass with higher ash content would have poor combustion.

The ultimate analysis results showed that the carbon content of the midrib samples are significantly low ranging between 36.3 to 37.9% as compared, for example with sugar cane bagasse (42%), oil palm frond (44.6%) and bituminous coal (73.1%). Concerning the hydrogen content, Sukariah showed the highest value of 5.70%, whereas Ajwah exhibited the lowest value of 5.38%. among all date palm midrib samples. For the nitrogen content, Jeddah has the highest value of 0.47% while Ajwah has the lowest value of 0.28%. Bituminous coal had the highest value of 1.4% nitrogen, which is the highest value compared to all the biomass [165, 168, 170].

Regarding sulfur content, Sukariah showed the highest value of 0.66%, whereas Ajwah had the lowest value of 0.33%. Higher content of the nitrogen and sulfur may result in harmful effect to the environment and thus additional treatment during the conversion process is needed.

Regarding the heating value Sukariah midribs showed the highest value of 16.8 kJ/kg, whereas Jeddah sample revealed the lowest value of 16.4 kJ/kg compared with the seed value of 18.97 kJ/kg and bituminous coal of 34 kJ/kg. The HHV difference of agricultural biomass wastes was not significantly high. Therefore it can be concluded that the date palm midribs can be used as an alternative feedstock for different energy conversion processes, such as gasification, pyrolysis and torrefaction. The presence of metalic elements can cause problems in the thermochemical processing systems and therefore require proper handling and treatment during process.

11.11 Fast Pyrolysis of Date Palm (Phoenix Dactylifera) Waste in a Bubbling Fluidized Bed Reactor

A study [171] has been conducted on the use of date palm residues as a source of energy. The date palms are predominantly grown in arid and semi-arid regions. The dates are either consumed as fruits or used in many foods processing industrials to produce date cookies, syrup, past, etc. [172, 173]. There are about 100–120 million date palms in the world, 70–90% of which are located in the Middle East and North African counties [172, 174]. In addition to dates the date palm residues have been historically used in building boats, shelters and shades as well as in the manufacture of a wide range of commodities such as food trays, rope, fish and animal traps, brushes and furniture. At present the date palm residues are either used for making compost, burnt in boilers to generate steam [175] or sent to landfills. In addition there are trials to use these materials for the manufacture of activated carbons as adsorbents [176, 177]. The date palm residues are mainly composed of cellulose, hemicellulose and lignin and can thus be used for biofuel production [154, 163]. In the past few years there has been a growing interest in the application of fast pyrolysis on solid biomass at temperatures above 500 °C to produce bio-oil and a permanent gas to be used for energy generation [178], as well as biochar that can be used as a soil conditioner [179–181]. Different studies have been conducted on pyrolysis of different biomass feedstocks for various applications (e.g. Heidari et al. [182], Mesa-Perez et al. [183], Fernandes et al. [184], Abu Baker and Titiloye [185], Madhu et al. [186], Acikalin et al. [187] and Choudhury et al. [63]). However, only a few examined the potential of date palm residues and its characteristics relevant to the fast pyrolysis process. The present study is devoted to the characterization of three date palm residues: leaves, empty fruit bunches and petioles and their fast pyrolysis in a bubbling fluidized bed reactor at 525 °C.

To conduct this study, the date palm residues were sourced from the American University of Sharjah campus (Sharjah, UAE), during the peak of pruning season (July, August). On the average each date palm produces 6–10 empty fruit bunches and 12–15 leaves. Each leave contains 120–240 leaflets [154, 163]. The products of pruning represent in the average 15–35 kg per date palm. The date palm residues in this study included leaves, petioles and empty fruit bunches. The samples were first air-dried for 2 days under direct sunlight with a peak temperature around 45 °C, then chopped into small pieces, milled using a cutting mill, type Retsch-SM 200 and finally sieved to a size 0.5–1.0 mm. The fast pyrolysis experiment was carried out at the European Bioenergy Research Institute in Aston University, Birmingham, UK.

The proximate analysis was carried out to determine the mass of moisture M, volatiles V, fixed carbon FC and ash in the biomass. The moisture and volatile mass content were determined by measuring the weight loss using a thermal analyser (TGA 4000 Instrument, PerkinElmer, USA). According to ASTMD7582 standard procedure, around 10–20 mg of biomass was placed in a crucible inside the thermal analyser furnace and subjected to programmed heating at the rate of 10 °C min^{-1} up to 900 °C. The obtained thermogravemetric (TG) curves represent the percentage weight loss, caused by decomposition of the surface functional groups during progressive heating. The differential thermogravimetric (DTG) curves were obtained from TG data. The estimated biomass moisture content was also confirmed by oven drying in air at 105 °C according to ASTMD3173 standard procedure [188]. All of the proximate analysis data were determined from triplicate measurements and expressed in mass percentage. The TG analysis was also applied to the produced bio-oil to gain an insight into the thermal decomposition behaviour of the product. The ash content of the feedstock was determined by combusting 1.5 g of biomass according to ASTM D3174-12 standard procedure [189]. The sample was placed in a porcelain crucible and combusted in a muffle furnace (Type: Vecstar furnace- model PF1) at a temperature of 550 ± 10 °C. Finally, once the mass percentage of moisture, volatiles and ash were determined, the percentage of fixed carbon was estimated by difference: FC% = 100–(M% + V% + Ash%).

The ultimate analysis was conducted to determine the mass of carbon, hydrogen, nitrogen, sulphur and oxygen in the biomass and the product bio-oil. The samples were analysed by an external company (MEDAC, UK) using a Carlo-Erba EA1108CHNS-O analyser employing complete oxidation followed by separation and quantification using chromatographic column and thermal conductivity detector (TCD), flushed by helium gas. The oxygen was determined by difference. The higher heating value (HHV) of the biomass and the fast pyrolysis products (bio-oil and biochar) were determined using a IKA-C1 static jacket oxygen bomb calorimeter [190].

The comparison of the proximate and ultimate analysis of the leaves, petioles and empty fruit bunches have shown great similarity, but with the leaves much higher ash was noticed. In comparison with other types of biomass and energy crops, the date palm residues can be classified as of high ash, high oxygen, low volatiles and average heating value. The fast pyrolysis product was found to consist of 38.65% bio-oil (including 10.39% reaction water), 37.23% biochar and 24.02% non-condensable gas. The GC–MS analysis revealed for the first time the detailed composition of the date palm pyrolysis oil consisting of at least 140 detectable compounds with the major ones being D-Allose (monosaccharide), phenols, catechol and apocynin. The latter two compounds are of particular interest due to their antioxidation characteristics. The bio-oil heating value was 20.88 MJ/kg, which falls within the low-intermediate range for most fast pyrolysis bio-oils. However the oxygen content was high, and this may have a negative impact on the oil stability and corrosivity. Future work will be directed to the application side of the pyrolysis products including the potentiality of use of bio-oil as a blend with biodiesel for combustion engines, as well as the long-term aging and stability of the bio-oil and potentiality of use of the biochar for soil amendment, especially in desert conditions.

11.12 Study on the Thermal Behavior of Different Date Palm Residues: Characterization and Devolatilization Kinetics Under Inert and Oxidative Atmospheres

An important study [191] has been conducted with the objective of investigation of the thermal behavior of date palm residues including: leaflets, rachises, trunk, spadix stems and date seeds. The use of agricultural residues as an energy source has attracted in the last decades the attention of researchers due to the advantages of economic feasibility and non-interference with the production of food [160]. The olive residues have received considerable attention of researchers [102, 192–197], whereas the date palm residues thermal behavior was not duely studied [198, 199]. The date palm is an important element of flora in arid and semi-arid regions of the world. A date palm may reach an age of 100 years and an average length of 20 m. An adult palm may have 100–125 green leaves, each composed of a rachis (53.4%) [200] and leaflets (46%) and 10–26 new leaves are being formed annually [201]. The number of date palms in the world can be estimated by 105 millions [161] and the annual production of dates moved from 6.5 million tons in 2004 to 7.4 million tons in 2009 [202]. Tunisia is currently among the 10 top dates producing countries and the first exporter of dates in the world possessing 4.5 million date palms [203]. The date palm residues include leaves, spadix stems, waste dates and date stones, in addition to trunks due to accidental death or renewal of date palm plantations. The date stones represent about 10% of weight of dates [204]. Assuming an average of 22 leaves per palm the annual pruning activity in Tunisia results in 198,000 tons of leaves [205]. Thus, there is a need for valorizing the date palm residues. In addition to the traditional forms of utilization of these residues [163, 206], there are new modern uses including preparing pulp and paper from leaves, as well as extraction of fibers for the reinforcement of composites [207, 208]. Date stones have been also used as biosorbents [209] and for the production of activated carbons by physical or chemical activation [210]. In a later paper the date stones have been used as a fuel [199].

To conduct this study date palm residues were obtained from a date palm oasis in Tozeur, Tunisia. The date palm leaflets were cut into small pieces and the date palm rachises, trunk samples, date seeds and spadix stems were air-dried for 2–3 days, milled and sieved to a particle size between 1 and 2 mm. The ultimate analysis corresponding to the elemental compositions of the samples was performed by Service Central d'Analyses (Vernaison, France). The proximate analysis was carried out using a thermogravemetric analyzer (CAHN121 thermobalance) with gas following upward through the furnace at 12 NL/h. The proximate TG method involved heating the samples under N_2 at a rate of 10 °C/min from room temperature to 110 °C with a hold uptime of 10 min to obtain the weight loss associated with moisture. Then temperature was ramped at 20 °C/min to 900 °C. The oxygen was then introduced into the furnace chamber to oxidize char formed and weight loss associated with this step was the fixed carbon. The remaining material after oxidation was the ash. The heating values were measured using a REKA adiabatic oxygen bomb calorimeter. The thermogravemetric analyses have been widely used to study the thermal degradation characteristics of agricultural residues [134, 211–220, 220, 221]. In this study the thermal behavior of the biomass has been studied under inert and oxidation conditions using a CAHN121 thermobalance. The TGA experiments have been performed at heating rates of 5 °C/min (gas flow of 12 NL/h) from room temperature to 900 °C. Slow heating rate was chosen to allow studying the thermal behavior under conditions where transport processes do not hide the study of chemistry effects [214]. Experimental tests were carried out in both inert and oxidative atmospheres to investigate different steps of thermal degradation of each biomass material and to determine reaction kinetic parameters.

The obtained results show that the tested samples have high content of volatiles, carbon, hydrogen and oxygen, whereas the relative contents of nitrogen and sulfur were low. As compared with biomass, these materials have typical composition [221]. The heating values (LHV) were found in the range from 15.2 to 19.0 MJ/kg, which are in the same order of magnitude as those for sawdust, olive solid waste, oil palm fruit bunches, Miscanthus, wood pellets and wood chips [214, 216, 217, 221]. It is interesting to note that among the date palm residues the date stones having the highest percentage of volatiles and fixed carbon and the lowest ash content enjoy the highest calorific values in terms of low heating value. The characteristic of bulk density is important in relation to the transportation and storage costs. The bulk density of date seeds was very high (656 kg/m^3), higher than that of wood chips (550 kg/m^3). Both the heating value and bulk density define the energetic density, which is the potential of energy available per unit volume of the biomass. The energetic density of the date seeds (11.4 GJ/m^3) was found much higher than other date palm residues and approaching that of wood pellets (12GJ/m^3). It can be thus concluded that the date palm seeds are the most attractive residue for energy production due to its high energetic density and thus low cost of transportation. But the heating values of the other date palm residues are high enough to overcome the problems associated with low energetic density. It was possible from TGA and DTG to determine the thermal behavior and to obtain information on reactivity. Thermal degradation of all samples has exhibited quite similar behavior except the date seeds. Thus the trunk and spadix stems were found to be the most reactive materials in inert conditions; while in oxidative conditions, the rachises are the highest. The activation energies corresponding to devolatilization regions under inert and oxidative conditions were 49.8, 45.2, 52.7, 50.9, 89.1 kJ/mol for date palm leaflets, rachises, trunk, date seeds and spadix stems respectively. Thus, the obtained results can be useful for the design of processing systems for the production of energy from date palm leaflets, rachises, trunks, date stones and spadix stems.

11.13 Evaluation of Date Palm Residues Combustion in Fixed Bed Laboratory Reactor: A Comparison with Sawdust Behavior

A research [205] has been conducted with the objective of assessing the combustion behavior of date palm residues. There are strong expectations that the fossil fuels such as oil, coal and natural gas will be depleted within the next 40–50 years [222]. Therefore, there is a growing interest among researchers to study the potentiality of use of biomass for energy production as a more sustainable substitute for fossil fuels, as well as for the rescue of the environment from CO_2 emissions [223]. In Tunisia huge masses of agricultural residues are annually available for use for energy purposes as a substitute for fossil fuels. The olive wastes have received considerable attention [194, 195, 197], whereas few studies were devoted to date palm residues.

The date palm plantations in Tunisia, as well as the whole date palm belt extending from Morocco in the far west to Iraq in the far east can be a source of different sustainable date palm residues including the leaflets, midribs, spadix stems, date seeds and trunks, available due to the substitution of palms with declined production yield. These residues have been successfully used in waste water treatment (e.g. removal of Cu (II) and Methylene Blue from aqueous solutions [209, 224]). Recently studies have been published on the thermogravimetric analysis under inert and oxidative atmospheres of date palm residues revealing that the date palm trunk was the most reactive material, whereas the date stones were the least reactive material [191]. Sait et al. [168] have also determined the pyrolysis and combustion kinetics of three date palm residues using thermogravemetric analysis. Kinetics has been also obtained for thermal degradation for seed, leaf and petioles. In addition the gasification of the date palm stones char has been investigated with CO_2 in a controlled environment using thermogravemetric analyzer (TGA) at temperatures ranging from 600 to 1000 °C [222]. Al-Omari has investigated the potential of date palm stones and stalks as an energy source [198, 199]. The date stones contained much volatile compounds as coal and the heat transfer rates per unit mass of the fuel in the same experimental conditions were in the same order of magnitude as tested coal [198]. However, few studies were concerned with quantifying the gaseous emissions from combustion of date palm residues [225–227] (see [136]).

To conduct this study samples of Deglet Nour date palm residues were sourced from Djerid region in Tunisia and air-dried during 2–3 days. The leaflets (DPL) were manually separated from the rachises (DPR) and cut to 1–2 mm in width and 4–5 mm in length. The rachises, trunk samples (DPT), spadix stems (FP) samples and date stones (DS) were ground and sieved to a size of 1–2 mm. As a reference for comparison natural pine sawdust 0.5–0.7 mm particle size was selected.

The ultimate analyses corresponding to the elemental composition of the date palm residues samples were carried out by Service Central d'Analyses (Vernaison, France) according to the relevant XP CEN/TS15014 standard method. Carbon, hydrogen and nitrogen compositions were performed by combustion at 1050 °C using an elemental analyzer. Oxygen and sulfur were determined by pyrolysis at 1080 °C and combustion at 1350 °C respectively. Proximate analysis was conducted using a Thermogravemetric analyser (CAHN 121 thermobalance). This involved heating the sample (under N_2) at a rate of 10 °C/min to 110 °C then holding for 10 min to obtain the weight loss associated with moisture. Then the temperature was ramped from 110 °C at a rate of 20 °C/min to obtain the weight loss associated with volatiles release. Air was then introduced to oxidize the carbon in the char and the weight loss is the fixed carbon. The remaining material after combustion is the ash. The high heating values (HHV) were measured following XPCEN/T5 15,103 standard methods using an adiabatic oxygen bomb calorimeter (IKA). The energetic potential for different residues was estimated basing on the calculation of the low heating values, bulk density, energetic density that is the potential of energy per unit of biomass volume. The combustion tests were carried out in a vertical laboratory scale furnace including a fixed bed fused silica reactor (internal diameter 37 mm) placed in an electrically heated oven. A mobile fused silica grid is placed in this reactor at the beginning of the isothermal zone which length is estimated to be 10 cm. A thermocouple is placed 5 mm under the grid to record the temperature. Samples are being introduced by means of a long stainless steel spoon. The reactor is heated until the selected temperature is reached, then the samples are overthrown on the grid simulating the biomass combustion in a real boiler. The device included also a gas cleaner (glass wool filter, dust filter) and a gas dryer. The gaseous emissions were made using gas analyzers. After leaving the reactor, a part of the exhaust gas flow was aspirated toward a flame ionization detector (COSMA graphite 655: range 1–10,000 ppm) to quantify the volatile organic compounds (VOC). The exhaust gases were cooled gradually when passing through the gas line. If the FID analyser is placed at the end of the gas analyzing system, VOC will be under estimated due to condensation and deposition phenomenon in the pipe. After being dried and cleaned, the rest of the exhaust gas flow passes through a set of analysers where mol fractions of CO_2, CO, NO_2 and SO_3 are measured by Rosemont infrared analyzers (BINOS 100 for CO and CO_2: range 0–6% and 0–10% respectively: and NGA 2000 per for NO, NO_2 and SO_2: range 0–1000 ppm). Paramagnetic analyser (OXYNOS 100: range: 0–25%) is used to quantify oxygen. The exhaust gases are diluted before analysis to have gas concentrations in the range of each analyzer. Combustion tests were performed at 600 °C under a flow rate of 30 and 60NL/h. Sample masses were close to 120 mg and each combustion test was triplicated to assure results reproducibility.

Referring to the results of ultimate and proximate analyses, the weight fractions of the different date palm residues were found of the same order as several biomasses [228, 229]. When placing the date palm residues in a Van Krevelen type diagram according to their H/C and O/C ratios, it was found that FP, DPT and DPR are located in the region of typical biomasses, whereas DS and DPL were located in the common region between refuse-derived fuels and typical biomasses [229]. The comparison between the date palm residues and conventional biomasses has shown higher chlorine content for DPR, FP and DPT samples with values above 1%. Thus future controls of this element in gas and particles in fumes are needed in order to reduce both corrosion impacts and emission factors of persistent organic pollutants as dioxins and furans [230, 231]. The low heating values (LHV) were found within the range from 15.2 to 19.0 MJ/kg, i.e. in the same order of magnitude as for olive solid waste, Miscantus, wood pellets and wood chips [214, 215]. Compared with data from literature concerning wood biomass and energy crops these materials have typical composition [134, 215, 218]. The date stones (DS) were found to have the highest percentage of volatiles (VM) and fixed carbon (FC) and the lowest ash content similar to sawdust.

The bulk density is an important characteristic of the biomass materials in relation to transport and storage cost. It was found that the bulk density value for DS is very high (656 kg/m^3); higher than that for wood chips (550 kg/m^3). In addition the energy density (ED) of DS was found 11.4 near to that for wood pellets (12 GJ/m3). Thus it can be concluded that the date seeds (DS) is the most attractive material for energy production because of its high energetic density and therefore low cost of transportation. The date seeds is one of the best biofuels enjoying the advantages of highest bulk density, calorific value and volatile matter content and the lowest ash content close to 1.2%. Its energy density (11.4 GJ/m^3) is much higher than other date palm residues. Although the highest values of LHV was obtained for date palm leaflets their high ash content (15.2%) represents a hindrance for their development as a biofuel since it may lead to corrosion problems in the combustion chambers. As biofuels the date palm rachis (DPR), date palm trunk (DPT) and spadix stem (FP) have very close energetic density and chemical composition. The high amounts of chlorine in DPR and DPT may introduce potential risks of corrosion in exchange boiler tubes and the formation of persistent organic pollutants as dioxin during combustion in district or domestic applications.

11.14 Chemical Analysis of Different Parts of Date Palm (Phoenix dactylifera L.) Using Ultimate, Proximate and Thermo-Gravimetric Techniques for Energy Production

A study [154] has been conducted with the objective of evaluating the potentiality of energy production from different date palm residues. There is a growing trend among researchers to discover new renewable resources as alternative materials to natural forest resources for energy production, such as corncobs [232], tobacco stems [120], rice waste [121], common reed [122, 233], vine pruning [123] and date palm midribs [119]. The date palm (Phoneix dactylifera L.) is one of the most important plantations in the Middle East known as a source of dates, as well as different residues used as construction materials [119, 234], lumber substitutes [130], particleboard [135, 235, 236], pulp and paper [237–239], in wood-cement composites [137], in wood-plastic composites [133, 240], oriented strandboard [241], pyrolysis applications [242], briquette production [243] and biochar production [89, 244].

The Kingdom of Saudi Arabia is considered as one of the main countries in date palm plantation producing 970,000 tons of dates annually from an area of about

150,000 ha [131, 245]. Each date palm has about 6–10 bunches and 12–15 new leaves are formed annually and thus big quantities of date palm residues are available. An average of 35 kg of date palm residues per tree is generated annually [246]. Thus the annual wastes resulting from annual pruning in Saudi Arabia are estimated by about one million tons that are either burnt or sent to landfills. As far as the use of date palm residues for energy production is concerned, lignin and higher extractives content contribute to a high heating value [126, 127], whereas ash is consided an undesirable material [126, 129]. The heating value and density of the raw material are very important factors affecting the over all energy production output [124–126, 132, 237]. The chemical energy of a solid fuel is stored in the forms of volatiles and fixed carbon [247]. Although there are few studies that investigated the suitability of date palm residues for energy production [119, 136], there is very little information on what components of date palm residues are more suitable for charcoal and energy production [136]. Abed et al. [248], tested the pyrolysis of date palm stones and stalks by thermogravimetric analysis. Babiker et al. [242] performed a thermogravimetric analysis of date palm stones (DPS) of 6 cultivars and concluded that DPS contain much volatile compounds. El May et al. [136] measured the gaseous and particulate matter obtained during the combustion of 4 parts of date palm residues and their energy recovery and concluded that DPS were the most convenient for energy production.

To conduct this study the Sukkari cultivar has been chosen as an important variety cultivated in most areas of Saudi Arabia. Thus five defect-free Sukkari date palms at the Experimental Station For Research and Agriculture of King Saudi University with ages ranging from 10–15 years were used for experimentation [119]. Acacia tortilis wood was used as a control sample. After air-drying, the residues were cut into small particles and ground using a Wiley mill and sieved to two sizes: 20–40 mesh for the determination of fuel characteristics and 40–60 mesh for the determination of the chemical and ultimate analysis of the specimens.

The total extractives of the date palm residues were determined using a Soxhlet apparatus according to ASTM D1037 [249]. The cellulose, hemicellulose and lignin were determined by the meal free-extractive method based on the oven-dry weights for each residue according to ASTM D1037 standard [249]. The ash contents were calculated according to a standard method developed by National Renewable Energy Laboratory (NREL, Golden, Co, USA) [250]. The ultimate analysis of the residues including carbon, hydrogen and nitrogen contents was carried out using a CHN Elemental Analyzer (model 2400 Series II Perkin Elmer, Waltham, MA, USA) at 925 °C [251]. The oxygen content was calculated by difference. To conduct the proximate analysis, the residues samples were oven-dried to a constant weight at a temperature of 103 $\pm$ 5 °C. Samples having 40–60 mesh particle size were used to determine the moisture content (MC), volatile matter content (VMC) and ash content, whereas the fixed carbon content (FCC) was calculated by difference [249]. The fuel characteristics of the residues including heating value (HV) based on dry basis (db) and dry ash-free (daf) as well as fuel value index (FVI) were determined. HV was determined according to ASTM D2015-85 [142]. Approximately one gram of oven-dry ground sample with a particle size ranging between 20 and 40 mesh was converted into pellets using a hydraulic pellet press and loaded into a model 6300 oxygen bomb calorimeter (Parr, Moline, IL USA). The calorimeter was calibrated before the analysis of samples using benzoic acid as a standard. No correction for acid formation was included in the heating value calculations. Six samples from each residue were combusted to estimate the heating value. The fuel value index (FVI) was calculated using a modified method of Bhatt and Todaria [143]: FVI = HVx density/ash content. Dry ash-free fuel (daf) was also calculated. The date palm residues in addition to Acacia tortilis wood were rated according to the positive and/or negative impacts on the bioenergy content and environmental impact according to Munalula and Meincken [252]. These ratings were calculated as the sum of all values in the column divided by the number of measured properties ($\Sigma/8$).

This study aimed to evaluate the potentiality of use for energy production of 8 Sukkari date palm residues, namely palm trunk (PT), palm frond base (petiole) (PFB), palm leaflets (PL), fruit stalk (FS), fruit empty bunch (FEB), date palm stone (DPS) and palm leaf sheaths fiber (LSF) taking Acacia tortilis wood (AT) as a reference. First of all the chemical composition of the 8 residues were found significantly different leading to significant differences in their fuel characteristics. These residues were found to have medium to high cellulose content (33–48%) and lignin (26–40%) and low to medium hemicellulose content (13–31%). The total extractives were (8–33%) and the ash content (1–15%). The volatile matter content ranged from 74.3% for PL to 87.5% for FEB; fixed carbon content ranged from 10.5% for PL to 17.6% for PT and the ash content ranged from 1.4%(DPS) to 15.2% for PL. The heat values of the residues varied from 15.47 MJ/kg for PFB to 19.93 MJ/kg for LSF. However, the heat values based on ash-free dry weigh had a wide range from 16.5 MJ/kg for FEB to 22.6 MJ/kg for PL due to the large variation in the ash content (1.3%-11.6%). The fuel value index of DPS was higher (2078) than the value for A.tortilis wood (1170) and other date palm residues. Concerning the ranking of date palm residues, the date palm stones (DPS) showed the best value (1.9), followed by leaf sheaths fiber (LSF) (2.5), while the palm frond base (PFB) showed the poorest rating (6.3). Thus it can be concluded that the date palm seeds and leaf sheaths fiber are the most suitable among the date palm residues for energy production.

11.15 Ultrasound Assisted Oil Extraction from Date Palm Kernels for Biodiesel Production

A research [253] has been conducted with the objective of determining the effects of ultrasonication and solvent types on oil extraction and biodiesel production from date palm kernels. In this research three solvent types were used: hexane, isopropyl alcohol and ethanol for oil extraction from date palm kernels and ultrasound was applied for 5 min to 25 min at 5 levels using an ultrasonic dismembrator. Biodiesel was produced using transesterification of oils with methanol and potassium hydroxide. The composition of the fatty acid methyl ester was determined using gas chromatography. It was found that the ultrasonic-assisted hexane oil extraction provided the highest oil yield by extracting 85% of the total available oil present in the date palm kernels. The biodiesel samples produced from oils extracted with and without ultrasonication had similar physical and chemical properties. Thus it can be concluded that ultrasonication has a potential to enhance the industrial processes by reducing the oil extraction time and energy.

11.16 Hydrothermal Pretreatment of Date Palm (Phoenix dactylifera L.) Leaflets and Rachis to Enhance Enzymatic Digestibility and Bioethanol Potential

A research [254] has been devoted to the hydrothermal pretreatment of date palm leaflets and midribs to evaluate their potential for bioethanol production. The increasing world demand for energy, together with the uncertainty of oil reserves and the associated climate change of their use has pushed the search for alternative energy resources as a transportation fuel [255, 256]. There is a growing evidence that the lignocellulosic biomass is the most sustainable alternative for fossil fuels [256]. The lignocellulosic materials in the form of grasses, wood and agricultural residues having the advantages of renewability and geographic availability possess great potentials as a feedstock for ethanol and other liquid fuels [257]. The past 7 years have witnessed a wide commertialization of cellulosic biofuels in North and South America, Europe and China [258–261]. In the Middle East, the researchers have focused on the use of marine (e.g. macroalgae) biomass [262], halophytes (e.g. Salicornia bigelovii) [263] and agricultural residues (e.g. date palm fruit and sap) [11, 264] for the production of bioethanol. This region is distinguished with wide date palm plantations possessing 70% of the 120 million world's date palms [174]. It can be estimated that each date palm renders 10–30 dried leaves annually, each (including leaflets and a midrib) of 2–3 kg mass [265]. Thus each date palm renders 50 kg of residues per year amounting to over 4 million tons annually on the world level. This treasure of renewable materials is most dominantly treated as waste being open-field burnt or sent to landfills. Thus, there is a necessity to explore avenues for valorization of date palm residues, as for example their use as a feedstock for bioethanol production. Within this context a pretreatment step is needed to reduce the recalcitrance of this lignocellulosic feedstock by removing hemicellulose from the microfibrils to expose the crystalline cellulose core to be hydrolyzed by cellulytic enzymes [255]. The hydrothermal pretreatment (also known as autohydrolysis or liquid hot water pretreatment) is much preferred as an eco-friendly green processing technology as compared with acid treatment and thus avoiding corrosion problems, acid recycling, and the formation of neutralization sludge. Another advantage is that hydrothermal pretreatment tends to result in lower inhibiting hydrolyzates which can decrease yield in the subsequent fermentation process [266]. An optimum hydrothermal condition (Peterson et al. [267]) has been reported on wheat straw at 195 °C for 6–12 min, yielding 70% hemicellulose recovery and 93–94% cellulose recovery in the fibers and approximately 89% of the cellulose was converted into ethanol by commercial cellulase mixture. Kumar et al. [268] reported that almost 100% cellulose and 92% hemicellulose were recovered from hydrothermally pretreated switch grass and more than 80% of glucan digestibility was achieved at 190 °C for 20 min. Arides Romani et al./ [269] reported that 94% of polysaccharides were recovered in the hydrolysis media as mono- or oligosaccharides when using Eucalyptus globulus wood pretreated at 220 °C. Despite the wide interest to convert biomass to energy the scaling up of experimental project to commercial levels is not easy [270]. A necessary starting point is to estimate the energy production potential (e.g. bioethanol) to reveal the specific dynamics and interrelationships between environmental and socioeconomic systems [271]. Wild-type S.cerevisiae strain readily ferment C6 sugars including glucose, mannose, fructose, and glactose as well as the disaccharides sucrose and maltose [272]. On the other hand, other of the most abundant sugar monomers from biomass D-xylose and L-arabinose (C5 sugars) require either extensive metabolic engineering of S.cerevisiae [272] or other fermentative organisms such as Kluyvermyes marxianus [273], Zymomonas mobilis [274], and Pichia stipitis [273].

To conduct this research, date palm leaves were sourced from Abu Dhabi in 2013. Leaflets were separated from the rachises, dried and stored before use. The dried material was milled using a knife mill (IKA, 10 MF Basic) to pass through a 1 mm screen. Sequential Soxhlet extractions with water and ethanol were performed based on National Renewable Energy Laboratory (NREL) protocol [275]. Structural carbohydrates and lignin of extractives-free date palm leaflets and rachis before and after hydrothermal pretreatment were subjected to two-step acid hydrolysis according to the analytical procedure of NREL [276]. The hydrolyzates were analyzed for sugars using High Performance Liquid Chromatography (Agilent 1260 Infinity Bio-Inert Binary LC). The Hi Plex-H column (Agilent) and refractive index detector (RID) were used to determine the concentrations of glucose, xylose and arabinose at 65 °C using 0.005 MH_2SO_4 as the mobile phase (eluent) with a flow rate 0.06 ml/min. The hydrothermal pretreatment were performed at 10% w/w dry matter loading at 4 temperature levels (180, 190, 200 and 210 °C). Processing time was maintained at 10 min. Combined severity factors corresponding to each condition were calculated [22]. The fiber composition analysis was conducted by subjecting the pretreated fibers to two-step acid hydrolysis [276] to determine sugar recovery. The process was carried out following the same protocol as in the case of the extractives-free raw leaflets and rachises. The enzymatic convertibility assay based on commercial Cellic CTec2 (117 FPU/mL, protein content 194 mg protein/L) and Cellic HTec2 enzymes (Novozymes A/S, Denmark) was used to determine the efficiency of the pretreatment. Protein concentration of enzymes was determined as described by Bradford [277]. The hydrolysis was performed according to National Renewable Energy Laboratory protocol [278] using 100 g/L dry biomass loading and 15 FPU cellulase/g dry matter of biomass (with cellulase-to-hemicellulose ratio of 1:9). The process was performed in the presence of 50 mM citric buffer (pH5) at 50 °C and samples were shaken at 150 rpm for 72 h. Glucose released during the enzymatic hydrolysis was quantified using HPLC at the same operating conditions as applied in the acid hydrolysis samples (described above).

The research results showed that high glucan (>90% for both leaflets and rachis) and high xylan (>75% for leaflets and > 79% for rachis) recovery were achieved. Under the optimal conditions of hydrothermal pretreatment (210 °C/10 min) highly digestible (glucan convertibility, 100% to leaflets, 78% to rachis) and fermentable (ethanol yield 96% to leaflets, 80% to rachis) solid fractions were obtained. The fermentability test of liquid fractions proved that no considerable inhibitors to saccharomyces cerevisisae were produced in hydrothermal pretreatment. Proceeding from the high sugar recovery, enzymatic digestibility, and ethanol yield it can be concluded that the production of bioethanoal by hydrothermal pretreatment from date palm residues is feasible.

11.17 Pyrolysis of Date Palm Waste in a Fixed-Bed Reactor: Characterization of Pyrolysis Products

A research [72] has been conducted on the characterization of pyrolytic products of date palm residues. Regarding the increasing demands on energy and the concern about the climate change associated with the fossil fuels there is a growing interest in the renewable sources of energy in the world and particularly in Tunisia [279]. Within this context, the agricultural residues are one of the feedstocks of primary importance as an alternative source of energy due to their advantages of high availability and renewability, low cost and sustainability [280]. In Tunisia one of the main countries in the world in date palm plantations the date palm residues amount to an annual quantity of 200,000 tons [281]. The traditional avenues of utilization of these residues (e.g. composting, incineration, animal feed and manufacture of decorative objects) are not satisfactory [282]. Therefore, it is necessary to discover new avenues of utilization of these renewable resources such as conversion to biofuels via the pyrolysis process. The pyrolysis process is a thermal cracking of the biomass in an inert atmosphere at temperatures ranging from 300 to 700 °C to produce useful liquid biofuel (bio-oil), solid biocombustible fuel (biochar) and renewable syngas. In the past few decades, a lot of research work has been directed to the pyrolysis technology [178]. This technology has been applied to different lignocellulosic biomasses including rice husk [282], switch grass [283], sunflower seed press cake [284], birch wood [285], Arundo donax [286], a-cellulose [287], sawdust [288] and corn cobs and corn stover [52]. The pyrolysis technology has been applied to the date palm residues mainly to obtain low cost biochar to be used for soil amendment and as an adsorbent for organic and non-organic pollutants. As examples Mahdi et al. [289] studied the use of date palm seeds in slow pyrolysis process at temperatures 350, 450, 550 and 650 °C, whereas Hamdoun et al. [290], studied the pyrolysis of date palm empty fruit bunches to produce activated carbon. Other studies investigated the thermal behavior of date palm residues via TGA and compared them with other lignocellulosic materials [205,

242]. Other researches focused on the preparation of activated carbon from date palm seeds at final temperature 600 °C; the obtained activated carbon was valorized for the Cu(II) and methylene blue removal from sewage water [291]. Few studies were devoted to the production of biofuels from date seeds [291]. Therefore, the objective of this study is the investigation of the main characteristics of the obtained products from the pyrolysis of the date palm residues to evaluate their potential as a feedstock for renewable energy and chemical industry.

To conduct this study, four date palm residues samples were obtained from the National Institute of Arid Zone(IRA-Kebili, Tunisia): midribs, leaflets, empty fruit bunches and spathes. The samples were finely crushed to small pieces with sizes from 2 to 4 mm and air-dried. The pyrolysis experiments were conducted on a laboratory scale pyrolysis pilot plant. The pyrolysis setup used in this study was described in detail in [292, 293], under the operational condition: 500 °C as final temperature, 15 °C/min and 300 g mass initial of the used sample. The percentages of bio-oil and biochar yields were determined by the equation:

$$Biofuel(\%) = \frac{weightofcharorliquid(g)}{weightofdryrawmaterial(g)} X100$$

Applying the principle of mass conservation, the gas production yield has been deducted as the difference between percentage of char and liquid yields from the total percentage of 100%

Syngas(%) = 100−(bio-oil)%−(bio-char)%.

The moisture content of date palm residues samples was measured with the method of (AFNOR XP CEN/TS 14,773). The volatile matter (VM) content was measured according to ASTM method D-1762–84, 1990. The ash content was obtained using the standard methods: AFNOR XP CEN/TS14775; ASTM method D-1762–84 and ASTM, 1990. The fixed carbon content was obtained as follows: FC = DM−VM + Ash). The lignocellulosic components (cellulose, hemicellulose and lignin of studied samples were determined according to the method used by Sun et al.[294]. The thermogravimetric analysis (TG-DTG) was applied to determine the thermal degradation behavior of biomass weight to the change in temperature by using thermogravimetric analyser (SIITG/DTA7200), under N_2 atmosphere in the temperature range 25–800 °C and heating rate of 10 °C/min. The elemental composition (CHN-O) of date palm residues, bio-oil and biochar were determined using an elemental analyser (LECOCHNSTRuSpec); the O content was determined by difference. The morphology of date palm residues and their biochars were determined by scanning electron microscopy (FEI Quanta 450 FEG apparatus). Fourier transform infrared (FTIR) analysis was conducted to identify the chemical functional groups present in the date palm residues, bio-oil and biochar. The FTIR spectra were collected in the spectral range of 4000–400 cm^{-1} using FTIR-ATR Perkin Elmer FTIR 100 spectrometer.

The proximate and ultimate analyses have shown that the date palm spathes had the least ash content of 2.4%, whereas the leaflets had the highest ash content (11.58%). But in general the date palm researched residues had high volatile matter content and ash fairly compared with those found in the literature for lignocellulosic materials converted into biofuels using pyrolysis (e.g. Lee et al. [295] found values of 72.88% and 11.105 for volatile matter and ash respectively for empty fruit bunches). The studied residues had slightly low amounts of carbon (from 39.95% to 43.65%), hydrogen (from 7.19% to 7.59%) and nitrogen (from 0.15% to 0.19%) and high amounts of oxygen (52.7–52.74%). The calculated high heat values ranged from 17.88 to 19.09 MJ/kg for all studied residues, which are in the same order as corn stover (HHV = 19.01 MJ/kg) [296], palm kernel shell (HHV = 19.72 MJ/kg) and empty fruit bunches (HHV = 19.65 MJ/kg) [295]). These values are low for a commercial fuel. The bio-oil yield ranged from 17.03 wt% for leaflets to 25.99 wt% for empty fruit bunches. Concerning the biochar, the highest yield 36.66 wt% was obtained for leaflets, whereas the lowest one (31.66wt%) was obtained for the spathes, while the syngas production varied from 39.1 wt% for midribs to 46.31 wt% for leaflets. As a conclusion of this study, the bio-oil which represents moderate amounts of carbon and hydrogen compared to petroleum-based fuels, could be used as biofuels after grading. The biochar could be used as biocoke in industrial applications. The presence of CH_4 and H_2 in significant proportions in the gaseous mixture gives the obtained syngas good combustion properties.

11.18 Seawater as Alternative to Freshwater in Pretreatment of Date Palm Residues for Bioethanol Production in Coastal and/or Rid Areas

A study [297] has been conducted with the objective of evaluating the technical feasibility of using seawater instead of freshwater in the pretreatment of date palm leaflets for bioethanol production. The lignocellulosic biorefineries are one of the most promising alternatives for fossil oil. But one of the obstacles of proliferation is the excessive utilization of fresh water (1.9–5.9 m^3 water per m^3 of biofuel [298–301]), which may be in shortage in arid and semi-arid regions, where the date palms are usually grown. Thus the assessment of the potentiality of use of nonpotable water resources as reaction media for biorefineries seems to be of great significance [302]. Within this context the research trials of use of seawater in succinic acid fermentation processes seem to be encouraging [78, 303–306]. In one of these researches fermentation was conducted on a wheat-derived medium and natural seawater and succinic acid was produced with a concentration of 49 gL^{-1} with a yield of 0.94gg^{-1} and a productivity of 1.12 gL^{-1} h^{-1} [78, 303–306]. In the process of depolymerization of several amorphous and crystalline celluloses using commercially available enzyme cocktail (Accellerase 1500), only slightly lower production rates (≈90%) were observed in seawater media relative to those in pure citrate buffer [78, 303–306]. A pretreatment is necessary to reduce lignocellulosic biomass recalcitrance by depolymerizing lignin and solubilizing hemicellulose. The removal of hemicellulose from the microfibrils is thought to expose the crystalline cellulose core, which can be then hydrolyzed by cellulytic enzymes [255]. The ionic liquids (ILs) have emerged as alternative solvents to overcome the challenges of recalcitrance of crystalline cellulose [307–310]. However, the implementation of ILs has not been commercially applied due to the high cost, energy-intensive recycling and environmental concerns of ILs toxicity and unbiodegradability [270, 311, 312]. Therefore new concepts (e.g., distillable ILs) or methodologies (e.g. spontaneous product segregation) have been developed to reduce the complexity of ILs recycling and product isolation [313–315]. The action of inorganic salt added during the strong inorganic acid catalysis of cellulose breakdown is analogous to that of ionic liquids. The addition of saline water (e.g. NaCL, 30 wt%) or concentrated (≈5 x) seawater, organic-acid-catalyzed cellulose depolymerization was able to proceed efficiently under mild reactions (T = 100–125 °C) [316]. A variety of inorganic salts have been directly used in hydrothermal pretreatment for the degradation of xylose monomer and xylotriose or for the decomposition of corn stover and sweet sorghum bagasse [317–319]. Chloride salts including KCl, NaCl, CaCl, $MgCl_2$, $FeCl_2$ and $CuCl_2$ have been tested individually [317–319]. Sulfate salts like $FeCl_3$ and $Fe_2(SO_4)_3$ have been also investigated [317–319]. The above inorganic salts showed significantly improved degradation of xylose monomer and xylotriose and decomposition of hemicellulose and even cellulose.

The date palm (Phoenix dactylifera L.) is an essential fruit crop in most of Middle East countries [320], historically connected with sustaining human life and traditions of people. The Middle East countries possess 70% of the 120 million date palms in the world [174]. The annual date palm residues in the world can be estimated by 4 million tons [321]. In this previous study it has been showed that hydrothermal pretreatment could be a promising method for valorizing the date palm residues by producing biofuels [321]. In this study artificial seawater has been used to replace fully the freshwater in the hydrothermal pretreatment of date palm leaflets to produce bioethanol. The physicochemical changes of the biomass were analysed by attenuated total reflection Fourier transform infrared (ATR-FTIR) spectroscopy and X-ray diffraction (XRD). Comparison of the digestibility and fermentability of pretreatment solids and the inhibition of pretreatment liquids were conducted to assess the possibility of utilizing seawater as a sustainable alternative to freshwater in biorefineries. The challenges of seawater application in lignocellulosic biorefineries and proposed solutions were also discussed. Comparisons between pretreatment with freshwater and seawater were examined under the same conditions in [321] (leaflets were pretreated at T = 200 °C/t = 10 min showing high enzymatic digestibility [(95.6 ± 2.9)% glucan-to-glucose conversion] and no inhibition was observed to saccharomyces cerevisiae in ethanol fermentation).

The results of this study confirm the feasibility of replacing freshwater with seawater in the hydrothermal pretreatment of date palm leaflets to produce bioethanol. However a lower crystallinity of cellulose has been observed after treatment with seawater rather than freshwater. Pretreatment by using seawater produced slightly lower digestibility of solids (glucan-to-glucose conversion) in enzymatic hydrolysis than pretreatment by using freshwater. But there was no significant difference in the bioethanol yield. Moreover, the fermentability test showed no significant difference in the ethanol yield between liquids from pretreatment by freshwater and seawater.

Thus it can be concluded that seawater could be a promising alternative to freshwater in biorefineries processing lignocellulosic materials. But the challenges of applying seawater in biorefineries cannot be ignored, such as the corrosion caused by seawater and the negative effect on biological conversion processes, which could bring inactivation to cellulolytic enzymes and inhibition to fermentation [303–306, 322, 323]. In addition the residual salts discharged into wastewater streams could be an issue for wastewater treatment.

11.19 Bioethanol Production from Date Palm Fruit Waste Fermentation Using Solar Energy

A study [324] has been devoted to the realization of an experimental solar batch fermenter for the production of bioethanol from date palm waste (DPW). The energy production and utilization is an issue of great significance for the development of countries [218]. The energy potentials in Algeria can be estimated by 3.7 million tons of oil equivalent (TOE) coming from forests and 1.3 million TOE coming from agriculture and urban wastes [325]. It has been estimated that 3–6 MW electricity can be obtained from the discharge of Qued-Smar in Algiers [326]. The increasingly use of biogas, particularly produced from landfills has commenced with national pilot research projects initiated by Renewable Energy Development Center (CDER, http:ilportial.cder.dz) studying methane recovery from cull dates and processing of fruits byproducts. Among the bioenergy alternatives including bioethanol, biodiesel and other biogases, the bioethanol has higher burning effect € 2.25/kWh and lower environmental effect [327]. According to the Renewable Fuels Association, USA, Brasil, EU and China are expected to share 50.3, 25.5, 4.5 and 2% of the world production of bioethanol in 2014 respectively [328]. It is evident that the countries with agronomic-based economy are more appropriate for bioethanol production [329]. The production of bioethanol captures the attention of researchers as having many advantages and a mean for realization of sustainable fuel for future [330]. The use of non-edible lignocellulosic materials for the production of biofuels, such as bioethanol is considered as the main sustainable alternative for satisfying the future demands on energy [331]. The bioethanol has various advantages [332]: favorability to mix with gasoline [331], prevention of engine knocking and premature explosion due to its high octane index of 110, wider flammability, higher heat vanishing and speed of flame [333]. In addition bioethanol has 35–40% lower energy content as compared to gasoline and 35% of higher oxygen content making the combustion cleaner and with lower emission of toxic substances [334]. Bioethanol helps as well to reduce CO_2 emission up 80% as compared with gasoline and thus promoting cleaner environment for the future [335]. In addition bioethanol is less unsafe than gasoline and its higher flash point (13 °C) gives better storage treatment ability and it is less flammable. The auto-ignition temperature is between 333 and 423 °C and ethanol relatively requires an elevated temperature than gasoline to be an auto-detonation fuel; hence ethanol vapor will be only combusted later than gasoline without a forced detonation[336].

Many microorganisms (yeasts) are used to convert biomass into bioethanol [337]. For example S.cerevisiae yeast has been used to produce 48.8 ml/kg of bioethanol from pineapple byproducts and 120.7 ml/kg from sorghum juice [338]. Within the expected increase of demand on energy for homes, industries and transportation the expected contribution of biofuels will grow from 50 EJ/year in 2012 to more than 160 EJ/year in 2050 [339]. The major cost factors in production of bioethanol are the biomass and energy costs [340]. After the oil crisis of 1970, biofuels were perceived in many countries as a realistic alternative to fossil-based fuels. Today, bioethanol is seen as the main biofuel for the future [341].

One of the most sustainable feedstocks for bioethanol production is the date palm residues. For example, the date palm grove area in Algeria has increased by 69% from 101000 ha in 2000 to 169361 ha in 2009 with a total of 18.7 million date palms [342]. The date palm is the main constituent of flora in Algeria creating a microclimate for other cultivars of cereals and vegetables. Algeria is one of the top ten countries in the world in production of dates including Egypt (17.2%); Saudi Arabia (13.7%); Iran (13%); UAR (9.8%); Pakistan (9.6%); Algeria (9%); Iraq (7.2%); Sudan (5.4); Oman (3.5%) and Libya (2%) [343]. The annual quantity of date palm wastes (DPW): during picking, storage, commercialization and conditioning, caused by the fungus, infestation by insects or simply due to their quality and thus unconsumed by humans can be estimated by ~ 30% of dates produced in Algeria [76]. There is not at the present industries to valorize these DPW, which are predominantly treated as waste. DPW can be considered as a sustainable material base for the production of different metabolites, biopolymers, organic acids, antibiotics, amino acids, enzymes, bakery yeast, as well as butanol and hydrogen [11, 344–346].

To conduct this study DPW of Hchef, Kacien and other varieties of date's scraps of the cattle food were sourced from Algerian Sahara, washed, plunged in a water bath, rubbed carefully and rinsed with pure water to eliminate sand pebbles, insects and reminder plants, petted to separate seeds, ground and imbibed in hot water at 90 °C to 95 °C to facilitate sugars extraction. About 200 g of DPW was diluted into 800 ml of tap water and sulfuric acid was simultaneously added and adjusted to ensure the pH between 4.3 and 4.7 to inhibit the bacterial growth and favor overgrowth of the yeast [347]. The fermentation medium was then inoculated with 1 g/L of S.cerevisiae and reactivated within 60 to 90 min under an ambient temperature of 25–30 °C in an aqueous solution in glucose with 12% V/V. A batch fermenter was designed and installed to operate efficiently by using a solar water heater in order to reduce the cost of the bioethanol generation process. The fermenter was realized within the South Society of Metallic Construction (ECOMES) [348], located in Adrar. The fermenter consisted of a tank of 50 L built in double walls of galvanized stainless steel and thermally insulted by fiberglass wool of 5 cm thickness. The heat exchanger was placed between the two walls of the tank. The lid of the tank was quite tight and contained a hole for evaluation of gasses and a second hole was provided with a copper pipe and contained a thermostat for adjusting the substrate temperature close to 30 °C. The fermenter was equipped with a manual agitator shaft and connected to a temperature data logger model Fluke 2635A with an internal memory card. The hot water was stored in a tank of 200L and its circulation was ensured by a hydraulic pump controlled by a thermostat. Experimentation was performed during the cold season of the year, the 1st week of January 2015. During the fermentation process, the density sugars consumed, pH and the alcohol degree of the DPW juice were controlled. The glucose evolution was controlled using Dubois method given in Michel DuBois et al. [349]. Reducing sugar (RS) and total sugar (TS) were assessed by titration using a spectrometer UV. Saccharose content was estimated as:

Saccharose (%) = (TS − 0.95RS)%

The pH was measured by a digital pH-meter model Mettle Toledo methods [350, 351] and the date's juice temperature during the alcoholic fermentation was recorded using thermocouples K type, connected to the date logger. After 72 h of alcoholic fermentation, the substrate juice to extract the bioethanol was used to filter. At the beginning of the distillation process, the degree of alcohol was measured every 30 min, and once the process got slowed, the alcohol was recorded every one hour. The process was stopped when the degree of alcohol became very weak. The distillation temperature was kept at about 78 °C.

The results of this study indicate that DPW constitute a favorable medium for S.cerevisiae growth, due to its sugar content and is thus considered an attractive feedstock. It is thus technically feasible to produce bioethanol using the solar batch fermenter at relatively moderate cost. The DPW distilled juice produced the highest bioethanol concentration of about 90° with an acceptable productivity of 3.47 ml/kg/h assessing a scale efficiency 33%. These results represent a strong support to continue R&D in the renewable energy field. It is thus necessary to start to build semi-pilot and pilot fermenters and investigate new methods, microorganisms and other byproducts to improve the quantity of ethanol produced and to reduce the energy consumption during the bioethanol process transformation to improve the economics of bioethanol production.

11.20 Biogas Production from Date Palm Trees Residues

A study [124] has been conducted with the objective of exploring the anaerobic digestive technology for processing of date palm residues. The rapid population growth and the associated increase of demand on energy have push the researchers to seek new sources of renewable energy. One of the most available and sustainable feedstocks for renewable energy is the date palm residues. These residues, annually available with huge quantities, are most dominantly treated as waste: being open-field burnt or sent to landfills. Therefore the exploitation of these residues for the production of methane gas should be seriously considered. Within this context the date palm residues should meet the nutritional requirements of the microorganisms in terms of energy sources and various components needed to build new cells, as well as the various components needed for the activity of microbial enzyme systems, such as trace elements and vitamins [62]. A carbon to nitrogen (C/N) ratio between

20 to 30 is regarded optimum for the anaerobic digestion process. If the C/N ratio is too high, methanogens will rapidly consume the nitrogen for meeting their protein requirments and will no longer react with the left over carbon content of the material resulting into low gas production. If conversely C/N ratio was very low, nitrogen will be liberated and accumulated in the form of ammonium ions (NH_4). The presence of excess NH_4 will increase the pH of the biodigestate in the digester and thus a PH higher than 8.5 will start showing toxic effect on methanogens population [352]. The bioenergy crops with high contents of cellulose, hemicellulose and lignin must be broken down into their corresponding monomers sugars, so that microorganisms can utilize them in the energy conversion process through biological route. Therefore it is beneficial to apply pre-treatment to break up the complex structure of cellulose and make it more accessible for digestion. The pretreatment processes are classified to mechanical or physical, chemical and physic-chemical and biological [352]. The effectiveness of hot water pretreatment grows with increasing processing temperature and time. Liquid hot water pretreatment increases the amount of biogas and decreases the residence time in the fermenter [233]. Mukhtar M.Ashur and Iman M.Bengharbia studied the effect of temperature and pH on biogas production from organic fraction of municipal waste experiments at 35, 45, 55, 65 °C and pH was controlled at 7. They have concluded that 35 °C gave the highest biogas yield, whereas the lowest yield was at 65 °C. The value of $pH < 6.0$ inhibited gas production and the methane gas percentage was 45%[353]. Lili Yang et al. investigated the effect of pH on biogas generation and concluded that a significant increase of methane yield could be achieved by pH adjustment for food waste and that a pH of 8.0 gives the maximum methane yield of 171.0 ml/g of total solids, which is 7.57 times higher than the pH uncontrolled group [354].

Z.Ismail and A.Talib studied the potentiality of anaerobic co- digestion for biogas production using the date palm residues. They showed that the volume of produced biogas was significantly affected by inoculum addition, pretreatment of waste materials and temperature conditions. The thermophilic conditions improved the biogas yield by ~ 23% [355]. Jaafar studied the potentiality of using an Iraqi date and fruit named Zahdi pulp waste from syrup production as a substrate for biogas production at thermophilic digestion with activated sludge as inoculum. Methane was produced with a yield of 570 ml/g volatile solids of substrate. Addition of 1% yeast extract solution as a nutrient increased methane yield by 5.9% [356]. Aljuhaini et al. studied the effect of pretreatment and operating conditions on the yield of biogas through anaerobic digestion of date palm residues using NaOH for pretreatment [198].

To conduct this study midribs, fruit bunches and rotten date were sourced from the southern oases of Libia from the area of Wadi Al Hayat valley, located in midland of Fizan region. Each date palm yields annually an average of 25–35 kg of rotten dates in addition to (12–15) leaves with an average weight of 20 kg, and a number of (7–10) fruit bunches, with on average weight of 15 kg. These huge quantities of residues are treated as waste. They are either open-field brunt or disposed by burying in the desert sand.

The midribs and fruit bunches were thermally pretreated to break the resistant layer of lignin. Water under high pressure and temperature were used for periods of

15 and 30 min and then introduced to an electrical mixer for further size reduction. The inoculum used was a mixture of 270 g of caw's manure and 135 g of food waste with a ratio 2:1. The mixture was then mixed with water to produce a solution of 1600 ml and introduced to a batch reactor under mesosphilic conditions (35–38 °C) until gas production started. In this research two-liters glass batch digester was used. It had three openings on the top. The middle was used for introducing raw materials and then for fixing a temperature sensor, the other was used for collecting generated gas and the third for fixing a pH sensor. The reactor was installed in a water bath equipped with an electric heater with a thermostat to control temperature. In the first experiment 3 batch reactors were used and named as R1, R2 and R3. To study the impact of thermal pretreatment and the pH, the three reactors had a total solid content of 10%. Reactors R1 and R2 contained dry waste thermally pretreated for 30 min. PH was controlled to be less than 6.5 and not higher than 8.0 for R1 and not to be less than 5.5 for R2, whereas the mixture in R3 was thermally treated for 15 min and the pH was controlled to be not less than 6.5. Concerning the second experiment fresh date palm wastes were tested in order to make comparison between the potentiality of yielding biogas from fresh wastes and dry wastes. Three reactors R1, R2 and R3 were used. R1 was filled with 10% of dry date palm residues that were thermally treated for 30 min with adding 1 g of NH_4CI in order to maintain a source of Nitrogen. R_2 was filled with 10% of fresh date palm residues without adding any nitrogen source. The three reactors were operated under mesophilic conditions.

Concerning the effects of thermal pretreatment and pH, the substrate treated for 30 min gave higher biogas production. When pH was maintained within 6.5–8 to avoid high acidic condition higher biogas production was obtained for the same pretreatment time. Retention time was 30 days. As far as the effects of using dry or fresh residues and the addition of nitrogen source are concerned, it has been observed that the production of biogas from fresh residues with nitrogen resource (NH_4CI) gave the best biogas production as compared to dry residues with nitrogen source and fresh residues without nitrogen source. In addition, biogas production from dry residues with nitrogen was better than from fresh residues without nitrogen. Regarding biogas analysis results, it was found that the biogas produced from dry residues had low concentration of methane due to the high percent of carbon to nitrogen. After the addition of chemical nitrogen source an improvement in the concentration of methane was observed. Conversely, the biogas produced from fresh residues showed high concentration of methane, which is attributed to the balance percentage of carbon to nitrogen ratio in the residues as compared to dry material. Based on the results of this study, it can be concluded that the heat treatment of date palm residues is very useful to improve the production of biogas and that the percentage of methane gas in the produced biogas has reached 48% for fresh residues with addition of a nitrogen source. PH in the digester should be kept in the range 6.5–8 for more efficient biogas production. It was found that the retention time for fresh residues inside the digester is less than the retention time for dry residues. Date palm residues contain a high ratio of carbon compared with nitrogen, especially dry residues and the addition of nitrogen source is required.

11.21 Biogas Production by Anaerobic Digestion of Date Palm Pulp Waste

A research [356] has been conducted with the objective of evaluating the potentiality of using Iraqi Zahdi date waste as a resource for biogas production. The date palm is an essential element of flora in the Middle East and North Africa region. The annual date palm residues can be estimated by about 3.3 million tons including midribs, leaflets, spadix stems and leaf sheaths fibers [357]. Iraq possesses ~ 9 million date palms, located in the middle and southern parts of the country. The dates are fruits with a high sugar content (55–70%wt) [358], traditionally used for food, production of sweets, sweet syrup (Dibs in Arabic), vinegar and alcoholic products. The date wastes, only partially blended in animal fodder, are most dominantly treated as waste and sent to landfills. Thus it is more rational to use these date wastes being mainly cellulosic compounds with sucrose sugar, fats and minerals as a feedstock for biogas or biofuel production via fermentation process. Many papers [50, 359–361], described the dates physiochemical characteristics. But very few research has been conducted on the date palm biomass as a biofuel resource.

This study has been devoted to investigate the usage of date palm biomass as a source of energy. The Zahdi date cultivar has been chosen because of its abundance in Iraq representing 60% of the country production [358]. Cellulosic biomass can be converted into biofuel by catalysis with dilute acid, concentrated acid, or enzymes known as cellulases. Anaerobic digestion of organic matter can be applied to both liquids and semi-solid wastes [362].

To conduct this research Zahdi date biomass was prepared to be in the same conditions as it is usually expelled from syrup (Dibs) industry in Iraq. 200 g of date fruits harvested in 2008 from the middle area of Iraq were pitted and boiled with one liter distilled water in a beaker, cooled, filtered with fabric mesh, rinsed with water and filtered again to get rid as much as possible from its water soluble sugar content. The resulting biomass weight was 51% of the original pulp weight and kept in 4 °C refrigeration for further testing. The Waste Management Lab/Corneal University (USA) methodology for anaerobic digestion, was used in this research work. The biomass from Zahdi dates was treated with anaerobic digestion in thermophilic conditions at 55 °C. 300 ml glass jars fitted with wireless computer controlled modules from Ankom were used as the small batch reactor. The remote sensing modules measured the total gas pressure produced during the period of fermentation experiment.

In order to keep the fermentation medium close to neutral pH (7.0–7.5), which is necessary for hydrolysis of fatty acids, 50 mM Beryllium Sulfide buffer solution (N, N-bis-2 hydroxyethyl-2-amino-ethane sulphonic acid) was used to prepare incubation medium. One ml of 5 g/L yeast extract solution was added as the only external nutrient for the inoculums. No trace elements were added to fermentation solution since the date biomass is normally rich with many minerals such as Potassium, Calcium, Boron, Iron, Manganese, Magnesium and Cobalt [358]. Blank samples of deionized water were placed under experiment conditions to observe pressure drop/increase due to temperature variation during sampling and adding substrate.

Methane yield was normalized to the volatile solids of biomass and original carbon added to the solution.

To conduct initial substrate testing, samples of substrate and inoculums were tested for total solid and volatiles solid contents to verify the best food/biomass ratio of mixing. Proportion of C:N was also measured with metal analyzer from CE Instruments, NC 2500 to calculate initial carbon and nitrogen added to the fermentation batch. Three grams of substrate were weighed in dried and cooled crucibles, then placed in 105 °C Fisher isotemp oven (0–200 °C) for 24 h and then cooled in the desiccators for 1 h. After weighing crucibles were placed in 550 °C lindberg 5B Muffle, furnace (0–600 °C) to maintain total solid, volatile solid and ash content of the substrate.

To follow the anaerobic digestion procedure Beryllium Sulfide solution (50 mM) was used to prepare two groups of triplicate samples for digestion. The first group included 100 ml Beryllium Sulfide plus 5 ml of activated sludge with addition of 1% yeast extract solution as the nutrient, whereas the second group was only Beryllium Sulfide solution with 5 ml of activated sludge. A third group of jars containing deionized water only were incubated as well and treated exactly as the samples. The jars were flushed with nitrogen to remove any traces of oxygen from the fermentation medium. After incubation for 48 h to exclude any biogas produced from the activated sludge, incubation jars were opened to add 0.5 g of substrate and flushed again with nitrogen. Gas samples for Gas Chromatography analysis were taken every 48 h through the experiment period. Liquid gas chromatography from Gow-Mac Series 580, dual detector for Methane/ CO_2/N_2 measurement with supleo 80/100 Porpak Q, 6' x ¼" column was used. Helium was used as the carrier gas with 35 ml/min flow rate. The hydrogen percentage in the total gas produced was measured with another Gow-Mac series 580, dual detector with Hydrogen Analysis Column: Supelco 60/80 Carboxen 1000, 5' x $\frac{1}{8}$". Samples from liquid phase were filtered with 0.2 μm paper filter then 1:1 with 2% hydrofluoric acid and tested for VFA with Hewleet-Packard 5890 Series 580, dual detector with Hydrogen Analysis Column: Supelco 60/80 Carboxen 100, 5' x $\frac{1}{8}$". Samples from liquid phase were filtered with 0.2 μm paper filter, then treated 1:1 with 2% Hydrofluoric acid and tested for VFA with Helweltt-Packard 5890 Series 11(with autosampler) and VFA column: Supelco Nukol, 15m X 0.53 mm X0.5 μm film thickness. The experiment took place until the total gas pressure was constant for 3 consecutive days, then jars were opened for final measurements of pH, VFA and SCOD.

The results of the study indicate that the volatile solids of substrate and inoculums was 39.82%, 2.37% respectively with a ratio of 16.8:1. The nitrogen content of the substrate was found 2.35 indicating the demand for extra amount of nutrients to provide nitrogen for bacteria growth in the fermentation batch. A total gas pressure with 67% Methane was produced from date pulp waste fermentation with a yield of 0.57 Lit for each gram volatile solid of the substrate. The addition of 1% yeast extract solution as a nutrient increased Methane yield in liters by 5.9%. The high volatile solids content in the date palm biomass compared to the inoculums indicates a high potential of biogas production from a small amount of biomass. Given the great abundance of date palms in the Middle East and North Africa there are great future potentials for biogas and biofuels in a commercial scale.

11.22 A Study of Biogas Production from Date Fruit Wastes

A study [363] has been devoted to the evaluation of potentiality of biogas production from date palm fruit wastes. Dates are of great significance in human nutrition due to their rich contents of essential nutrients including carbohydrates, salts, minerals, dietary fiber, vitamins, fatty acids and protein [264]. Carbohydrates are the major chemical element of the date mainly including glucose, fructose and small amounts of cellulose and starch [364]. In addition the date wastes could be a sustainable material base for microbial fermentation potential for bioenergy production [365]. In Iraq, only a small portion of date wastes are used in animal feed, whereas most of the date wastes of 9 million date palms are treated as waste and discarded [356]. To produce biogas from digestible materials by anaerobic digestion, the decomposition of organic matter occurs in 4 processes simultaneously: hydrolysis, acidogenesis, acetogenesis and methanogenesis [354]. The maximization of gas production can be attained by controlling these stages in a suitable manner of equilibrium between their rates. Many researchers have proposed different methods for biogas production from agricultural and industrial wastes [366–370]. Assirey [366] reported that dates contain moisture ranging from 10 to 22%, total sugars 71.2–81.4%, protein 1.72–4.73%, lipid 0.12–0.72% and ash 1.89–3.94% on dry weight basis. Methane with a yield of 570 L/kg volatile solids from Zahdi date biomass as a byproduct from syrup industry in Iraq was reported by Jaafar [356]. She claimed that the addition of 1% of yeast extract solution as a nutrient, at a thermophilic temperature of 55 °C, increased methane yield in liters by 5.9%. Ismail and Talib [371] examined the recycling of date palm residues as a feedstock for biogas production. They proved that the effect of inoculums addition was more significanant than alkaline pretreatment of date palm residues (petiole, rachis, leaflets) and the biogas recovery from inoculated residues exceeded its recovery without inoculation by 140% at the mesophilic conditions. Biogas production with a value of 141.66 L/kg volatile solids was obtained by the co-digestion of inoculated date wastes. Shafiei et al. [79] studied the effect of the date palm fruits composition on the fermentation process. His results indicated that successful hydrolysis and removal of date palm fruits fibers has led to fewer problems in fermentation processes.

The present study investigated the production of biogas from date palm fruit wastes by measuring the gas volumetric flow rates directly. Samples of Digal date fruit wastes in their final stage of maturity (with hard texture) were sourced from stores of Diyala province in Iraq. The measured values of particle size, moisture content, total solid, volatile solids and pH were determined according to the standard method 1684 [372]. The results were found: 2–5 mm, 8.8%, 91.2%, 88.75% and 7 ± 0.5 respectively. Removal processes of pits have been carried out before the samples being weighted, shredded, and loaded inside the reactor. In each run of experiments, 1 kg of wastes was diluted with distilled water to prepare substrates mixing ratios of 0.5, 0.25, 0.15 and 0.1 (weight of fruit waste/weight of water (w/w). To study the effect of recycled digestate addition on the biogas recovery, substrate mixture with maximum biogas production selected from the experiments was fed with digested materials. Residual liquid wastes (effluents from digestion process), which were collected from previous digestion processes were adjusted to be 25% of the substrate content.

A lab-scale digestion system was used in this research. The bioreactor is integrated with a control unit that contains temperature control, a pH measuring system that ranges between 2 and 10, and gas cylinder and regulator. The main components of the bioreactor are: jacket heating system, temperature sensor and pH probe (inchtrode N75F, Hamilton Robotics, Reno, USA). To remove its water vapor, the produced biogas is passed through a closed container filled with formaldehyde solution. In addition, there are 5 filters installed on the biogas outlet pipe to ensure the complete removal of water vapor before the gas enters the flow meter. All the parameters of the bioreactor operation conditions such as set point and measured values for temperature, pressure and pH were displayed on a touch screen installed on the face of the control unit and is used to input all required parameters for testing samples. All recorded date for measuring biogas parameters such as volumetric flow rate, temperature, and pressure were recorded and transformed from the gas flow meter to a laptop. Alicat's flow vision software program was installed to get the required data out of the flow meter readings when it is connected to the laptop. The software was programmed to record readings every 3S. Determination of the major constituent, methanol gas, of the produced biogas was carried out by using Gas Chromatography (GC14B, Shimadzu corp, Kyoto, Japan), with SUS Packed Column Porapak Q FID detector. The carrier gas was argon with the flow rate of 30 ml/min.

After the samples were pitted, weighted, mixed with a proper amount of water and placed in the digester, eight mixtures of substrates were divided into two groups. Each group contained 4 samples of substrates having ratios of 0.5, 0.2, 0.15 and 0.1 (w/w). They were subjected to anaerobic digestion at 37 °C for mesophilic conditions and 55 °C for thermophilic conditions. The discharge process of biogas was carried out every 3–5 days and the data was saved on a laptop. The pH readings were recorded after and before each anaerobic digestion process. The experiment was stopped at the end of the biogas production to measure the amount of produced gas. For analyzing the total volume of produced biogas, Alicat Scientific M-series digital mass flow meter (O-200 SCCM) was used to measure volumetric flow rate, pressure and temperature of the produced biogas. All volumes of the produced biogas were corrected at the standard conditions of 25 °C and 1 at.

Concerning the effect of solid mixing ratio the results of the study showed that under mesophilic temperature of 37 °C, the highest biogas yield was achieved in the case of 0.15 w/w (182 L/kg volatile solids mass), whereas the lowest biogas yield was in the case of 0.5 w/w (84 L/kg volatile solids mass). These results, compared with the results of Jaafar [356], showed a lower amount of biogas production due to using different types of date fruit wastes and inoculum materials. During anaerobic digestion of 0.5w/w, the response of the reactor to produce biogas was slower than those of other samples. The pH values for all experimental tests ranged between 7 ± 0.5 before the digestion process and 3.2 ± 0.6 after it. To study the effect of temperature the same experiments of mesophilic conditions (37 °C) have been reported under the thermophilic conditions (55 °C). The highest biogas yield was in the case of 0.15 w/w (133 L/kg volatile solids); whereas the lowest biogas yield was in the case of 0.5 w/w (16 L/kg volatile solids). It was observed that the increase of temperature above mesophilic temperature decreases the biogas production. As far as the effect of recycled digestate is concerned, the research results showed that the use of recycled digestate has improved the production of biogas by 12%. This can be explained by the increase of active bacteria since the recycled materials are a rich source of microorganisms. These results are in agreement with other results of previous works [354, 373]. Thus it can be concluded that the date palm fruit wastes are a suitable source for biogas production and that a mesophilic system is the best option for producing biogas from date wastes. A maximum biogas production of 203L/kg volatile solids was achieved for a solid concentration of 0.15 (w/w), when the substrate was mixed with recycled digestate at 25% of the substrate content.

11.23 Biogas Production from Raw and Oil-Spent Date Palm Seeds Mixed with Wastewater Treatment Sludge

A study [374] has been conducted with the objective of assessing the potential impact of date seeds on biogas production from date seeds and wastewater treatment sludge mixtures. The date palm is one of the most essential elements of flora in the Middle East and North Africa. UAE is among the top date fruit producers in the world [375]. The date fruit seed constitutes about 10% of the fruit weight [376]. These seeds are rich in biodegradable organic matter, including protein, carbohydatres, fibers and lipids [359, 377]. These seeds are most dominantly treated as waste. However a small proption of these seeds is used as an animal feed supplement, coffee substitute, a source of edible date seed oil and fiber, pharmaceuticals, cosmetics, oil for biofuel production and production of activated carbon [46, 49, 359, 377, 378]. There is strong evidenece that date seeds possess antibacterial activity against various Gram negative and Gram positive pathogenic bacteria [379–381]. The antibiotic active ingredients in seeds, roots and leaves originate from their antioxidant contents [382–384] and date seeds are known to be rich in antioxidants, especially phenolics [385–388].

The locally available waste products in UAR include organic solids waste, wastewater treatment sludge and date palm (Phoenix dactylifera L.) residues. Previous studies [16, 389] were devoted to produce oil from the date seeds for the production of biodiesel as an alternative to fossil fuels. Only one study [390] was concerned with using date seeds in a mixture with cattle manure for the production of biogas, where the ratio of 10% was found optimum.

To conduct this study seeds of Khalas and Khudari cultivars were selected. After deseeding the date fruits were washed to remove fruit flesh, air dried and then oven dried at 50 °C for 24 h and crushed using a steel hammer and converted into powder using a household blender with a 2000 rpm rotational speed. However, the blender was unable to crush the seeds and the ground seed material was sieved and sorted into three sizes: 1:18–3.75, 0.6–1.18 and 0.425–0.6 mm. Samples of the smallest size range were analysed in the laboratories of Sharjah Municipality for fat, carbohydates, proteins and volatile and ash contents. The ash and volatile solids contents were measured following standard methods. The fat contents were also verified through oil extraction from the date seeds. Oil was extracted from the date seeds using an automatic soxhlet extractor (Buchi B-811 LSV, Flawil Switzerland) with a solvent mixture consisting of methanol, chloroform and water (volume ratios of 2:1:0.8). The oil-spent date seeds were recovered from the extractor, dried and then used for biogas production. The primary wastewater treatment sludge was sourced from a nearby domestic wastewater treatment plant. The total solids of the primary sludge was adjusted to approximately 2% and stored in a freezer in small containers until used. The volatile solids after total solids adjustment reached 1.64%. The preparation of the date seeds/sludge mixtures was conducted by thawing the frozen sludge from containers, mixing and adding the predetermined date seed quantity and then adjusting the pH if necessary to approximately 7.3. Samples (30 ml) of prepared date seed/sludge mixtures were then placed in 50 ml serum bottles using a wide mouth pipette, then capped using rubber stoppers and aluminium sealing rings. The serum bottles were incubated at 35 °C and total and methane biogas production were measured weekly for 14 weeks. Seven date seed to sludge total solids dry weight ratios were prepared: 0, 2.5, 5, 7.5, 10, 20 and 40%. For example, 0.2 g date seeds added to 100 ml sludge with 2% solids (i.e. 2 g total solids in 100 mL sludge) amounts to 10% date seeds/sludge ratio. The total biogas was measured by displacing distilled water filling an inverted capped bottle with an outlet tube leading to a graduated cylinder. The collected biogas in the bottle was then passed through a second inverted displacement bottle containing NaOH solution to remove CO_2 from the biogas and measure methane. Ammonia was measured by using Hack Method 4500-NH3 and Hack Spectrophotometer DR 2800 (Loveland Colorodo, USA). The pH was measured using a pH Meter (10N Meter Model #3345, Keison Products, Chelmsford, Essex, England). Four identical serum bottles were prepared for each date seed ratio of each type of the 4 types of date seeds (raw Khalas, oil-extracted Khalas, raw Khudari, oil-extracted Khudari). The smallest size ground date seeds, 0.425–0.6 mm were used in all experiments except those that were meant to determine the impact of size on biogas production.

After 14 weeks of incubation the biogas production, expressed in terms of the date seed/sludge ratios, was in the following order: 10% > 7.5% ≈5%≈2.5%≈0% > 20% > 40%. The size of date seed particles did not significantly affect biogas production. The specific biogas production was in the range of 370–390 mlg^{-1} volatile solids for the 0–10% seed/sludge ratios, 245 mlg^{-1} at 20% and 120mlg^{-1} for volatile solids at 40%. The relatively low biogas production from the 20 and 40% seed/sludge mixtures indicated inhibition, which was also shown by the low pH in the mixtures following digestion. Oil extraction from the date seeds reduced, but did not fully overcome, inhibition of biogas production from the 20 and 40% mixtures. The inhibition reduction due to oil extraction was indicated by observed discrepancy between biogas production and digestion efficiency, combined with observed low pH after digestion. The discrepancy between biogas production and solids digestion, suggest that oil extraction reduced inhibition of the acid forming but not that of the more sensitive methane forming microorganisms.

References

1. Ramzy Abu Ayana (2017) Palm by-products. Al Rajhi Endowment Administration (in Arabic).
2. Hamed El-Mously, Mohamed Saber (2019, May 20) medium density fiberboards from date palm residues a strategic industry in the Arab world. By-Products of Palm Trees and Their Applications.
3. Al-Zuhair S, Ahmed K, Abdulrazak A, El-Naas MH (2013) Synergistic effect of pretreatment and hydrolysis enzymes on the production of fermentable sugars from date palm lignocellulosic waste. J Ind Eng Chem 19(2):413–415. https://doi.org/10.1016/j.jiec.2012.09.022
4. Cybulska I, Brudecki GP, Zembrzuska J, Schmidt JE, Lopez CG-B, Thomsen MH (2017) Organosolv delignification of agricultural residues (date palm fronds, Phoenix dactylifera L.) of the United Arab Emirates. Appl Energy 185:1040–1050. https://doi.org/10.1016/j.apenergy.2016.01.094
5. Murphy JD, McCarthy K (2005) Ethanol production from energy crops and wastes for use as a transport fuel in Ireland. Appl Energy 82(2):148–166. https://doi.org/10.1016/j.apenergy.2004.10.004
6. Dürre P (2007) Biobutanol: an attractive biofuel. Biotechnol J 2(12):1525–1534. https://doi.org/10.1002/biot.200700168
7. Zhang Y, Li Q, Su J, Lin Y, Huang Z, Lu Y, Sun G, Yang M, Huang A, Hu H, Zhu Y (2015) A green and efficient technology for the degradation of cellulosic materials: structure changes and enhanced enzymatic hydrolysis of natural cellulose pretreated by synergistic interaction of mechanical activation and metal salt. Biores Technol 177:176–181. https://doi.org/10.1016/j.biortech.2014.11.085
8. Edhilvia J. Campos, Nasib Qureshi, Hans P. Blaschek (2001) Production of acetone butanol ethanol from degermed corn using clostridium beijerinckii BA101. In Biotechnology for fuels and chemicals.
9. Qureshi N, Saha BC, Dien B, Hector RE, Cotta MA (2010) Production of butanol (a biofuel) from agricultural residues: part I—use of barley straw hydrolysate☆. Biomass Bioenerg 34(4):559–565. https://doi.org/10.1016/j.biombioe.2009.12.024
10. *Annual Statistical Agricultural Data for Arab Republic of Egypt*. (2017). FAOSTAT.
11. Abd-Alla MH, Elsadek El-Enany A-W (2012) Production of acetone-butanol-ethanol from spoilage date palm (Phoenix dactylifera L.) fruits by mixed culture of clostridium acetobutylicum and Bacillus subtilis. Biomass Bioenerg 42:172–178. https://doi.org/10.1016/j.biombioe.2012.03.006
12. Baghlaf AO, Abou-Zeid AA, Yassein M (1980) Influence of nitrogen and phosphorus sources on the fermentative production of acetone and butanol by clostridium acetobutylicum. Zentralblatt Für Bakteriologie, Parasitenkunde, Infektionskrankheiten Und Hygiene. Zweite Naturwissenschaftliche Abteilung: Mikrobiologie Der Landwirtschaft, Der Technologie Und Des Umweltschutzes 135(6):515–522. https://doi.org/10.1016/S0323-6056(80)80071-1
13. Foda MI, Joun H, Li Y (2010) Study the suitability of cheese whey for bio-butanol production by Clostridia. J American Sci:39–46.

14. Macrelli S, Mogensen J, Zacchi G (2012) Techno-economic evaluation of 2 and generation bioethanol production from sugar cane bagasse and leaves integrated with the sugar-based ethanol process, Biotechnol. Biofuels:1–18.
15. Abd-Alla MH, Zohri A-NA, El-Enany A-WE, Ali SM (2015) Acetone–butanol–ethanol production from substandard and surplus dates by Egyptian native clostridium strains. Anaerobe 32:77–86. https://doi.org/10.1016/j.anaerobe.2014.12.008
16. Amani MA, Davoudi MS, Tahvildari K, Nabavi SM, Davoudi MS (2013) Biodiesel production from Phoenix dactylifera as a new feedstock. Ind Crops Prod 43:40–43. https://doi.org/10.1016/j.indcrop.2012.06.024
17. Pearl GC (2002, September 22) Animal fat potential for bioenergy use In: Bioenergy. The Biennial Bioenergy Conference, Boise.
18. Peterson CL (1986) Vegetable oil as a diesel fuel: Status and research priorities. Trans, ASAE.
19. Hu J, Du Z, Li C, Min E (2005) Study on the lubrication properties of biodiesel as fuel lubricity enhancers. Fuel S0016236105000700. https://doi.org/10.1016/j.fuel.2005.02.009
20. Junior JS, Mariano AP, Angelis DF (2009) Biodegradation of biodiesel/diesel blends by Candida viswanathii. African J Biotechnology.
21. Xia S, Baker GA, Li H, Ravula S, Zhao H (2014) Aqueous ionic liquids and deep eutectic solvents for cellulosic biomass pretreatment and saccharification. RSC Adv 4(21):10586. https://doi.org/10.1039/c3ra46149a
22. Pedersen M, Meyer AS (2010) Lignocellulose pretreatment severity—relating pH to biomatrix opening. New Biotechnol 27(6):739–750. https://doi.org/10.1016/j.nbt.2010.05.003
23. Al-Farsi MA, Lee CY (2008) Optimization of phenolics and dietary fibre extraction from date seeds. Food Chem 108(3):977–985. https://doi.org/10.1016/j.foodchem.2007.12.009
24. Besbes S, Blecker C, Deroanne C, Lognay G, Drira N-E, Attia H (2005) Heating effects on some quality characteristics of date seed oil. Food Chem 91(3):469–476. https://doi.org/10.1016/j.foodchem.2004.04.037
25. Al-Farsi M, Alasalvar C, Al-Abid M, Al-Shoaily K, Al-Amry M, Al-Rawahy F (2007) Compositional and functional characteristics of dates, syrups, and their by-products. Food Chem 104(3):943–947. https://doi.org/10.1016/j.foodchem.2006.12.051
26. Al-Hooti S, Sidhu JS, Qabazard H (1998) Chemical composition of seeds of date fruit cultivars of United Arab Emirates. J Food Sci Technol 35(1):44–46
27. Al-shahib W, Marshall RJ (2003) The fruit of the date palm: Its possible use as the best food for the future? Int J Food Sci Nutr 54(4):247–259. https://doi.org/10.1080/09637480120091982
28. Al-Showiman SS (n.d.) Chemical composition of some date palm seeds (Phaenix dactylifera L) in Saudi Arabia. Arab GulfJ Sci:15–24.
29. FAO (2010) FAO statistics (Food and Agriculture Organization of the United Nations), 2010 [Statistics]. FAO.
30. Campo P, Sorial GA, Suidan MT, Venosa AD (2006) Statistical evaluation of an analytical GC/MS method for the determination of long chain fatty acids. Talanta 68(3):888–894. https://doi.org/10.1016/j.talanta.2005.06.031
31. Kalayasiri P, Jayashike N, Krisnangkura K (1996) Survey ofseed oils for use asdiesel fuels. JAOCS 73(4).
32. Krisnangkura K (1986) A simple method for estimation of cetane index of vegetable oil methyl esters. Journal of the American Oil Chemists' Society 63(4):552–553. https://doi.org/10.1007/BF02645752
33. Kamil M, Ramadan K, Olabi AG, Ghenai C, Inayat A, Rajab MH (2019) Desert palm date seeds as a biodiesel feedstock: extraction, characterization, and engine testing. Energies 12(16):3147. https://doi.org/10.3390/en12163147
34. Kontorovich AE (2009) Estimate of global oil resource and the forecast for global oil production in the 21st century. Russ Geol Geophys 50(4):237–242. https://doi.org/10.1016/j.rgg.2009.03.001
35. Labeckas G, Slavinskas S, Kanapkienė I (2019) Study of the effects of biofuel-oxygen of various origins on a CRDI diesel engine combustion and emissions. Energies 12(7):1241. https://doi.org/10.3390/en12071241

36. Arunkumar M, Kannan M, Murali G (2019) Experimental studies on engine performance and emission characteristics using castor biodiesel as fuel in CI engine. Renewable Energy 131:737–744. https://doi.org/10.1016/j.renene.2018.07.096
37. Hirner FS, Hwang J, Bae C, Patel C, Gupta T, Agarwal AK (2019) Performance and emission evaluation of a small-bore biodiesel compression-ignition engine. Energy 183:971–982. https://doi.org/10.1016/j.energy.2019.07.015
38. Schatowitz B, Brandt G, Gafner F, Schlumpf E, Bühler R, Hasler P, Nussbaumer T (1994) Dioxin emissions from wood combustion. Chemosphere 29(9–11):2005–2013. https://doi.org/10.1016/0045-6535(94)90367-0
39. Luque R, Lin CSK, Wilson K, Clark JH (2016) Biofuels production: processes and technologies (2nd ed.). Woodhead.
40. Kamil M, Ramadan KM, Awad OI, Ibrahim TK, Inayat A, Ma X (2019) Environmental impacts of biodiesel production from waste spent coffee grounds and its implementation in a compression ignition engine. Sci Total Environ 675:13–30. https://doi.org/10.1016/j.scitotenv.2019.04.156
41. Lee M, Yang M, Choi S, Shin J, Park C, Cho S-K, Kim YM (2019) Sequential production of lignin, fatty acid methyl esters and biogas from spent coffee grounds via an integrated physicochemical and biological process. Energies 12(12):2360. https://doi.org/10.3390/en12122360
42. Azeem MW, Hanif MA, Al-Sabahi JN, Khan AA, Naz S, Ijaz A (2016) Production of biodiesel from low priced, renewable and abundant date seed oil. Renewable Energy 86:124–132. https://doi.org/10.1016/j.renene.2015.08.006
43. Al-Yahyai R, Manickavasagan (2012) A an overview of date palm production. Dates production, processing, food, and medicinal values. CRC Press, Boca Raton, pp 3–12
44. Mrabet A, Ferchichi A, Chaira N, Salah Moha B, Baaziz M, Mrabet Pen T (2008) Physico-chemical characteristics and total quality of date palm varieties grown in the southern of Tunisia. Pak J Biol Sci 11(7):1003–1008. https://doi.org/10.3923/pjbs.2008.1003.1008
45. Akasha IAM (2004) Extraction and characterisation of protein fraction from date palm (Phoenix dactylifera L.) seeds [Ph.D]. Heriot-Watt University, Edinburgh, Scotland.
46. Afiq A, Rahman A, Man C (2013) Date seed and date seed oil. Int Food Res 20(5):2035–2043
47. Besbes S, Drira L, Blecker C, Deroanne C, Attia H (2009) Adding value to hard date (Phoenix dactylifera L.): compositional, functional and sensory characteristics of date jam. Food Chemistry 112(2):406–411. https://doi.org/10.1016/j.foodchem.2008.05.093
48. Besbes S, Blecker C, Deroanne C, Drira N-E, Attia H (2004) Date seeds: Chemical composition and characteristic profiles of the lipid fraction. Food Chem 84(4):577–584. https://doi.org/10.1016/S0308-8146(03)00281-4
49. Al-Farsi MA, Lee CY (2011) Usage of date (Phoenix dactylifera L.) seeds in human health and animal feed. In: Nuts and seeds in health and disease prevention, Elsevier, pp 447–452. https://doi.org/10.1016/B978-0-12-375688-6.10053-2
50. van Maris AJA, Abbott DA, Bellissimi E, van den Brink J, Kuyper M, Luttik MAH, Wisselink HW, Scheffers WA, van Dijken JP, Pronk JT (2006) Alcoholic fermentation of carbon sources in biomass hydrolysates by Saccharomyces cerevisiae: current status. Antonie Van Leeuwenhoek 90(4):391–418. https://doi.org/10.1007/s10482-006-9085-7
51. Ali-Mohamed AY, Khamis ASH (2004) Mineral ion content of the seeds of six cultivars of Bahraini date palm (*Phoenix dactylifera*). J Agric Food Chem 52(21):6522–6525. https://doi.org/10.1021/jf030518x
52. Rahman MS, Kasapis S, Al-Kharusi NSZ, Al-Marhubi IM, Khan AJ (2007) Composition characterisation and thermal transition of date pits powders. J Food Eng 80(1):1–10.https://doi.org/10.1016/j.jfoodeng.2006.04.030
53. Devshony S, Eteshola E, Shani A (1992) Characteristics and some potential applications of date palm (*Phoenix dactylifera* L.) seeds and seed oil. J American Oil Chemists' Society 69(6):595–597. https://doi.org/10.1007/BF02636115
54. Al-Zuhair S, Taher H, Al Dhaheri S, Wajeeh S, Nour M, El-Najjar E (2017) Biodiesel production from oils extracted from date pits. Green Sustain Chem 07(01):48–56. https://doi.org/10.4236/gsc.2017.71004

55. Jamil F, Al-Muhtaseb AH, Al-Haj L, Al-Hinai MA, Hellier P, Rashid U (2016) Optimization of oil extraction from waste "Date pits" for biodiesel production. Energy Convers Manage 117:264–272. https://doi.org/10.1016/j.enconman.2016.03.025
56. Taghizadeh-Alisaraei A, Ghobadian B, Tavakoli-Hashjin T, Mohtasebi SS, Rezaei-asl A, Azadbakht M (2016) Characterization of engine's combustion-vibration using diesel and biodiesel fuel blends by time-frequency methods: a case study. Renewable Energy 95:422–432. https://doi.org/10.1016/j.renene.2016.04.054
57. Aydin H, Bayindir H (2010) Performance and emission analysis of cottonseed oil methyl ester in a diesel engine. Renewable Energy 35(3):588–592. https://doi.org/10.1016/j.renene.2009.08.009
58. Hill J, Nelson E, Tilman D, Polasky S, Tiffany D (2006) Environmental, economic, and energetic costs and benefits of biodiesel and ethanol biofuels. Proc Natl Acad Sci 103(30):11206–11210. https://doi.org/10.1073/pnas.0604600103
59. Murugesan A, Umarani C, Subramanian R, Nedunchezhian N (2009) Bio-diesel as an alternative fuel for diesel engines—a review. Renew Sustain Energy Rev 13(3):653–662. https://doi.org/10.1016/j.rser.2007.10.007
60. Tadesse H, Luque R (2011) Advances on biomass pretreatment using ionic liquids: an overview. Energy Environ Sci 4(10):3913. https://doi.org/10.1039/c0ee00667j
61. Taghizadeh-Alisaraei A, Ghobadian B, Tavakoli-Hashjin T, Mohtasebi SS (2012) Vibration analysis of a diesel engine using biodiesel and petrodiesel fuel blends. Fuel 102:414–422. https://doi.org/10.1016/j.fuel.2012.06.109
62. Brown RC (Ed) (2011) Thermochemical processing of biomass: conversion into fuels, chemicals and power. Wiley. https://doi.org/10.1002/9781119990840
63. Choi IS, Lee YG, Khanal SK, Park BJ, Bae H-J (2015) A low-energy, cost-effective approach to fruit and citrus peel waste processing for bioethanol production. Appl Energy 140:65–74. https://doi.org/10.1016/j.apenergy.2014.11.070
64. Deublein D, Steinhauser A (2010) Biogas from waste and renewable resources: an introduction. Wiley-VCH Verlag GmbH & Co KGaA. https://doi.org/10.1002/9783527632794
65. Maghanaki MM, Ghobadian B, Najafi G, Galogah RJ (2013) Potential of biogas production in Iran. Renew Sustain Energy Rev 28:702–714. https://doi.org/10.1016/j.rser.2013.08.021
66. Gunaseelan VN (2004) Biochemical methane potential of fruits and vegetable solid waste feedstocks. Biomass Bioenerg 26(4):389–399. https://doi.org/10.1016/j.biombioe.2003.08.006
67. Basu P (2010) Working Together to grow libraries in developing countries. Elsevier
68. Shapouri H, Salassi M (2006) The economic feasibility of ethanol production from sugar in the United States. USDA Rep.
69. Chtourou H, Bouallagui Z, Sayadi S, Dhouib A (2012) Bioethanol production from cull dates by *Candida kefyr*. J Biobased Mater Bioenergy 6(5):588–593. https://doi.org/10.1166/jbmb.2012.1249
70. Louhichi B, Belgaib J, Benamor H, Hajji N (2013) Production of bio-ethanol from three varieties of dates. Renewable Energy 51:170–174. https://doi.org/10.1016/j.renene.2012.07.028
71. Zeinelabdeen MA, Abasaeed AE, Gaily MH, Sulieman AK, Putra MD (2013) Coproduction of fructose and ethanol from dates by S. Cerevisiae Atcc 36859. https://doi.org/10.5281/ZENODO.1088526
72. Bensidhom G, Ben Hassen-Trabelsi A, Alper K, Sghairoun M, Zaafouri K, Trabelsi I (2018) Pyrolysis of date palm waste in a fixed-bed reactor: characterization of pyrolytic products. Biores Technol 247:363–369. https://doi.org/10.1016/j.biortech.2017.09.066
73. Fadhil AB, Alhayali MA, Saeed LI (2017) Date (Phoenix dactylifera L.) palm stones as a potential new feedstock for liquid bio-fuels production. Fuel 210:165–176. https://doi.org/10.1016/j.fuel.2017.08.059
74. Alamri SA, Hashem M, Alrumman SA, Qahtani MSAA (2015) Enhancement of bio-ethanol production from date molasses by non-conventional yeasts. Res J Microbiol 10(3):114–125. https://doi.org/10.3923/jm.2015.114.125

75. Ghanim, AN (2013) Bioethanol production from Iraqi date palm resources. J Babylon Univ/Eng Sci 21(1).
76. Chniti S, Djelal H, Hassouna M, Amrane A (2014) Residue of dates from the food industry as a new cheap feedstock for ethanol production. Biomass Bioenerg 69:66–70. https://doi.org/10.1016/j.biombioe.2014.07.011
77. Alzain MA (2012) Bioethanol production from date fruit waste and its usage in petrol engine [Master, University of Malaya, Kualalumpur]. http://studentsrepo.um.edu.my/id/eprint/3834
78. AL Bassam R (850) Industrial ethanol production using juice of dates in a fixed cell process. Emirates Dates Factory -AI Saad.
79. Senneca O (2007) Kinetics of pyrolysis, combustion and gasification of three biomass fuels. Fuel Process Technol 88(1):87–97. https://doi.org/10.1016/j.fuproc.2006.09.002
80. Dong T, Knoshaug EP, Pienkos PT, Laurens LML (2016) Lipid recovery from wet oleaginous microbial biomass for biofuel production: A critical review. Appl Energy 177:879–895. https://doi.org/10.1016/j.apenergy.2016.06.002
81. Alrumman SA (2016) Enzymatic saccharification and fermentation of cellulosic date palm wastes to glucose and lactic acid. Braz J Microbiol 47(1):110–119. https://doi.org/10.1016/j.bjm.2015.11.015
82. Afkari-sayah A, Said Minaei (2009) The principles for assessing the mechanical waste in agricultural products. (SID) Publication, Iran.
83. Mousavi A (n.d.) The use of palm lignocellulosic wastes. Livest Agro-Ind Mag
84. Alananbeh KM, Bouqellah NA, Al Kaff NS (2014) Cultivation of oyster mushroom Pleurotus ostreatus on date-palm leaves mixed with other agro-wastes in Saudi Arabia. Saudi J Biolog Sci 21(6):616–625. https://doi.org/10.1016/j.sjbs.2014.08.001
85. Al-Faifi SA, Migdadi HM, Algamdi SS, Khan MA, Ammar MH, Al-Obeed RS, Al-Thamra MI, EL-Harty EH (2016) Development, characterization and use of genomic SSR markers for assessment of genetic diversity in some Saudi date palm (Phoenix dactylifera L.) cultivars. Electron J Biotechnol 21:18–25. https://doi.org/10.1016/j.ejbt.2016.01.006
86. Hoddle MS, Al-Abbad AH, El-Shafie HAF, Faleiro JR, Sallam AA, Hoddle CD (2013) Assessing the impact of areawide pheromone trapping, pesticide applications, and eradication of infested date palms for Rhynchophorus ferrugineus (Coleoptera: Curculionidae) management in Al Ghowaybah, Saudi Arabia. Crop Prot 53:152–160. https://doi.org/10.1016/j.cropro.2013.07.010
87. ISIRI, Iran National (2002) Standard 5998: properties of packaging dates.
88. Sait HH, Hussain A, Salema AA, Ani FN (2012) Pyrolysis and combustion kinetics of date palm biomass using thermogravimetric analysis. Biores Technol 118:382–389. https://doi.org/10.1016/j.biortech.2012.04.081
89. vomStein T, Grande PM, Leitner W, deMa-ría P (2011) ChemSusChem 4:1592–1594.
90. Jonker JGG, van der Hilst F, Junginger HM, Cavalett O, Chagas MF, Faaij APC (2015) Outlook for ethanol production costs in Brazil up to 2030, for different biomass crops and industrial technologies. Appl Energy 147:593–610. https://doi.org/10.1016/j.apenergy.2015.01.090
91. Rubin EM (2008) Genomics of cellulosic biofuels. Nature 454(7206):841–845. https://doi.org/10.1038/nature07190
92. Aditiya HB, Mahlia TMI, Chong WT, Nur H, Sebayang AH (2016) Second generation bioethanol production: a critical review. Renew Sustain Energy Rev 66:631–653. https://doi.org/10.1016/j.rser.2016.07.015
93. Chovau S, Degrauwe D, Van der Bruggen B (2013) Critical analysis of techno-economic estimates for the production cost of lignocellulosic bio-ethanol. Renew Sustain Energy Rev 26:307–321. https://doi.org/10.1016/j.rser.2013.05.064
94. Hamelinck CN, van Hooijdonk G, Faaij AP (2005) Ethanol from lignocellulosic biomass: techno-economic performance in short-, middle- and long-term. Biomass Bioenerg 28(4):384–410. https://doi.org/10.1016/j.biombioe.2004.09.002
95. Limayem A, Ricke SC (2012) Lignocellulosic biomass for bioethanol production: current perspectives, potential issues and future prospects. Prog Energy Combust Sci 38(4):449–467. https://doi.org/10.1016/j.pecs.2012.03.002

96. Li S-Z, Chan-Halbrendt C (2009) Ethanol production in (the) people's Republic of China: potential and technologies. Appl Energy 86:S162–S169. https://doi.org/10.1016/j.apenergy.2009.04.047
97. Kim S, Kim CH (2013) Bioethanol production using the sequential acid/alkali-pretreated empty palm fruit bunch fiber. Renewable Energy 54:150–155. https://doi.org/10.1016/j.renene.2012.08.032
98. Scheffran J (2010) The global demand for biofuels: technologies, markets and policies. In: Verts AA, Qureshi N, Blaschek HP, Yukawa H (Eds) Biomass to biofuels, pp 27–54. Blackwell Publishing Ltd. https://doi.org/10.1002/9780470750025.ch2
99. Taha I, Steuernagel L, Ziegmann G (2006) Chemical modification of date palm mesh fibres for reinforcement of polymeric materials. part I: examination of different cleaning methods. polymers and polymer composites 14(8):767–778. https://doi.org/10.1177/096739110601400802
100. Yu Q, Zhuang X, Yuan Z, Qi W, Wang Q, Tan X (2011) The effect of metal salts on the decomposition of sweet sorghum bagasse in flow-through liquid hot water. Biores Technol 102(3):3445–3450. https://doi.org/10.1016/j.biortech.2010.10.084
101. Chen H, Fu X (2016) Industrial technologies for bioethanol production from lignocellulosic biomass. Renew Sustain Energy Rev 57:468–478. https://doi.org/10.1016/j.rser.2015.12.069
102. Chouchene A, Jeguirim M, Favre-Reguillon A, Trouvé G, Le Buzit G, Khiari B, Zagrouba F (2012) Energetic valorisation of olive mill wastewater impregnated on low cost absorbent: Sawdust versus olive solid waste. Energy 39(1):74–81. https://doi.org/10.1016/j.energy.2011.03.044
103. Shanableh A, Radeef W (2020) Biogas production from raw and oil-spent date palm seeds mixed with wastewater treatment sludge. Biofuels 11(6):707–714. https://doi.org/10.1080/17597269.2017.1398954
104. Alvira P, Tomás-Pejó E, Ballesteros M, Negro MJ (2010) Pretreatment technologies for an efficient bioethanol production process based on enzymatic hydrolysis: a review. Biores Technol 101(13):4851–4861. https://doi.org/10.1016/j.biortech.2009.11.093
105. Dwivedi P, Alavalapati JRR, Lal P (2009) Cellulosic ethanol production in the United States: conversion technologies, current production status, economics, and emerging developments. Energy Sustain Dev 13(3):174–182. https://doi.org/10.1016/j.esd.2009.06.003
106. Ahring BK, Albernas Y, Antunes FAF, Azuma J-I, Ballesteros I, Ballesteros M, Berni M, Bikaki N, Bisaria VS, Biswas R, Biswas R, Bychkov A, Carneiro TF, Carrier M, Carvalheiro F, Castro E, Coca M, Collard F-X, Corsano G, … Zhu JY (2016) List of contributors. In: Biomass fractionation technologies for a lignocellulosic feedstock based biorefinery, Elsevier, pp xvii–xix. https://doi.org/10.1016/B978-0-12-802323-5.01002-1
107. Hendriks ATWM, Zeeman G (2009) Pretreatments to enhance the digestibility of lignocellulosic biomass. Biores Technol 100(1):10–18. https://doi.org/10.1016/j.biortech.2008.05.027
108. Ismail ZZ, Talib AR (2014) Assessment of anacrobic co-digestion of agro wastes for biogas recovery: a bench scale application to date palm Wastes. Int J Energy Environment 5(5):591–600
109. Taherzadeh MJ, Karimi K (2015) Acid-based hydrolysis processes for ethanol from lignocellulosic materials: a review. BioResources 2.
110. Sun XF, Sun RC, Tomkinson J, Baird MS (2003) Preparation of sugarcane bagasse hemicellulosic succinates using NBS as a catalyst. Carbohyd Polym 53(4):483–495. https://doi.org/10.1016/S0144-8617(03)00150-4
111. WEA-World Energy Assessment (2000) United Nations development program, pp 3–4. Bureau for Development Policy.
112. Ismail AZ, Ali W, Aziz KK, Syahadah AAH, Kelvin SCW, Ali IW, Aziz KK, Syahadah AAH, Kelvin SCW, Arifin Z, Ismail AZ, Ali W, Aziz KK, Syahadah AAH, Kelvin SCW, Ali IW, Aziz KK, Syahadah AAH, Kelvin SCW, Arifin Z (2014) Bioethanol production from acid hydrolysates of date palm fronds using a co- culture of Saccharomyces cerevisiae and Pichia stipitis. International Journal of Enhanced Research in Science Technology & Engineering 3(5):35–44

113. Ziouti A, El Modafar C, Fleuriet A et al (1996) Phenolic compounds in date palm cultivars sensitive and resistant to Fusarium oxysporum. Biol Plant 38(3):451–457
114. Al-Muhtaseb AH, Jamil F, Al-Haj L, Al-Hinai MA, Baawain M, Myint MTZ, Rooney D (2016) Efficient utilization of waste date pits for the synthesis of green diesel and jet fuel fractions. Energy Convers Manage 127:226–232. https://doi.org/10.1016/j.enconman.2016.09.004
115. de Sousa FP, Cardoso CC, Pasa VMD (2016) Producing hydrocarbons for green diesel and jet fuel formulation from palm kernel fat over Pd/C. Fuel Process Technol 143:35–42. https://doi.org/10.1016/j.fuproc.2015.10.024
116. Liu Q, Zuo H, Zhang Q, Wang T, Ma L (2014) Hydrodeoxygenation of palm oil to hydrocarbon fuels over Ni/SAPO-11 catalysts. Chin J Catal 35(5):748–756. https://doi.org/10.1016/S1872-2067(12)60710-4
117. Wang S, Huffman J, Littell R (1981) Characterization of Melaleuca biomass as a fuel for direct combustion. Wood Sci 13(4):216–219
118. Duan J, Han J, Sun H, Chen P, Lou H, Zheng X (2012) Diesel-like hydrocarbons obtained by direct hydrodeoxygenation of sunflower oil over Pd/Al-SBA-15 catalysts. Catal Commun 17:76–80. https://doi.org/10.1016/j.catcom.2011.10.009
119. Nasser RA-S (2014) An evaluation of the use of midribs from common date palm cultivars grown in saudi arabia for energy production. BioResources 9(3):4343–4357. https://doi.org/10.15376/biores.9.3.4343-4357
120. Pesevski M, Iliev B, Zivkovic D, Jakimovska-Popovska V, Srbinoska M, Filiposki B (2010) Possibilities for utilization of tobacco stems for production of energetic briquettes. J Agricultural Sciences, Belgrade 55(1):45–54. https://doi.org/10.2298/JAS1001045P
121. Sakai S, Nakashimada Y, Inokuma K, Kita M, Okada H, Nishio N (2005) Acetate and ethanol production from H2 and CO2by Moorella sp. using a repeated batch culture. J Bioscience Bioengineering 99(3):252–258. https://doi.org/10.1263/jbb.99.252
122. Kitzler H, Pfeifer C, Hofbauer H (2012) Combustion of reeds in a 3 MW district heating plant. Int J Environmental Sci Development 407–411. https://doi.org/10.7763/IJESD.2012.V3.256
123. Nasser RA, Salem MZM, Al-Mefarrej HA, Abdel-Aal MA, Soliman SS (2013) Fuel characteristics of vine prunings (Vitis vinifera L.) as a potential source for energy production. BioResources 9(1):482–496. https://doi.org/10.15376/biores.9.1.482-496
124. Mukhtar M. Ashur, Salh M. Aldulaimi, Iman M. Bengharbia (n.d.) Biogas production from date palm trees residues. 2019.
125. Feldman D (1985) Wood—chemistry, ultrastructure, reactions, by Fengel D, Wegener G, Walter de Gruyter, Berlin and New York, 1984, 613 pp. Price: 245 DM. Journal of Polymer Science: Polymer Letters Edition 23(11):601–602. https://doi.org/10.1002/pol.1985.130231112
126. Nasser RA-S, Aref IM (2014) fuelwood characteristics of six acacia species growing wild in the southwest of saudi arabia as affected by geographical location. BioResources 9(1):1212–1224. https://doi.org/10.15376/biores.9.1.1212-1224
127. Scholl AL, Menegol D, Pitarelo AP, Fontana RC, Filho AZ, Ramos LP, Dillon AJP, Camassola M (2015) Ethanol production from sugars obtained during enzymatic hydrolysis of elephant grass (Pennisetum purpureum, Schum.) pretreated by steam explosion. Biores Technol 192:228–237. https://doi.org/10.1016/j.biortech.2015.05.065
128. Baker A (1983) Wood fuel properties and fuel properties from wood. Michigan State University, East Lansing, MI, Fuelwood Management and Utilization Seminar
129. Kataki R, Konwer D (2002) Fuelwood characteristics of indigenous tree species of North-east India. Biomass Bioenerg 22(6):433–437. https://doi.org/10.1016/S0961-9534(02)00026-0
130. El-Mousty HI (1997) The rediscovery of local raw materials: New opportunities for developing countries. Industry Environ 20(1–2):17–20
131. Al-Fuhaid KM (2006) The famous date varieties in the kingdom of Saudi Arabia," Ministry of Agriculture, 1st edition, Public Relation and Agriculture Information, FAO, p 245
132. Nasser RA (2012) Physical and mechanical properties of three-layer particleboard manufactured from the tree pruning of seven wood species. World Appl Sci J 19:741–753

133. Aref I, Nasser R, Ali I, Al-Mefarrej H, Al-Zahrani S (2013) Effects of aqueous extraction on the performance and properties of polypropylene/wood composites from *Phoenix dactylifera* and *Acacia tortilis* wood. J Reinf Plast Compos 32(7):476–489. https://doi.org/10.1177/0731684412454462
134. Dorge S, Jeguirim M, Trouvé G (2011) Thermal degradation of Miscanthus pellets: kinetics and aerosols characterization. Waste and Biomass Valorization 2(2):149–155. https://doi.org/10.1007/s12649-010-9060-4
135. Hegazy SS, Aref IM (2010) Suitability of some fast-growing trees and date palm fronds for particleboard production. Forest Products J 60(7):599–604. https://doi.org/10.13073/0015-7473-60.7.599
136. May YE, Dorge S, Jeguirim M, Trouvé G, Said R (2012) Measurement of gaseous and particulate pollutants during combustion of date palm wastes for energy recovery. Aerosol and Air Quality Research 12(5):814–825. https://doi.org/10.4209/aaqr.2012.03.0056
137. Nasser RA, Al-Mefarrej HA (2011) Midribs of date palm as a raw material for wood-cement composite industry in Saudi Arabia. J Appl Sci Res 15(12):1651–1658
138. ASTM D 1037 (1989) Standard methods of evaluating the properties of wood-base fiber and particle panel materials; ASTM: Philadelphia. PA, USA
139. Nikitin VM (1960) Himia drevesini I telliuloze Goslesbumiz. Cimia Lemnului SI A Celuloze, 1.
140. Romaní A, Garrote G, Alonso JL, Parajó JC (2010) Bioethanol production from hydrothermally pretreated eucalyptus globulus wood. Biores Technol 101(22):8706–8712. https://doi.org/10.1016/j.biortech.2010.06.093
141. NREL (2005) NREL laboratory analytical procedure. "Standard method for the determination ash in biomass.
142. ASTM D-2015 (1987) Standard test method for gross heating value of coal and coke by adiabatic bomb calorimeter. American Society for Testing and Materials, Philadelphia, PA, USA
143. Bhatt BP, Todaria NP (1992) Fuelwood characteristics of some Indian mountain species. For Ecol Manage 47(1–4):363–366. https://doi.org/10.1016/0378-1127(92)90285-H
144. Kollman F, Côte W (1968) Principles of wood science and technology (Vol. 1). Springer.
145. Panshin A, DeZeeuw C (1980) Textbook of wood technology. McGraw-Hill
146. Haygreen JC, Bowyer JI (1996) Forest products and wood science (3rd ed.). Iowa State University Press.
147. Bakhashwain A, El-Feel A, Hindi S, Bakhashwain A, Hindi S, Bakhashwain A, Hindi S, Bakhashwain A, Hindi S, Bakhashwain A (2010) Physico-chemical characterization of some Saudi lignocellulosic natural resources and their suitability for fiber production. Journal of King Abdulaziz University-Meteorology, Environment and Arid Land Agriculture Sciences 21(2):45–55. https://doi.org/10.4197/met.21-2.4
148. Sabra W, Groeger C, Sharma PN, Zeng A-P (2014) Improved n-butanol production by a non-acetone producing Clostridium pasteurianum DSMZ 525 in mixed substrate fermentation. Appl Microbiol Biotechnol 98(9):4267–4276. https://doi.org/10.1007/s00253-014-5588-8
149. Chao CCT, Krueger RR (2007) The date palm (Phoenix dactylifera L.): overview of biology, uses, and cultivation. Hort Sci 42(5):1077–1082.
150. Chehma A, Longo HF (2001) Valorisation des Sous-Produits du Palmier Dattier en Vue de leur Utilisation en Alimentation du Bétail. Rev Energ Ren : Production et Valorisation –Biomasse:59–64
151. Sulaiman A (2017) Characterization of date palm fronds as a fuel for energy production. Bull Chem Soc Ethiop 30(3):465. https://doi.org/10.4314/bcse.v30i3.15
152. Pascual JJ, Fernández C, Díaz, J. R., Garcés, C., & Rubert-Alemán, J. (2000) Voluntary intake and in vivo digestibility of different date-palm fractions by Murciano-Granadina (Capra hircus). J Arid Environ 45(2):183–189. https://doi.org/10.1006/jare.1999.0622
153. Motghare KA, Rathod AP, Wasewar KL, Labhsetwar NK (2016) Comparative study of different waste biomass for energy application. Waste Manage 47:40–45. https://doi.org/10.1016/j.wasman.2015.07.032

154. Nasser R, Salem M, Hiziroglu S, Al-Mefarrej H, Mohareb A, Alam M., & Aref, I. (2016). Chemical analysis of different parts of date palm (phoenix dactylifera l.) using ultimate, proximate and thermo-gravimetric techniques for energy production. Energies 9(5):374. https://doi.org/10.3390/en9050374
155. Jenkins BM, Baxter LL, Miles TR, Miles TR (1998) Combustion properties of biomass. Fuel Process Technol 54(1–3):17–46. https://doi.org/10.1016/S0378-3820(97)00059-3
156. Ibrahim AEDMM (2014) Nox and Sox emissions and climate changes. World Appl Sci J 31(8):1422–1426
157. Bibbiani C, Fantozzi F, Gargari C, Campiotti CA, Schettini E, Vox G (2016) Wood biomass as sustainable energy for greenhouses heating in Italy. Agriculture and Agricultural Science Procedia 8:637–645. https://doi.org/10.1016/j.aaspro.2016.02.086
158. Tang Z-X, Shi L-E, Aleid SM (2014) Date and their processing byproducts as substrates for bioactive compounds production. Braz Arch Biol Technol 57(5):706–713. https://doi.org/10.1590/S1516-89132014005000017
159. Telliard WA (2001) Method 1684: total, fixed, and volatile solids in water, solids, and biosolids. US Environmental Protection Agency.
160. Lee S, Speight JG, Loyalka SK (2007) Handbook of alternative fuel technologies. Taylor & Francis Group.
161. Agoudjil B, Benchabane A, Boudenne A, Ibos L, Fois M (2011) Renewable materials to reduce building heat loss: characterization of date palm wood. Energy and Buildings 43(2–3):491–497. https://doi.org/10.1016/j.enbuild.2010.10.014
162. Bhansali RR (2010) Date palm cultivation in the changing scenario of indian arid zones: challenges and prospects. In: KG Ramawat (Ed) Desert plants. Springer, Berlin Heidelberg, pp 423–459. https://doi.org/10.1007/978-3-642-02550-1_20
163. Barreveld WH (1993) Date palm products. FAO Agricultural Services Bulletin (FAO).
164. Faghihi K, Asghari A, Hajibeygi M (2012) Synthesis and characterization of thermally stable poly(amide-imide)-montmorillonite nanocomposites based on bis(4-carboxyphenyl)-N, N'-pyromellitimide acid. Bull Chem Soc Ethiop 27(1):95–104. https://doi.org/10.4314/bcse.v27i1.10
165. Guangul FM, Sulaiman SA, Raghavan VR (2012) The 4th international Meeting of Advances in thermofluids (IMAT 2011) [Personal communication].
166. Njoya D, Hajjaji M, Nkoumbou C, Elimbi A, Kwekam M, Njoya A, Yvon J, Njopwouo D (2010) Chemical and mineralogical characterization and ceramic suitability of raw feldspathic materials from Dschang (Cameroon). Bulletin of the Chemical Society of Ethiopia 24(1). https://doi.org/10.4314/bcse.v24i1.52959
167. Ravikumar C, Senthil Kumar P, Subhashni SK, Tejaswini PV, Varshini V (2017) Microwave assisted fast pyrolysis of corn cob, corn stover, saw dust and rice straw: experimental investigation on bio-oil yield and high heating values. Sustain Mater Technol 11:19–27. https://doi.org/10.1016/j.susmat.2016.12.003
168. Saidur R, Abdelaziz EA, Demirbas A, Hossain MS, Mekhilef S (2011) A review on biomass as a fuel for boilers. Renew Sustain Energy Rev 15(5):2262–2289. https://doi.org/10.1016/j.rser.2011.02.015
169. Demirbas A (2004) Combustion characteristics of different biomass fuels. Prog Energy Combust Sci 30(2):219–230. https://doi.org/10.1016/j.pecs.2003.10.004
170. Inayat M, Sulaiman SA, Hussain SB, Fiseha MG, Atnaw SM, Jamil AA (2015) Proceedings of the International Conference on Global Sustainability and Flare in Cogasification of Biomass Feedstock in Throated Downdraft. 203.
171. Makkawi Y, El Sayed Y, Salih M, Nancarrow P, Banks S, Bridgwater T (2019) Fast pyrolysis of date palm (Phoenix dactylifera) waste in a bubbling fluidized bed reactor. Renewable Energy 143:719–730. https://doi.org/10.1016/j.renene.2019.05.028
172. Ghnimi S, Umer S, Karim A, Kamal-Eldin A (2017) Date fruit (Phoenix dactylifera L.): an underutilized food seeking industrial valorization. NFS Journal 6:1–10. https://doi.org/10.1016/j.nfs.2016.12.001

173. Kamal-Eldin A, Hashim BI, Mohamed OI (2012) Processing and utilization of palm date fruits for edible applications. Recent Patents on Food, Nutrition & Agriculturee 4(1):78–86. https://doi.org/10.2174/2212798411204010078
174. El-Juhany LI (2010) Degradation of date palm trees and date production in Arab countries: causes and potential rehabilitation. Aust J Basic Appl Sci 4(8):1991–8178
175. Galiwango E, Al-Marzouqi AH, Abu-Omar MM., Khaleel AA, AbdelRahman NS (2017) Estimating combustion kinetics of UAE date palm tree biomass using thermogravimetric analysis. J Natural Sci Res 7(12).
176. Houache O, Al-Maamari R, Al-Rashidi B, Jibril B (2008) Study of date palm stem as raw material in preparation of activated carbon. J Eng Res [TJER] 5(1):47. https://doi.org/10.24200/tjer.vol5iss1pp47-54
177. Shen J, Agblevor FA (2010) Modeling semi-simultaneous saccharification and fermentation of ethanol production from cellulose. Biomass Bioenerg 34(8):1098–1107. https://doi.org/10.1016/j.biombioe.2010.02.014
178. Jahirul M, Rasul M, Chowdhury A, Ashwath N (2012) Biofuels production through biomass pyrolysis—a technological review. Energies 5(12):4952–5001. https://doi.org/10.3390/en5124952
179. Al-Wabel MI, Usman ARA, El-Naggar AH, Aly AA, Ibrahim HM, Elmaghraby S, Al-Omran A (2015) Conocarpus biochar as a soil amendment for reducing heavy metal availability and uptake by maize plants. Saudi J Biol Sci 22(4):503–511. https://doi.org/10.1016/j.sjbs.2014.12.003
180. Jeffery S, Verheijen FGA, van der Velde M, Bastos AC (2011) A quantitative review of the effects of biochar application to soils on crop productivity using meta-analysis. Agr Ecosyst Environ 144(1):175–187. https://doi.org/10.1016/j.agee.2011.08.015
181. Laird DA (2008) The Charcoal vision: a win–win–win scenario for simultaneously producing bioenergy, permanently sequestering carbon, while improving soil and water quality. Agron J 100(1):178–181. https://doi.org/10.2134/agronj2007.0161
182. Heidari A, Stahl R, Younesi H, Rashidi A, Troeger N, Ghoreyshi AA (2014) Effect of process conditions on product yield and composition of fast pyrolysis of Eucalyptus grandis in fluidized bed reactor. J Ind Eng Chem 20(4):2594–2602. https://doi.org/10.1016/j.jiec.2013.10.046
183. Mesa-Pérez JM, Rocha JD, Barbosa-Cortez LA, Penedo-Medina M, Luengo CA, Cascarosa E (2013) Fast oxidative pyrolysis of sugar cane straw in a fluidized bed reactor. Appl Therm Eng 56(1–2):167–175. https://doi.org/10.1016/j.applthermaleng.2013.03.017
184. Fernandes ERK, Marangoni C, Souza O, Sellin N (2013) Thermochemical characterization of banana leaves as a potential energy source. Energy Convers Manage 75:603–608. https://doi.org/10.1016/j.enconman.2013.08.008
185. Abu Bakar MS, Titiloye JO (2013) Catalytic pyrolysis of rice husk for bio-oil production. J Anal Appl Pyrol 103:362–368. https://doi.org/10.1016/j.jaap.2012.09.005
186. Madhu P, Kanagasabapathy H, Neethi Manickam I (2016) Cotton shell utilization as a source of biomass energy for bio-oil by flash pyrolysis on electrically heated fluidized bed reactor. J Mater Cycles Waste Manage 18(1):146–155. https://doi.org/10.1007/s10163-014-0318-y
187. Açıkalın K, Karaca F, Bolat E (2012) Pyrolysis of pistachio shell: effects of pyrolysis conditions and analysis of products. Fuel 95:169–177. https://doi.org/10.1016/j.fuel.2011.09.037
188. ASTM D 3173/D 3173–17a (2017) Standard test method for moisture in the analysis sample of coal and coke, ASTM International, West Conshohocken, PA. ASTM International.
189. ASTM D 3174–12 (2012) Standard test method for ash in the analysis sample of coal and coke from coal, ASTM International, West Conshohocken, PA. ASTM International.
190. Waldheim L, Nilsson T (2001) Heating value of gases from biomass gasification.
191. Elmay Y, Jeguirim M, Dorge S, Trouvé G, Said R (2012) Study on the thermal behavior of different date palm residues: characterization and devolatilization kinetics under inert and oxidative atmospheres. Energy 44(1):702–709. https://doi.org/10.1016/j.energy.2012.05.022

192. Blanco López MC, Blanco CG, Martínez-Alonso, A., & Tascón, J. M. D. (2002) Composition of gases released during olive stones pyrolysis. J Anal Appl Pyrol 65(2):313–322. https://doi.org/10.1016/S0165-2370(02)00008-6
193. Caputo AC, Scacchia F, Pelagagge PM (2003) Disposal of by-products in olive oil industry: waste-to-energy solutions. Appl Therm Eng 23(2):197–214. https://doi.org/10.1016/S1359-4311(02)00173-4
194. Chouchene A, Jeguirim M, Khiari B, Trouvé G, Zagrouba F (2010) Study on the emission mechanism during devolatilization/char oxidation and direct oxidation of olive solid waste in a fixed bed reactor. J Anal Appl Pyrol 87(1):168–174. https://doi.org/10.1016/j.jaap.2009.11.008
195. Chouchene A, Jeguirim M, Trouvé G, Favre-Reguillon A, Le Buzit G (2010) Combined process for the treatment of olive oil mill wastewater: absorption on sawdust and combustion of the impregnated sawdust. Biores Technol 101(18):6962–6971. https://doi.org/10.1016/j.biortech.2010.04.017
196. García-Ibañez P, Sánchez M, Cabanillas A (2006) Thermogravimetric analysis of olive-oil residue in air atmosphere. Fuel Process Technol 87(2):103–107. https://doi.org/10.1016/j.fuproc.2005.08.005
197. Masghouni M (2000) Energy applications of olive-oil industry by-products: — I. The exhaust foot cake. Biomass and Bioenergy 18(3):257–262. https://doi.org/10.1016/S0961-9534(99)00100-2
198. Al-Omari S-AB (2006) Experimental investigation on combustion and heat transfer characteristics in a furnace fueled with unconventional biomass fuels (date stones and palm stalks). Energy Convers Manage 47(6):778–790. https://doi.org/10.1016/j.enconman.2005.06.012
199. Al-Omari S-AB (2009) Evaluation of the biomass "date stones" as a fuel in furnaces: a comparison with coal combustion. Int Commun Heat Mass Transfer 36(9):956–961. https://doi.org/10.1016/j.icheatmasstransfer.2009.06.016
200. Bendahou A, Dufresne A, Kaddami H, Habibi Y (2007) Isolation and structural characterization of hemicelluloses from palm of Phoenix dactylifera L. Carbohyd Polym 68(3):601–608. https://doi.org/10.1016/j.carbpol.2006.10.016
201. Zaafouri K, Trabelsi BHA, Krichah S, Ouerghi A (2016) Enhancement of biofuels production by means of co-pyrolysis of Posidonia oceanica (L.) and frying oil wastes: experimental study and process modeling. Biores Technol 207:387–398. https://doi.org/10.1016/j.biortech.2016.02.004
202. FAOSTAT (2010) On-line statistical database of the food and agricultural organization of the United Nations [Statistics]. FAOSTAT.
203. REN21 (2014) Renewable energy networks. global status report [Global Status Report]. Renewable Energy Networks.
204. Haimour NM, Emeish S (2006) Utilization of date stones for production of activated carbon using phosphoric acid. Waste Manage 26(6):651–660. https://doi.org/10.1016/j.wasman.2005.08.004
205. Elmay Y, Jeguirim M, Dorge S, Trouvé G, Said R (2014) Evaluation of date palm residues combustion in fixed bed laboratory reactor: a comparison with sawdust behaviour. Renewable Energy 62:209–215. https://doi.org/10.1016/j.renene.2013.07.007
206. Bchini H, Hsayoui S, Aloui S (2002) Gestion de la matière organique et compostage des palmes sèches dans le milieu oasien. Annales de l'INRAT.
207. El-Morsy MMS (1980) Studies on the rachises of the Egyptian date palm leaves for hardboard production. Fibre Science and Technology 13(4):317–321. https://doi.org/10.1016/0015-0568(80)90018-4
208. Khiari R, Mhenni MF, Belgacem MN, Mauret E (2010) Chemical composition and pulping of date palm rachis and Posidonia oceanica—a comparison with other wood and non-wood fibre sources. Biores Technol 101(2):775–780. https://doi.org/10.1016/j.biortech.2009.08.079
209. Belala Z, Jeguirim M, Belhachemi M, Addoun F, Trouvé G (2011) Biosorption of basic dye from aqueous solutions by date stones and palm-trees waste: kinetic, equilibrium and thermodynamic studies. Desalination 271(1–3):80–87. https://doi.org/10.1016/j.desal.2010.12.009

210. Bouchelta C, Medjram MS, Bertrand O, Bellat J-P (2008) Preparation and characterization of activated carbon from date stones by physical activation with steam. J Anal Appl Pyrol 82(1):70–77. https://doi.org/10.1016/j.jaap.2007.12.009
211. Chen W-H, Kuo P-C (2011) Isothermal torrefaction kinetics of hemicellulose, cellulose, lignin and xylan using thermogravimetric analysis. Energy 36(11):6451–6460. https://doi.org/10.1016/j.energy.2011.09.022
212. Chouchene A, Jeguirim M, Khiari B, Zagrouba F, Trouvé G (2010) Thermal degradation of olive solid waste: Influence of particle size and oxygen concentration. Resour Conserv Recycl 54(5):271–277. https://doi.org/10.1016/j.resconrec.2009.04.010
213. Ghetti P, Ricca L, Angelini L (1996) Thermal analysis of biomass and corresponding pyrolysis products. Fuel 75(5):565–573. https://doi.org/10.1016/0016-2361(95)00296-0
214. Grønli MG, Várhegyi G, Di Blasi C (2002) Thermogravimetric analysis and devolatilization kinetics of wood. Ind Eng Chem Res 41(17):4201–4208. https://doi.org/10.1021/ie0201157
215. Jeguirim M, Dorge S, Loth A, Trouvé G (2010) Devolatilization kinetics of miscanthus straw from thermogravimetric analysis. Int J Green Energy 7(2):164–173. https://doi.org/10.1080/15435071003673641
216. Jeguirim M, Dorge S, Trouvé G (2010) Thermogravimetric analysis and emission characteristics of two energy crops in air atmosphere: Arundo donax and Miscanthus giganthus. Biores Technol 101(2):788–793. https://doi.org/10.1016/j.biortech.2009.05.063
217. Jeguirim M, Trouvé G (2009) Pyrolysis characteristics and kinetics of Arundo donax using thermogravimetric analysis. Biores Technol 100(17):4026–4031. https://doi.org/10.1016/j.biortech.2009.03.033
218. Munir S, Daood SS, Nimmo W, Cunliffe AM, Gibbs BM (2009) Thermal analysis and devolatilization kinetics of cotton stalk, sugar cane bagasse and shea meal under nitrogen and air atmospheres. Biores Technol 100(3):1413–1418. https://doi.org/10.1016/j.biortech.2008.07.065
219. Orfão JJM, Antunes FJA, Figueiredo JL (1999) Pyrolysis kinetics of lignocellulosic materials—three independent reactions model. Fuel 78(3):349–358. https://doi.org/10.1016/S0016-2361(98)00156-2
220. Stephanopoulos G (2007) Challenges in engineering microbes for biofuels production. Science 315(5813):801–804. https://doi.org/10.1126/science.1139612
221. Wilson DB (2011) Microbial diversity of cellulose hydrolysis. Curr Opin Microbiol 14(3):259–263. https://doi.org/10.1016/j.mib.2011.04.004
222. Saha BC, Yoshida T, Cotta MA, Sonomoto K (2013) Hydrothermal pretreatment and enzymatic saccharification of corn stover for efficient ethanol production. Ind Crops Prod 44:367–372. https://doi.org/10.1016/j.indcrop.2012.11.025
223. Klass DL (1998) Biomass for renewable energy, fuels, and chemicals. Academic Press
224. Belala Z, Jeguirim M, Belhachemi M, Addoun F, Trouvé G (2011) Biosorption of copper from aqueous solutions by date stones and palm-trees waste. Environ Chem Lett 9(1):65–69. https://doi.org/10.1007/s10311-009-0247-5
225. Alves CA, Gonçalves C, Pio CA, Mirante F, Caseiro A, Tarelho L, Freitas MC, Viegas DX (2010) Smoke emissions from biomass burning in a mediterranean shrubland. Atmos Environ 44(25):3024–3033. https://doi.org/10.1016/j.atmosenv.2010.05.010
226. Giuntoli J, de Jong W, Verkooijen AHM, Piotrowska P, Zevenhoven M, Hupa M (2010) Combustion characteristics of biomass residues and biowastes: fate of fuel nitrogen. Energy Fuels 24(10):5309–5319. https://doi.org/10.1021/ef100571n
227. Kamalak Kannan G, Gupta M, Chandra Kapoor J (2004) Estimation of gaseous products and particulate matter emission from garden biomass combustion in a simulation fire test chamber. Atmos Environ 38(39):6701–6710. https://doi.org/10.1016/j.atmosenv.2004.09.021
228. Miranda T, Esteban A, Rojas S, Montero I, Ruiz A (2008) Combustion analysis of different olive residues. Int J Mol Sci 9(4):512–525. https://doi.org/10.3390/ijms9040512
229. Senelwa K, Sims REH (1999) Fuel characteristics of short rotation forest biomass. Biomass Bioenerg 17(2):127–140. https://doi.org/10.1016/S0961-9534(99)00035-5

230. Jeguirim M, Chouchène A, Réguillon AF, Trouvé G, Le Buzit G (2012) A new valorisation strategy of olive mill wastewater: impregnation on sawdust and combustion. Resour Conserv Recycl 59:4–8. https://doi.org/10.1016/j.resconrec.2011.03.006
231. Sánchez ÓJ, Cardona CA (2008) Trends in biotechnological production of fuel ethanol from different feedstocks. Biores Technol 99(13):5270–5295. https://doi.org/10.1016/j.biortech.2007.11.013
232. Demirbaş A (2003) Relationships between lignin contents and fixed carbon contents of biomass samples. Energy Convers Manage 44(9):1481–1486. https://doi.org/10.1016/S0196-8904(02)00168-1
233. Komulainen MP, Simi E, Hagelberg I, Ikonen S, Lyytinen S (2008) Possibilities of Using the common reed for energy generation in southern Finland. Turku University of Applied Sciences, Turku, Finland, p 21
234. Tchapda A, Pisupati S (2014) A review of thermal co-conversion of coal and biomass/waste. Energies 7(3):1098–1148. https://doi.org/10.3390/en7031098
235. Amirou S, Zerizer A, Pizzi A, Haddadou I, Zhou X (2013) Particleboards production from date palm biomass. Eur J Wood Wood Prod 71(6):717–723. https://doi.org/10.1007/s00107-013-0730-3
236. Hegazy S, Ahmed K (2015) Effect of date palm cultivar, particle size, panel density and hot water extraction on particleboards manufactured from date palm fronds. Agriculture 5(2):267–285. https://doi.org/10.3390/agriculture5020267
237. Khiari R, Mauret E, Belgacem MN, Mhemmi F (2011) Tunisian date palm rachis used as an alternative source of fibres for papermaking applications. BioResource 265–281.
238. Khristova P (2005) Alkaline pulping with additives of date palm rachis and leaves from Sudan. Biores Technol 96(1):79–85. https://doi.org/10.1016/j.biortech.2003.05.005
239. Nasser RA, Hiziroglu S, Abdel-Aal MA, Al-Mefarrej HA, Shetta ND, Aref IM (2015) Measurement of some properties of pulp and paper made from date palm midribs and wheat straw by soda-AQ pulping process. Measurement 62:179–186. https://doi.org/10.1016/j.measurement.2014.10.051
240. Abu-Sharkh BF, Hamid H (2004) Degradation study of date palm fibre/polypropylene composites in natural and artificial weathering: Mechanical and thermal analysis. Polym Degrad Stab 85(3):967–973. https://doi.org/10.1016/j.polymdegradstab.2003.10.022
241. Hegazy S, Ahmed K, Hiziroglu S (2014) Oriented strand board production from water-treated date palm fronds. BioResources 10(1):448–456. https://doi.org/10.15376/biores.10.1.448-456
242. Babiker ME, Aziz ARA, Heikal M, Yusup S (2013) Pyrolysis characteristics of phoenix dactylifera date palm seeds using thermo-gravimetric analysis (TGA). International Journal of Environmental Science and Development 521–524. https://doi.org/10.7763/IJESD.2013.V4.406
243. Eissa AHA (2013, June 5) Quality characteristics for agricultural residues to produce briquettes. The International Conference on Agricultural, Biotechnology, Biological and Biosystems Engineering, New York, NY, USA.
244. Alhamed Y (2006) Activated carbon from dates' stone by ZnCl2 activation. J King Abdulaziz Univ-Eng Sci 17(2):75–98. https://doi.org/10.4197/Eng.17-2.4
245. Sluiter A, Hames B, Ruiz R, Scarlata C, Sluiter J, Templeton D (n.d.) Determination of Ash in Biomass. *NREL/TP-510–42622* [Technical Report]. National Renewable Energy Laboratories (NREL).
246. Al-Juhaimi FY, Hamad SH, Al-Ahaideb IS, Al-Otaibi MM, Ghafoor K, Abbasi T, Abbasi SA (2014) Biogas production through the anaerobic digestion of date palm tree wastes—process optimization. BioResources 9(2):3323–3333. https://doi.org/10.15376/biores.9.2.3323-3333
247. McKendry P (2002) Energy production from biomass (part 1): Overview of biomass. Biores Technol 83(1):37–46. https://doi.org/10.1016/S0960-8524(01)00118-3
248. Abed I, Paraschiv M, Loubar K, Zagrouba F, Tazerout M (2012) Thermogravimetric investigation and thermal conversion kinetics of typical North African and Middle Eastern lignocellulosic wastes. BioResource 7:1200–1220

249. *ASTM D-1037"Standard test method of evaluating the properties of wood-based fiber and particleboard panel materials* (1989) American Society for Testing and Materials.
250. Sindhu R, Binod P, Pandey A (2016) Biological pretreatment of lignocellulosic biomass—an overview. Biores Technol 199:76–82. https://doi.org/10.1016/j.biortech.2015.08.030
251. ASTM D 3176–89 (1989) Standard practice for ultimate analysis of coal and coke; ASTM: Philadelphia. PA, USA
252. Munalula F, Meincken M (2009) An evaluation of South African fuelwood with regards to calorific value and environmental impact. Biomass Bioenerg 33(3):415–420. https://doi.org/10.1016/j.biombioe.2008.08.011
253. Mudhafar Kareem Abdullah, A Bulent Koc. (2010) Ultrasound Assisted oil extraction from date palm kernels for biodiesel production. 2010 Pittsburgh, Pennsylvania, June 20 - June 23, 2010. https://doi.org/10.13031/2013.29928
254. Fang C, Schmidt JE, Cybulska I, Brudecki GP, Frankær CG, Thomsen MH (2015) Hydrothermal pretreatment of date palm (*Phoenix dactylifera* L.) leaflets and rachis to enhance enzymatic digestibility and bioethanol potential. Biomed Res Int 2015:1–13. https://doi.org/10.1155/2015/216454
255. Himmel ME, Ding S-Y, Johnson DK, Adney WS, Nimlos MR, Brady JW, Foust TD (2007) Biomass recalcitrance: engineering plants and enzymes for biofuels production. Science 315(5813):804–807. https://doi.org/10.1126/science.1137016
256. Sosa-Villalobos A, Rustri an E, Houbron E (2014) Anaerobic digestion of vinasse cane alcohol: the influence of OLR by a UASB reactor. IJMER 4(6):37–42
257. Rozmarin G, Simionescu C (1973) Determining hemicellulose content. 392.
258. Babcock BA, Marette S, Tréguer D (2011) Opportunity for profitable investments in cellulosic biofuels. Energy Policy 39(2):714–719. https://doi.org/10.1016/j.enpol.2010.10.044
259. Brown TR, Brown RC (2013) A review of cellulosic biofuel commercial-scale projects in the United States. Biofuels, Bioprod Biorefin 7(3):235–245. https://doi.org/10.1002/bbb.1387
260. Harris PV, Xu F, Kreel NE, Kang C, Fukuyama S (2014) New enzyme insights drive advances in commercial ethanol production. Curr Opin Chem Biol 19:162–170. https://doi.org/10.1016/j.cbpa.2014.02.015
261. Qiu H, Sun L, Huang J, Rozelle S (2012) Liquid biofuels in China: current status, government policies, and future opportunities and challenges. Renew Sustain Energy Rev 16(5):3095–3104. https://doi.org/10.1016/j.rser.2012.02.036
262. John RP, Anisha GS, Nampoothiri KM, Pandey A (2011) Micro and macroalgal biomass: a renewable source for bioethanol. Biores Technol 102(1):186–193. https://doi.org/10.1016/j.biortech.2010.06.139
263. Cybulska I, Chaturvedi T, Brudecki GP, Kádár Z, Meyer AS, Baldwin RM, Thomsen MH (2014) Chemical characterization and hydrothermal pretreatment of Salicornia bigelovii straw for enhanced enzymatic hydrolysis and bioethanol potential. Biores Technol 153:165–172. https://doi.org/10.1016/j.biortech.2013.11.071
264. Chandrasekaran M, Bahkali AH (2013) Valorization of date palm (Phoenix dactylifera) fruit processing by-products and wastes using bioprocess technology—review. Saudi J Biological Sci 20(2):105–120. https://doi.org/10.1016/j.sjbs.2012.12.004
265. Mallaki M, Fatehi R (2014) Design of a biomass power plant for burning date palm waste to cogenerate electricity and distilled water. Renewable Energy 63:286–291. https://doi.org/10.1016/j.renene.2013.09.036
266. Sadig H, Sulaiman SA, Zaidi Moni MN, Anbealagan LD (2017) Characterization of date palm frond as a fuel for thermal conversion processes. MATEC Web of Conferences 131:01002. https://doi.org/10.1051/matecconf/201713101002
267. Petersen MØ, Larsen J, Thomsen MH (2009) Optimization of hydrothermal pretreatment of wheat straw for production of bioethanol at low water consumption without addition of chemicals. Biomass Bioenerg 33(5):834–840. https://doi.org/10.1016/j.biombioe.2009.01.004
268. Kumar S, Kothari U, Kong L, Lee YY, Gupta RB (2011) Hydrothermal pretreatment of switchgrass and corn stover for production of ethanol and carbon microspheres. Biomass Bioenerg 35(2):956–968. https://doi.org/10.1016/j.biombioe.2010.11.023

269. Rinaldi R, Palkovits R, Schüth F (2008) Depolymerization of cellulose using solid catalysts in ionic liquids. Angew Chem 120(42):8167–8170. https://doi.org/10.1002/ange.200802879
270. Wu W-H, Hung W-C, Lo K-Y, Chen Y-H, Wan H-P, Cheng K-C (2016) Bioethanol production from taro waste using thermo-tolerant yeast Kluyveromyces marxianus K21. Biores Technol 201:27–32. https://doi.org/10.1016/j.biortech.2015.11.015
271. Tengberg M, Lombard P (2001) Environnement et économie végétale à Qal'at al-Bahreïn aux périodes Dilmoun et Tylos. Premiers éléments d'archéobotanique. *Paléorient* 27(1):167–181. https://doi.org/10.3406/paleo.2001.4727
272. Van Doorslaer C, Schellekens Y, Mertens P, Binnemans K, De Vos D (2010) Spontaneous product segregation from reactions in ionic liquids: application in Pd-catalyzed aliphatic alcohol oxidation. Phys Chem Chem Phys 12(8):1741–1749. https://doi.org/10.1039/B920813P
273. Rouhollah H, Iraj N, Giti E, Sorah A (2007) Mixed sugar fermentation by Pichia stipitis, Sacharomyces cerevisiaea, and an isolated xylosefermenting Kluyveromyces marxianus and their cocultures. African Journal of Biotechnolog 6(9):1110–1114
274. Pinilla L, Torres R, Ortiz C (2011) Bioethanol production in batch mode by a native strain of Zymomonas mobilis. World J Microbiol Biotechnol 27(11):2521–2528. https://doi.org/10.1007/s11274-011-0721-7
275. Sluiter A, Ruiz R, Scarlata C, Sluiter J, Templeton D (2008) Determination of extractives in biomass, NREL/TP-510–42619. National Renewable Energy Laboratory.
276. Sluiter A, Hames B, Ruiz R et al (2008) Determination of structural carbohydrates and lignin in biomass, NREL/TP-510–42618. National Renewable Laboratory.
277. Bradford MM (1976) A rapid and sensitive method for the quantitation of microgram quantities of protein utilizing the principle of protein-dye binding. Anal Biochem 72(1–2):248–254. https://doi.org/10.1016/0003-2697(76)90527-3
278. Selig M, Weiss N, Ji Y (2008) Enzymatic saccharification of lignocellulosic biomass: laboratory analytical procedure. National Renewable Energy Laboratory NREL/TP-510–42629.
279. Bonelli P (2001) Effect of pyrolysis temperature on composition, surface properties and thermal degradation rates of Brazil Nut shells. Biores Technol 76(1):15–22. https://doi.org/10.1016/S0960-8524(00)00085-7
280. Bertero M, Gorostegui HA, Orrabalis CJ, Guzmán CA, Calandri EL, Sedran U (2014) Characterization of the liquid products in the pyrolysis of residual chañar and palm fruit biomasses. Fuel 116:409–414. https://doi.org/10.1016/j.fuel.2013.08.027
281. CRDA Kébili (2011) Inventaire sur les Oasis.
282. Thomsen ST, Kádár Z, Schmidt JE (2014) Compositional analysis and projected biofuel potentials from common West African agricultural residues. Biomass Bioenerg 63:210–217. https://doi.org/10.1016/j.biombioe.2014.01.045
283. Imam T, Capareda S (2012) Characterization of bio-oil, syn-gas and bio-char from switchgrass pyrolysis at various temperatures. J Anal Appl Pyrol 93:170–177. https://doi.org/10.1016/j.jaap.2011.11.010
284. Angın D (2013) Effect of pyrolysis temperature and heating rate on biochar obtained from pyrolysis of safflower seed press cake. Biores Technol 128:593–597. https://doi.org/10.1016/j.biortech.2012.10.150
285. Khelfa A, Bensakhria A, Weber JV (2013) Investigations into the pyrolytic behaviour of birch wood and its main components: Primary degradation mechanisms, additivity and metallic salt effects. J Anal Appl Pyrol 101:111–121. https://doi.org/10.1016/j.jaap.2013.02.004
286. Bartoli M, Rosi L, Giovannelli A, Frediani P, Frediani M (2016) Production of bio-oils and bio-char from Arundo donax through microwave assisted pyrolysis in a multimode batch reactor. J Anal Appl Pyrol 122:479–489. https://doi.org/10.1016/j.jaap.2016.10.016
287. Bartoli M, Rosi L, Giovannelli A, Frediani P, Frediani M (2016) Pyrolysis of α-cellulose using a multimode microwave oven. J Anal Appl Pyrol 120:284–296. https://doi.org/10.1016/j.jaap.2016.05.016
288. Kumar P, Kumar P, Rao PVC, Choudary NV, Sriganesh G (2017) Saw dust pyrolysis: effect of temperature and catalysts. Fuel 199:339–345. https://doi.org/10.1016/j.fuel.2017.02.099

289. Mahdi et al., Mahdi Z, ElHanandeh A, Yu J (2015) Qimimg Date palm (Phoenix Dactylifera L.) seed characterization for bio-char preparation Fuels. 130–136.
290. Hadoun H, Sadaoui Z, Souami N, Sahel D, Toumert I (2013) Characterization of mesoporous carbon prepared from date stems by H3PO4 chemical activation. Appl Surf Sci 280:1–7. https://doi.org/10.1016/j.apsusc.2013.04.054
291. Joardder et al, Joardder MUH, Shazib Uddin M, Islam MN (2012) The utilization of waste dateseed as bio-oil and activated carbon by pyrolysis process Adv. Mech Eng:1–6.
292. Ben Hassen-Trabelsi A, Kraiem T, Naoui S, Belayouni H (2014) Pyrolysis of waste animal fats in a fixed-bed reactor: production and characterization of bio-oil and bio-char. Waste Manage 34(1):210–218. https://doi.org/10.1016/j.wasman.2013.09.019
293. Yu YH, Kim SD, Lee JM, Lee KH (2002) Kinetic studies of dehydration, pyrolysis and combustion of paper sludge. Energy 27(5):457–469. https://doi.org/10.1016/S0360-5442(01)00097-4
294. Sulaiman F, Abdullah N (2011) Optimum conditions for maximising pyrolysis liquids of oil palm empty fruit bunches. Energy 36(5):2352–2359. https://doi.org/10.1016/j.energy.2010.12.067
295. Lee XJ, Lee LY, Gan S, Thangalazhy-Gopakumar S, Ng HK (2017) Biochar potential evaluation of palm oil wastes through slow pyrolysis: Thermochemical characterization and pyrolytic kinetic studies. Biores Technol 236:155–163. https://doi.org/10.1016/j.biortech.2017.03.105
296. Shagufta Perveen, Muhammad Shahbaz, Muhammad Ashraf (2012) Changes in mineral composition, uptake and use efficiency of salt stressed wheat (triticum aestivum l.) plants raised from seed treated with triacontanol. Pak J Bot:27–35.
297. Fang C, Thomsen MH, Brudecki GP, Cybulska I, Frankaer CG, Bastidas-Oyanedel J-R, Schmidt JE (2015) Seawater as alternative to freshwater in pretreatment of date palm residues for bioethanol production in coastal and/or arid areas. Chemsuschem 8(22):3823–3831. https://doi.org/10.1002/cssc.201501116
298. Aden A (2007) SouthwestHydrol 6(4).
299. Dutta A, Phillips D (2009) Thermochemical ethanolvia direct gasification and mixed alcohol synthesis of lignocellulosic biomas, NREL/TP-51045913. National Renewable Energy Laboratory, Golden, CO.
300. Humbird D, Davis R, Tao L, Kinchin C, Hsu D, Aden A, Schoen P, Lukas J, Olthof B, Worley M, Sexton D, Dudgeon D (2011) Process design and economics for biochemical conversion of lignocellulosic biomass to ethanol: dilute-acid pretreatment and enzymatic hydrolysis of corn stover (NREL/TP-5100-47764, 1013269; p. NREL/TP-5100-47764, 1013269). https://doi.org/10.2172/1013269
301. Wu M, Chiu Y (2011) Consumptive water use in the production of ethanol and petroleum gasoline 2011 update. Argonne National Laboratory.
302. deMaría PD (2013) ChemCatChem 5:1643–1648
303. Grande PM, Domínguez de María P (2012) Enzymatic hydrolysis of microcrystalline cellulose in concentrated seawater. Biores Technol 104:799–802. https://doi.org/10.1016/j.biortech.2011.10.071
304. Lehmann C, Sibilla F, Maugeri Z, Streit WR, Domínguez de María P, Martinez R, Schwaneberg U (2012) Reengineering CelA2 cellulase for hydrolysis in aqueous solutions of deep eutectic solvents and concentrated seawater. Green Chem 14(10):2719. https://doi.org/10.1039/c2gc35790a
305. Lin CSK, Luque R, Clark JH, Webb C, Du C (2011) A seawater-based biorefining strategy for fermentative production and chemical transformations of succinic acid. Energy Environ Sci 4(4):1471. https://doi.org/10.1039/c0ee00666a
306. Grande PM, Bergs C, deMaría P (n.d.) Chem-SusChem 5, 2012.
307. Binder JB, Raines RT (2010) Fermentable sugars by chemical hydrolysis of biomass. Proc Natl Acad Sci 107(10):4516–4521. https://doi.org/10.1073/pnas.0912073107
308. Rhouma A (2005) Le palmier dattier en Tunisie e Le patrimoine génétique (Vol. 2). IPGRI.
309. Suresh Kumar Reddy K, Ahmed AS, Srinivasakannan C (2013) Gasification kinetics of date palm seed using carbon dioxide. Int J Environment Res 7(1).

310. Zheng Y-N, Li L-Z, Xian M, Ma Y-J, Yang J-M, Xu X, He D-Z (2009) Problems with the microbial production of butanol. J Ind Microbiol Biotechnol 36(9):1127–1138. https://doi.org/10.1007/s10295-009-0609-9
311. deMaría PDJ (2014) ChemTechnolBiotechnol:11–18.
312. Swatloski RP, Spear SK, Holbrey JD, Rogers RD (2002) Dissolution of cellose with ionic liquids. J Am Chem Soc 124(18):4974–4975. https://doi.org/10.1021/ja025790m
313. Ignatyev IA, VanDoorslaer C, Mertens PGN, Binnemans K, DeVos DE. Holzforschung (2012)
314. King AWT, Asikkala J, Mutikainen I, Järvi P, Kilpeläinen I (2011) Distillable acid-base conjugate ionic liquids for cellulose dissolution and processing. Angew Chem Int Ed 50(28):6301–6305. https://doi.org/10.1002/anie.201100274
315. Usman ARA, Abduljabbar A, Vithanage M, Ok YS, Ahmad M, Ahmad M, Elfaki J, Abdulazeem SS, Al-Wabel MI (2015) Biochar production from date palm waste: charring temperature induced changes in composition and surface chemistry. J Anal Appl Pyrol 115:392–400. https://doi.org/10.1016/j.jaap.2015.08.016
316. Vandepopuliere JM, Al-Yousef Y, Lyons JJ (1995) Dates and date pits as ingredients in broiler starting and coturnix quail breeder diets. Poult Sci 74(7):1134–1142. https://doi.org/10.3382/ps.0741134
317. Liu C, Wyman CE (2006) The enhancement of xylose monomer and xylotriose degradation by inorganic salts in aqueous solutions at 180 °C. Carbohyd Res 341(15):2550–2556. https://doi.org/10.1016/j.carres.2006.07.017
318. Liu L, Sun J, Cai C, Wang S, Pei H, Zhang J (2009) Corn stover pretreatment by inorganic salts and its effects on hemicellulose and cellulose degradation. Biores Technol 100(23):5865–5871. https://doi.org/10.1016/j.biortech.2009.06.048
319. Yong Z, Dong Y, Zhang X, Tan T (2015) Anaerobic co-digestion of food waste and straw for biogas production. Renewable Energy 78:527–530. https://doi.org/10.1016/j.renene.2015.01.033
320. Al-Dous EK, George B, Al-Mahmoud ME, Al-Jaber MY, Wang H, Salameh YM, Al-Azwani EK, Chaluvadi S, Pontaroli AC, DeBarry J, Arondel V, Ohlrogge J, Saie IJ, Suliman-Elmeer KM, Bennetzen JL, Kruegger RR, Malek JA (2011) De novo genome sequencing and comparative genomics of date palm (Phoenix dactylifera). Nat Biotechnol 29(6):521–527. https://doi.org/10.1038/nbt.1860
321. Fang C, Schmidt JE, Cybulska I, Brudecki GP, Fran ær CG, Thomsen MH (2015) BiomedRes.Int.
322. Casey E, Mosier NS, Adamec J, Stockdale Z, Ho N, Sedlak M (2013) Effect of salts on the co-fermentation of glucose and xylose by a genetically engineered strain of Saccharomyces cerevisiae. Biotechnol Biofuels 6(1):83. https://doi.org/10.1186/1754-6834-6-83
323. Tsai W, Lee M, Chang Y (2007) Fast pyrolysis of rice husk: product yields and compositions. Biores Technol 98(1):22–28. https://doi.org/10.1016/j.biortech.2005.12.005
324. Ahmed B, Mabrouk K, Cherif K, Boudjemaa B (2016) Bioethanol production from date palm fruit waste fermentation using solar energy. Afr J Biotech 15(30):1621–1627. https://doi.org/10.5897/AJB2016.15368
325. Hasni T (2006) President of New Energy Algeria (NEAL). Développement des Energies Renouvelables en Algérie. National Conference on Renewable Energies and Sustainable Development, Laghouat, Algeria.
326. Himri Y, Malik AS, Boudghene Stambouli A, Himri S, Draoui B (2009) Review and use of the Algerian renewable energy for sustainable development. Renew Sustain Energy Rev 13(6–7):1584–1591. https://doi.org/10.1016/j.rser.2008.09.007
327. Shoaib M, Al-Swaidan HM (2014) Ramp rate influence on synthesis of sliced porous activated carbon from date palm tree by physical activation method. Asian Journal of Chemistry 26(16):5295–5297. https://doi.org/10.14233/ajchem.2014.17449
328. Salleh F, Samsuddin R, Husin M (2011) Bio-fuel source from combination feed of sewage and rice waste:68–72.
329. Mielenz JR (2001) Ethanol production from biomass: technology and commercialization status. Curr Opin Microbiol 4(3):324–329. https://doi.org/10.1016/S1369-5274(00)00211-3

330. Oh TH, Hasanuzzaman M, Selvaraj J, Teo SC, Chua SC (2018) Energy policy and alternative energy in Malaysia: issues and challenges for sustainable growth—an update. Renew Sustain Energy Rev 81:3021–3031. https://doi.org/10.1016/j.rser.2017.06.112
331. Balat M, Balat H, Öz C (2008) Progress in bioethanol processing. Prog Energy Combust Sci 34(5):551–573. https://doi.org/10.1016/j.pecs.2007.11.001
332. Wang W-C, Thapaliya N, Campos A, Stikeleather LF, Roberts WL (2012) Hydrocarbon fuels from vegetable oils via hydrolysis and thermo-catalytic decarboxylation. Fuel 95:622–629. https://doi.org/10.1016/j.fuel.2011.12.041
333. Lashinsky A, Schwartz ND (2006) How to beat the high cost of gasoline. Forever! Fortune 153(2):40–47.
334. Li Y, Horsman M, Wang B, Wu N, Lan CQ (2008) Effects of nitrogen sources on cell growth and lipid accumulation of green alga Neochloris oleoabundans. Appl Microbiol Biotechnol 81(4):629–636. https://doi.org/10.1007/s00253-008-1681-1
335. Reed T, Reed TB, Das A (1988) Handbook of biomass downdraft gasifier engine systems, biomass energy foundation. Biomass Energy Foundation.
336. Sait HH, Salema AA (2015) Microwave dielectric characterization of Saudi Arabian date palm biomass during pyrolysis and at industrial frequencies. Fuel 161:239–247. https://doi.org/10.1016/j.fuel.2015.08.058
337. Nigam J (2000) Continuous ethanol production from pineapple cannery waste using immobilized yeast cells. J Biotechnol 80(2):189–193. https://doi.org/10.1016/S0168-1656(00)00246-7
338. Johansson TB, Patwardhan AP, Nakićenović N, Gomez-Echeverri L (2012) Global energy assessment: toward a sustainable future. Cambridge University Press.
339. Kwiatkowski JR, McAloon AJ, Taylor F, Johnston DB (2006) Modeling the process and costs of fuel ethanol production by the corn dry-grind process. Ind Crops Prod 23(3):288–296. https://doi.org/10.1016/j.indcrop.2005.08.004
340. vom Stein T, Grande P, Sibilla F, Commandeur U, Fischer R, Leitner W, Domínguez de María P (2010) Salt-assisted organic-acid-catalyzed depolymerization of cellulose. Green Chem 12(10):1844. https://doi.org/10.1039/c0gc00262c
341. FAO (2015) FAO statistics (Food and Agriculture Organization of the United Nations) [Statistics]. FAO.
342. Elsanhoty RM, Al-Turki IA, Ramadan MF (2012) Screening of medium components by Plackett-Burman design for carotenoid production using date (Phoenix dactylifera) wastes. Ind Crops Prod 36(1):313–320. https://doi.org/10.1016/j.indcrop.2011.10.013
343. FAO (2012) FAO statistics (Food and Agriculture Organization of the United Nations) [Statistics]. FAO.
344. Abdul Sattar Qureshi (2012) Production of pectinase by Bacillus subtilis EFRL 01 in a date syrup medium. Afr J Biotechnol 11(62):Article 62. https://doi.org/10.5897/AJB11.4236
345. Aditiya HB, Chong WT, Mahlia TMI, Sebayang AH, Berawi MA, Nur H (2016) Second generation bioethanol potential from selected Malaysia's biodiversity biomasses: a review. Waste Manage 47:46–61. https://doi.org/10.1016/j.wasman.2015.07.031
346. Shah A, Darr MJ, Dalluge D, Medic D, Webster K, Brown RC (2012) Physicochemical properties of bio-oil and biochar produced by fast pyrolysis of stored single-pass corn stover and cobs. Biores Technol 125:348–352. https://doi.org/10.1016/j.biortech.2012.09.061
347. Wilson L, Yang W, Blasiak W, John GR, Mhilu CF (2011) Thermal characterization of tropical biomass feedstocks. Energy Convers Manage 52(1):191–198. https://doi.org/10.1016/j.enconman.2010.06.058
348. ECOMES EURL (2015) Entreprise de Construction Métalliques du Sud. http://www.algerie3.com/index.phpoption=com_mtree&task=searchby&cf_id=6&value=01-Adrarwww.groupebakhadda.com
349. DuBois M, Gilles KA, Hamilton JK, Rebers PA, Smith F (1956) Colorimetric method for determination of sugars and related substances. Anal Chem 28(3):350–356. https://doi.org/10.1021/ac60111a017
350. AOAC (1998) Association of Official Agricultural Chemists.

351. ISO 11289:1993 (n.d.) Heat-processed foods in hermetically sealed containers—Determination of pH.
352. Chandra R, Takeuchi H, Hasegawa T (2012) Methane production from lignocellulosic agricultural crop wastes: a review in context to second generation of biofuel production. Renew Sustain Energy Rev 16(3):1462–1476. https://doi.org/10.1016/j.rser.2011.11.035
353. Ashur M, Bengharbia I (2015) Effect of temperature and pH on biogas production from organic fraction-MSW. Solar Energy Sustainable Development J 4(1):22–28. https://doi.org/10.51646/jsesd.v4i1.80
354. Xue J, Grift TE, Hansen AC (2011) Effect of biodiesel on engine performances and emissions. Renew Sustain Energy Rev 15(2):1098–1116. https://doi.org/10.1016/j.rser.2010.11.016
355. Zaid A, de Wet PF (2002) Origin, geographical, distribution and nutritional values of date palm. FAO Corporate Document Repository.
356. Jaffer KA (2014) Biogas production by anaerobic digestion of date palm pulp waste. Al-Khwarizmi Eng. J 5(5):591–600
357. Taghizadeh-Alisaraei A, Motevali A, Ghobadian B (2019) Ethanol production from date wastes: adapted technologies, challenges, and global potential. Renewable Energy 143:1094–1110. https://doi.org/10.1016/j.renene.2019.05.048
358. Al-Niaimi JH, Jaafer AA (1980) Physiology, anatomy and morphology of date palm tree. Basarh University Press, pp 185–192.
359. Hamada JS, Hashim IB, Sharif FA (2002) Preliminary analysis and potential uses of date pits in foods. Food Chem 76(2):135–137. https://doi.org/10.1016/S0308-8146(01)00253-9
360. Ismail B, Haffar I, Baalbaki R, Henry J (2008) Physico-chemical characteristics and sensory quality of two date varieties under commercial and industrial storage conditions. LWT Food Sci Technol 41(5):896–904. https://doi.org/10.1016/j.lwt.2007.06.009
361. Pillay AE, Williams JR, El Mardi MO, Hassan SM, Al-Hamdi A (2005) Boron and the alternate-bearing phenomenon in the date palm (Phoenix dactylifera). J Arid Environ 62(2):199–207. https://doi.org/10.1016/j.jaridenv.2004.11.007
362. Grover VI, Grover VK, Hogland W (2002) Recovering energy from waste various aspects. Science Publishers, Inc., Enfield (NH), USA, pp 44–45
363. Lattieff FA (2016) A study of biogas production from date palm fruit wastes. J Clean Prod 139:1191–1195. https://doi.org/10.1016/j.jclepro.2016.08.139
364. Tan L, Yu Y, Li X, Zhao J, Qu Y, Choo YM, Loh SK (2013) Pretreatment of empty fruit bunch from oil palm for fuel ethanol production and proposed biorefinery process. Biores Technol 135:275–282. https://doi.org/10.1016/j.biortech.2012.10.134
365. Gupta N, Kushwaha H (2011) Date palm as a source of bioethanol producing microorganisms. Date Palm Biotech. Springer, Netherlands, pp 711–727
366. de Souza SNM, Santos RF, Fracaro GPM (2011) Potential for the production of biogas in alcohol and sugar cane plants for use in urban buses in the Brazil. 418–424. https://doi.org/10.3384/ecp11057418
367. Janke L, Leite A, Nikolausz M, Schmidt T, Liebetrau J, Nelles M, Stinner W (2015) Biogas production from sugarcane waste: assessment on kinetic challenges for process designing. Int J Mol Sci 16(9):20685–20703. https://doi.org/10.3390/ijms160920685
368. Poh PE, Chong MF (2009) Development of anaerobic digestion methods for palm oil mill effluent (POME) treatment. Biores Technol 100(1):1–9. https://doi.org/10.1016/j.biortech.2008.06.022
369. Soliman SS (2012) Effects of strands thinning on yield and fruit quality of Succary date palm. African J Biotechnology 11(11):Article 11. https://doi.org/10.5897/AJB11.3277
370. Song Y-C, Kwon S-J, Woo J-H (2004) Mesophilic and thermophilic temperature co-phase anaerobic digestion compared with single-stage mesophilic- and thermophilic digestion of sewage sludge. Water Res 38(7):1653–1662. https://doi.org/10.1016/j.watres.2003.12.019
371. Ismail Z, Talib A (2014) Assessment of anaerobic co-digestion of agro wastes for biogas recovery: a bench scale application to date palm wastes. Energy and Environment 5(5):591–600

372. Vučurović DG, Dodić SN, Popov SD, Dodić JM, Grahovac JA (2012) Process model and economic analysis of ethanol production from sugar beet raw juice as part of the cleaner production concept. Biores Technol 104:367–372. https://doi.org/10.1016/j.biortech.2011.10.085
373. Alkanok G, Demirel B, Onay TT (2014) Determination of biogas generation potential as a renewable energy source from supermarket wastes. Waste Manage 34(1):134–140. https://doi.org/10.1016/j.wasman.2013.09.015
374. Shahravy A, Tabandeh F, Bambai B, Zamanizadeh HR, Mizani M (2012) Optimization of probiotic lactobacillus casei ATCC 334 production using date powder as carbon source. Chem Ind Chem Eng Q 18(2):273–282. https://doi.org/10.2298/CICEQ110709004S
375. Dangoggo SM, Fernando EC (1986) A simple biogas plant with additional gas storage system. Nigerian J Sol Energ:138–141.
376. Yassein N (n.d.) Antibacterial effect of date palm (Phoenix Dactylifera L.) pit aqueous extract on some bacteria cause urinary tract infection. Diyala J Pure Sci 8(3):112–120.
377. Hossain M, Waly M, Singh V, Sequeira V, Rahman M (2014) Chemical Compositions of date-pits and its potential for developing value-added product—a review. Polish J Food Nutrition Sci 64(4):215–226. https://doi.org/10.2478/pjfns-2013-0018
378. Basuny AMM, AL-Marzooq MA (2011) Production of mayonnaise from date pit oil. Food Nutr Sci 02(09):938–943. https://doi.org/10.4236/fns.2011.29128
379. Bhat RS, Al-Daihan S (2012) Antibacterial properties of different cultivars of Phoenix dactylifera L and their corresponding protein content. Scholars Research Library 3(10):4751–4757
380. Shafiei M, Karimi K, Taherzadeh MJ (2010) Palm date fibers: analysis and enzymatic hydrolysis. Int J Mol Sci 11(11):4285–4296. https://doi.org/10.3390/ijms11114285
381. Yang L, Huang Y, Zhao M, Huang Z, Miao H, Xu Z, Ruan W (2015) Enhancing biogas generation performance from food wastes by high-solids thermophilic anaerobic digestion: effect of pH adjustment. Int Biodeterior Biodegradation 105:153–159. https://doi.org/10.1016/j.ibiod.2015.09.005
382. Cragg GM, Simon JE, Jato JG, et al (1996) Drug discovery and development at the National Cancer Institute: Potential for new pharmaceutical crops. Progress in New Crops. Arlington. VA: ASHS, 554–560.
383. Cutter CN (2000) Antimicrobial effect of herb extracts against escherichia coli O157:H7, listeria monocytogenes, and salmonella typhimurium associated with Beef†. J Food Prot 63(5):601–607. https://doi.org/10.4315/0362-028X-63.5.601
384. Puupponen-Pimia R, Nohynek L, Meier C, Kahkonen M, Heinonen M, Hopia A, Oksman-Caldentey K-M (2001) Antimicrobial properties of phenolic compounds from berries. J Appl Microbiol 90(4):494–507. https://doi.org/10.1046/j.1365-2672.2001.01271.x
385. Biglari F, AlKarkhi AFM, Easa AM (2008) Antioxidant activity and phenolic content of various date palm (Phoenix dactylifera) fruits from Iran. Food Chem 107(4):1636–1641. https://doi.org/10.1016/j.foodchem.2007.10.033
386. Hong YJ, Tomas-Barberan FA, Kader AA, Mitchell AE (2006) The Flavonoid Glycosides and Procyanidin Composition of Deglet Noor Dates (*Phoenix dactylifera*). J Agric Food Chem 54(6):2405–2411. https://doi.org/10.1021/jf0581776
387. Mosa JS, Hifnawy MS, Mekkawi AG (1986) Phytochemical and biological investigations on date palm seeds (Phoenix dactylifera L.) produced in Saudi Arabia. Arab Gulf J Sci 4:495–507
388. Zhu S, Wu Y, Chen Q, Yu Z, Wang C, Jin S, Ding Y, Wu G (2006) Dissolution of cellulose with ionic liquids and its application: a mini-review. Green Chem 8(4):325. https://doi.org/10.1039/b601395c
389. Adam N, Shanableh A (2015) Comparison of biodiesel production from a variety of feedstocks. Energy Environ (ICWEE):362–368.
390. Al-Turki A, El-Hadidi Y, Al-Yahya S (2004) Utilization of date pits as a potential source of biogas and organic fertilizer. Food, Agric Environ 2(2):369–374

Chapter 12
Date Palm Byproducts in Other Fields of Applications

Abstract An important research has been conducted to investigate the feasibility of production of furfural from date palm midribs using a sulfuric acid-catalyzed hydrolysis process. In this research, the effect of reaction temperature, liquid-to-solid ratio, acid concentration and reaction time were investigated, and their optimum values determined to attain the maximum furfural yield. Furfural and its derivatives are considered as important chemicals due to their various applications. For example, furfural alcohol is employed in chemical industry as an additive or solvent in the manufacture of different resins. In addition, furfural is used as a selective solvent for the separation of saturated from unsaturated compounds in petroleum refining, gas oil and diesel fuel industries. In addition, furfural is also used as a fungicide, weedkiller and as a feedstock for the production of tetrahydrofuran. In most of the date-producing countries in the Arab world the products of pruning of date palms (e.g., leaflets, midribs, spadix stems, petioles and leaf sheaths fibers) are being treated as waste. i.e., open-field burnt or sent to landfills. This represents a great loss of such an important treasure of locally available renewable resources. As a response to this situation, an important research has been conducted with the objective of isolation and structural characterization of hemicellulose-type polysaccharides from date palm leaflets and midribs. The hemicellulose-type polysaccharides were successfully extracted from the leaflets and midribs of the date palms. The sugar analysis and nuclear magnetic resonance measurements indicate that they belong to xylans family. This may open new potentialities for the economic utilization of date palm products of pruning. A study has been conducted to investigate the potentiality of transforming the date palm midrib into sizing agent for use in the textile industry as a substitute for expensive commercial products imported from aboard. The date palm midribs were collected from Monastir (Tunisia). To obtain sodium cellulose carboxyl methylate (NaCMC), the cellulose was transformed into alkali-cellulose, and then etherification agent (monochloroacetic acid) was then introduced. The performance of NaCMCs was then compared to that of the imported commercial product. To conduct yarn sizing, 100% cotton yarns were utilized. To evaluate the performance of the sized yarn, the yarn hairiness and the load and elongation at break were measured. Hairiness is defined as the number of fibrils outside the main axis of the yarn. The results of this research prove the potentiality of use of date palm

H. EL-Mously et al., *Date Palm Byproducts: A Springboard for Circular Bio Economy*, Materials Horizons: From Nature to Nanomaterials,
https://doi.org/10.1007/978-981-99-0475-4_12

midrib in the manufacture of sizing materials for yarn as a substitute for expensive imported commercial products. A research has been conducted with the purpose of extracting cellulose fibers from date palm petioles. The date palm petioles were collected in Monastir, Tunisia. The samples were air-dried, milled and sieved to a grain size between 200 μm and 1 mm. The extraction of fibers was conducted under alkaline conditions to remove hemicellulose and break the hydrogen bonds and the hydrolyzed ester groups. The chemical composition of the date palm petioles was determined and found: 1,4% extractives, 27.5% lignin and 67.7% holocellhouse. In this work, it was possible to extract and purify cellulose fiber from date palm petioles with a yield of 42%. Moreover, the solubility of cellulose in ionic liquid (N-buthyl-N-methylpyrrolidium buthyl-dibutthylphosphate) allowed the dissolving of 11 g of cellulose per 100 g of this ionic liquid at 80 °C. The results of this study provide an environment-friendly method of dissolving cellulose from date palm petioles. It also represents an eloquent example of valorization of date palm products of pruning generally treated as waste. A research has been conducted to evaluate the potentiality of use of date palm seeds powder in sealing of fractures in oil wells. The expenses of drilling of oil wells represent 25% of the total oilfield exploitation cost. The drilling fluids represent 15 to 18% of the total cost of petroleum well drilling operations. One of the problems in well drilling is the loss of drilling liquids into drilling-induced fractures or natural fractures in wells. Therefore, it is necessary to find appropriate materials that can efficiently seal these fractures during drilling operations. In this research two superior fracture seal materials made from crushed date palm seeds and shredded waste tires were tested in laboratory conditions to seal artificially fractured holes under high temperature and pressure conditions. Mixtures of either crushed date palm seeds or shredded waste car tires of different grain sizes proved its ability to completely seal the samples at pressures up to 1000 psi and temperature up to 90 °C. In addition to its superior ability to seal the fractured formations, the date palm seeds are cheap, locally available in commercial quantities, environmentally friendly and easy to crush into various required grain sizes. A research has been conducted to study the effect off using a perforated plate on the sound absorption of date palm leaf sheaths fibers. To conduct this research, fibers were collected from the sheathing leaf bases and dried in shade at room temperature for 2 days. The pulp (paranchyma) was removed from fibers by combing and the fibers were then scrapped to remove the pulp completely. The average diameter and density of the fibers were 0.408 mm and 919 kg/m^3 respectively. The plastic molds were fabricated with diameters 28 mm and 100 mm to suit the two impedance tubes used in this research. The thickness and density of prepared samples were 30 mm and 77 kg/m^3, respectively. The frequency span of the experiment was 100–5000 Hz with 3 Hz resolution. An aluminum performed plate was used to enhance the sound absorption. The experiment was conducted for the panel without air gap, with air gap of 10, 20 and 30 mm between the date palm fiber sample and the rigid backing of the impedance tube. The use of the perforated plate has led to the increase of the absorption coefficient between 1000 and 3000 Hz by shifting the peak toward the low frequency range. However, the sound absorption coefficient decreased above 3000 Hz and the peak decreased by 4%. The results show that the best performance for improving the sound absorption at low frequency range can be achieved using the date palm leaf sheaths fibers combined with the perforated plate facing and the 10 mm air gap backing. The performance of the date fibers can be improved by increasing the samples density and using plates with different perforations.

Keyword Furfural · Date palm midribs · Hemicellulose · Date palm leaflets and midribs · Sizing agent for use in textile industry · Extracting cellulose fibers · Date palm petioles · Use of date palm seeds powder · Sealing of fractures in oil wells · Perforated plate · Date palm leaf sheaths fibers · Sound absorption · Air gap backing · Low frequency range

12.1 Furfural from Date Palm Midribs

An important research [1] has been conducted to investigate the feasibility of production of furfural from date palm midribs using a sulfuric acid catalyzed hydrolysis process. In this research, the effect of reaction temperature, liquid–to-solid ratio, acid concentration and reaction time were investigated, and their optimum values were determined to attain the maximum furfural yield.

Furfural and its derivates are considered as important chemicals due to their various applications. For example, furfural alcohol is employed in chemical industry as an additive or solvent in the manufacture of different resins. In addition, furfural is used as a selective solvent for the separation of saturated from unsaturated compounds in petroleum refining, gas oil and diesel fuel industries. In addition, furfural is also used as a fungicide, weed killer and as a feedstock for the production of tetrahydrofuran.

Many researchers have investigated the potentiality of production of furfural from different lignocellulosic waste materials, such as sugar cane bagasse, timber residues, olive stones, rice hulls, etc. The date palm plantations should be pruned every year rendering enormous quantities of leaves (e.g., 250,000 t in Saudi Arabia alone). These renewable resources, conventionally treated as a waste, represent a challenge to be used as a feedstock for the manufacture of furfural.

In this work, fresh midribs from Sukkaria date palms were thoroughly washed, cleaned and dried. After manual chopping, they were ground using a disc mill. Samples were then sieved to particle sizes in the range of 0.25–0.5 mm. The pentosan content of these samples was determined according to AOAC Phloroglucinal method. Hydrolysis reaction was conducted in a Parr batch reactor, model number 5112. It is made of a 1500 ml stirred type glass jacketed reactor with a maximum pressure of 10 bars. At the start of each experiment, the required volume of acid of a given concentration was placed in the reactor. The required amount of raw material was also placed. The experiments have been conducted at three different acid concentrations of 5, 10 and 15 wt.%, three levels of liquid-to-solid weight ratios of 50,70 and 100 ml/g, and 140 °C.

The research results have proven that the content of pentosan in date palm midribs is about 14.52%, which compares well with the corresponding values in wheat hulls, cotton seeds and nut shells, commonly used for furfural production. The maximum furfural yield has been obtained (53%) at 140 °C, 15 wt % acid concentration and liquid-to-solid ratio 50. The research indicates that all the formed side products are valuable chemicals having wide industrial applications. They include: furfural alcohol, 5 Hydroxymethyl furfural (5HMF), and 5 MF is also a valuable and important chemical intermediate [2]. It is also used for the production of tetrahydrofurfuryl alcohol and various synthetic fibers, rubbers, resins resistant to acids and bases and resins used for strengthening ceramics [3]. It is used, as well, as a chemical intermediate in the manufacture of lysine, vitamin C, lubricants and dispersing agents [4]. 5HMF is an important substance for plastic manufacture [5]. In addition, it is also a catalyst in biofuel chemistry and petrochemical industry. 5HMF is also an important chemical intermediate. However, it is desirable to limit the concentration of these byproducts to less than 1% of the final furfural product.

12.2 Isolation and Structural Characterization of Hemicellulose from Date Palm Leaflets and Rachis

In most of the date-producing countries in the Arab world, the products of pruning of date palms (e.g., leaflets, rachises, spadix stems, petioles and leaf sheaths fibers) are being treated as waste, i.e., open-field burnt or sent to landfills. This represents a great loss of such an important treasure of locally available renewable resources. As a response to this situation, an important research work [6] has been conducted with the objective of isolation and structural characterization of hemicellulose-type polysaccharides from date palm leaflets and rachis. This may open new potentialities for the economic utilization of such available and neglected resources.

The research results show that the water-soluble hemicelluloses are more highly branched than water-insoluble hemicelluloses. The hemicellulose-type polysaccharides were successfully extracted from the leaflets and rachises of the date palm. The sugar analysis and nuclear magnetic resonance measurements indicate that they belong to xylans family. As far as the leaflets are concerned the water-soluble fractions are arabinoglucuronoxylans which are mono-substituted at C-3 with arabinose, while the water-non-soluble fraction is 4–0-methyl-glucuronoxylans. As far as the rachis is concerned: both water soluble and water non-soluble polysaccharides are 4–0 methyl-glucuronoxylans.

12.3 Cellulose Derivatives from Date Palm Rachis as a Sizing Agent for Cotton Yarn

The products of pruning of date palms represent in most of cases a huge burden on the environment. That is why a research [7] has been devoted for finding an approach for rational utilization of date palm rachises as a suitable solution for their disposal. The main idea of this research was to investigate the potentiality of use of the date palm rachises as a starting raw material to prepare cellulose derivatives.

Sodium carboxyl methylate cellulose has wide applications: in textile, paper, agro-food, adhesive, cosmetic and pharmaceutical industries, as well as in manufacture of absorbent materials. The purpose of this research was the study of the potentiality of transforming date palm rachis into sizing agent for use in the textile industry as a substitute for expensive commercial products imported from aboard.

The date palm rachises were collected from Monastir (Tunisia). The samples were air-dried, milled and sieved to produce grains of size 1 mm to 200 μm. The samples were treated with caustic soda solution (15%) and stirred. The obtained fibers were washed and bleached with sodium hypochlorite solution (12% active chlorine) under a basic pH (Ca.12). To obtain sodium cellulose carboxyl methylate (NaCMC), the cellulose was transformed into alkali-cellulose, and then the etherification agent (monochloroacetic acid) was then introduced. The performance of NaCMCs was then compared to that of the imported commercial product. To conduct yarn sizing, 100% cotton yarns were utilized. To evaluate the performance of the sized yarn, the yarn hairiness and the load and elongation at break were measured. Hairiness is defined as the number of fibrils outside the main axis of the yarn. In this study, NaCMC was prepared both from virgin date palm rachises and from extracted bleached cellulose. The hairiness of the sized yarn, treated with NaCMC obtained from bleached cellulose, was found similar to the commercial NaCMC sized yarn. The NaCMC, prepared from non-purified raw material, had lower sizing performance. The results of this research prove the potentiality of use of date palm rachises in the manufacture of sizing materials for yarns as a substitute for expensive imported commercial products.

12.4 Cellulose Fibers from Date Palm Petioles

The limited reserves of petroleum, in addition to the greenhouse consequences of use of petroleum in recent years, have motivated scientific research to find more environment-friendly alternatives. Cellulose is the most common natural polymer in the world. The date palm residues are generally treated as waste. In this research, a trial has been made to valorize date palm resides via the extraction of cellulose from them. Date palm petioles were collected in Monastir, Tunisia. The samples were air-dried, milled and sieved to obtain grains of size between 200 μm and 1 mm. The extraction of fibers was conducted under alkaline conditions to remove hemicellulose and break the hydrogen bonds and the hydrolyzed ester groups. The chemical composition of date palm petioles was determined and found: 1, 4% extractives, 27.5% lignin and 67.7% holocellulose. In this work, it was possible to extract and purify cellulose fibers from date palm petioles with a yield of 42%. Moreover, the solubility of cellulose in the ionic liquid (N-buthyl-N-methylpyrrolidium buthyl-dibuthylphosphate) allowed the dissolving of 11 g of cellulose per 100 g of this ionic liquid at 80 °C. The results of this study provide an environment-friendly method of dissolving cellulose from date palm petioles. It also represents an eloquent example of valorization of date palm products of pruning, generally treated as waste.

12.5 Innovative Wellbore Strengthening Using Crushed Date Palm Seeds and Shredded Waste Car Tyres

A research [8] has been conducted to evaluate the potentiality of date palm seeds powder in sealing of fractures in oil wells. The expenses of drilling of oil wells represent 25% of the total oilfield exploitation cost. The drilling fluids represent 15 to 18% of the total cost of petroleum well drilling operations. One of the problems in well drilling is the loss of drilling liquids into drilling-induced fractures or natural fractures in wells. Therefore, it is necessary to find appropriate materials that can efficiently seal these fractures during drilling operations. In this research, two superior fracture seal materials made from crushed date palm seeds and shredded waste tires were tested in laboratory conditions to seal artificially fractured holes under high temperature and pressure conditions. For this purpose, a conventional 500 ml high temperature and high pressure press was modified to accommodate an artificially fractured core plug of length and diameter equal to 38.1 mm instead of the ceramic disc. Mixtures of either crushed date palm seeds or shredded waste car tires of different grain sizes proved its ability to completely seal the artificially made fracture in the test core samples at pressures up to 1000psi and temperatures up to 90 °C. In addition to its superior ability to seal the fractured formations, the date palm seeds are cheap, locally available in commercial quantities, environmentally friendly and easy to crush into various required grain sizes.

12.6 Experimental Investigation of Sound Absorption Properties of Date Palm Leaf Sheaths Fibers Panel

A research [9] has been conducted to study the effect of using a perforated plate on the sound absorption of date palm leaf sheaths fibers. The sources of vegetable fibers include leaves, roots, fruits and seeds of plants. But proceeding from the economic and technological points of view, the most significant fibers are those of cotton, kenaf, sisal, flax, palm, coconut coir, arecanut and banana fibers [10]. The date palm is an essential element of flora in wide regions of the world extending from the Atlantic coast of Mauritania to India and from the Mediterranean shoreline to about 15° in Africa [11]. The date palm has different sources of fibers: leaf fibers in the peduncle, bast fibers in the stem, wood fibers in the trunk and leaf sheaths fibers around the trunk [12]. Riahi et al. [13] investigated the use of date palm fibers as filters and porous medium for tertiary domestic wastewater treatment, whereas Al-Sulaian [14] evaluated the performance of date palm fibers as wetted pats in evaporative cooling.

The glass and mineral fibers are most dominantly used in the building construction industry. But the growing health concerns, associated with the use of these non-renewable materials have urged researchers to search for alternative renewable feedstocks [15, 16]. Yang et al. [16] reported that the sound absorption coefficient of rice straw particle composite boards are higher than other wood materials in the 500–8000 Hz frequency range due to the low specific gravity and high porosity of these boards. The impedance tube measurements of bamboo fiber samples revealed similar properties to that of glass wool and superior sound absorption as compared with plywood material of similar density [17]. Erosy and Kucuk [18] experimentally investigated the sound absorption properties of tea leaf fiber as an industrial waste material. Coir fibers from coconut husk are one of the hardest natural fibers having high lignin content. The sound absorption property of coir fibers has been investigated in a reverberation room [19] and using impedance tube [20]. The acoustic properties of panels can be improved by using perforated plate design. Davern [21] investigated the effect of the airspace layers, perforated plate and porosity on the acoustic properties of materials. He showed that the porosity of the perforated plate, and the density of the porous material significantly affected the acoustic impedance and sound absorption of the panel. Lee and Chen [22] found that the acoustic absorption of multilayer materials is better with a perforated plate backed with airspaces. Fouladi et al. [23] investigated the enhancement of coir fiber acoustical absorption using perforated plates and air gap layers. The research on the acoustical properties of the date palm fibers was started at the Khartoum University [24]. The sound absorption coefficient of date palm fibers was simulated using Delany and Bazley model based on the flow resistivity of the fibers. Elwaleed et al. [25] studied experimentally the variation of sound absorption coefficient against frequency for date palm fiber sample at normal incidence as measured by the impedance tube.

To conduct this study, fibers were collected from the sheathing leaf bases and dried in shade at room temperature for 2 days. The pulp (parenchyma) was removed from fibers by combing and the fibers were then scrapped to remove the pulp completely. The average diameter and density of the fibers were 0.408 mm and 919 kg/m^3, respectively. Two plastic molds were fabricated with diameters 28 mm and 100 mm to suit the two impedance tubes used in this research. The thickness and density of the prepared samples were 30 mm and 77 kg/m^3 respectively. The experiment was conducted using the two impedance tubes, noise generator, two-channel data acquisition system 01 dB, two ¼ in microphones type GRAS-40BP in each tube and software package SCS8100. The measurements were made based on 150 10,534-2 standard [26]. The microphones' sensitivity was calibrated using calibrator type

GRAS-42AB at 114 dB level and 1 kHz. The noise generator transmitted a random noise into the tubes. The interior sound pressure spectrum was measured by the two microphones and transfer functions between them were calculated. The acoustical absorption coefficient was calculated from these transfer functions and distances between the microphones and date palm fiber sample. The frequency span of the experiment was 100–5000 Hz with 3 Hz resolution. Before experimentation, the two impedance tube microphones were calibrated relative to each other using the standard switching technique. This was based on mounting the sample in the sample holder and conducting the measurement to make sure that the sound field inside the tube was well defined. An aluminum perforated plate (PP) was used to enhance the sound absorption. The experiment was conducted for the panel without air gap, with air gap of 10, 20 and 30 mm between the date palm fiber sample and the rigid backing of the impedance tube.

The use of the perforated plate has led to the increase of the absorption coefficient between 1000 and 3000 Hz by shifting the peak toward the lower frequency range. However, the sound absorption coefficient decreased above 3000 Hz and the peak decreased by 4%. The results show that the best performance for improving the sound absorption at the low frequency range can be achieved using the date palm leaf sheaths fibers combined with the PP facing and the 10 mm air gap backing. However, this coincides with a decrease in the sound absorption at the medium frequency range. The further increase of the gap thickness moves the peaks toward lower frequencies and improves the adsorption: at low frequencies and above 4000 Hz. The performance of the date palm leaf sheaths fibers can be improved by increasing the sample density and using plates with different perforations.

References

1. Bamufleh HS, Alhamed YA, Daous MA (2013) Furfural from midribs of date-palm trees by sulfuric acid hydrolysis. Ind Crops Prod 42:421–428. https://doi.org/10.1016/j.indcrop.2012.06.008
2. Yang W, Sen A (2011) Direct catalytic synthesis of 5-methylfurfural from biomass-derived carbohydrates. Chemsuschem 4(3):349–352. https://doi.org/10.1002/cssc.201000369
3. Huang W, Li H, Zhu B, Feng Y, Wang S, Zhang S (2007) Selective hydrogenation of furfural to furfuryl alcohol over catalysts prepared via sonochemistry. Ultrason Sonochem 14(1):67–74. https://doi.org/10.1016/j.ultsonch.2006.03.002

4. Merlo AB, Vetere V, Ruggera JF, Casella ML (2009) Bimetallic PtSn catalyst for the selective hydrogenation of furfural to furfuryl alcohol in liquid-phase. Catal Commun 10(13):1665–1669. https://doi.org/10.1016/j.catcom.2009.05.005
5. Fujiki J, Fan H-J, Hattori H, Tajima K, Tsai Y-S, Furuya E (2008) Computer-aided design of surface modified adsorbent for adsorption of 5-hydroxy-methyl-furfural. Sep Purif Technol 60(3):223–229. https://doi.org/10.1016/j.seppur.2007.08.019
6. Bendahou A, Dufresne A, Kaddami H, Habibi Y (2007) Isolation and structural characterization of hemicelluloses from palm of Phoenix dactylifera L. Carbohyd Polym 68(3):601–608. https://doi.org/10.1016/j.carbpol.2006.10.016
7. Ramzi K, Ikhlass F, Farouk MM (2010) Study of liquids absorption and retention capacities of new cellulosic materials and sodium cellulose carboxylmethylate prepared from Posidonia. Fibers and Polymers 11(4):593–597. https://doi.org/10.1007/s12221-010-0593-x
8. Musaed NJ, Khaled Elshreef A, Ahmed Algobany (2018, November) Innovative Wellbore strengthening using crushed date palm seeds and shredded waste car tyres. The 10th ISRM International Asian Rock Mechanics SymposiumAt: Singapore.
9. Elwaleed AK, Nikabdullah N, Nor MJM, Tahir MFM, Zulkifli R (2013) Experimental investigation of sound absorption properties of perforated date palm fibers panel. IOP Conference Series: Materials Science and Engineering 46:012027. https://doi.org/10.1088/1757-899X/46/1/012027
10. Rao KMM, Rao KM (2007) Extraction and tensile properties of natural fibers: Vakka, date and bamboo. Compos Struct 77(3):288–295. https://doi.org/10.1016/j.compstruct.2005.07.023
11. Khristova P (2005) Alkaline pulping with additives of date palm rachis and leaves from Sudan. Biores Technol 96(1):79–85. https://doi.org/10.1016/j.biortech.2003.05.005
12. Kriker A, Debicki G, Bali A, Khenfer MM, Chabannet M (2005) Mechanical properties of date palm fibres and concrete reinforced with date palm fibres in hot-dry climate. Cement Concr Compos 27(5):554–564. https://doi.org/10.1016/j.cemconcomp.2004.09.015
13. Riahi K, Mammou AB, Thayer BB (2009) Date-palm fibers media filters as a potential technology for tertiary domestic wastewater treatment. J Hazard Mater 161(2–3):608–613. https://doi.org/10.1016/j.jhazmat.2008.04.013
14. Al-Sulaiman F (2002) Evaluation of the performance of local fibers in evaporative cooling. Energy Convers Manage 43(16):2267–2273. https://doi.org/10.1016/S0196-8904(01)00121-2
15. Khedari J, Nankongnab N, Hirunlabh J, Teekasap S (2004) New low-cost insulation particleboards from mixture of durian peel and coconut coir. Build Environ 39(1):59–65. https://doi.org/10.1016/j.buildenv.2003.08.001
16. Yang H-S, Kim D-J, Kim H-J (2003) Rice straw–wood particle composite for sound absorbing wooden construction materials. Biores Technol 86(2):117–121. https://doi.org/10.1016/S0960-8524(02)00163-3
17. Koizumi T, Tsujiuchi N, Adachi A (2002) The development of sound absorbing materials using natural bamboo fibres. In: Brebbia CA, De Wilde WP (eds) High performance structures and composites 4. High performance structures and materials.
18. Ersoy S, Küçük H (2009) Investigation of industrial tea-leaf-fibre waste material for its sound absorption properties. Appl Acoust 70(1):215–220. https://doi.org/10.1016/j.apacoust.2007.12.005
19. Zulkifli R, Nor MJM, Tahir MFM, Ismail AR, Nuawi MZ (2008) Acoustic properties of multilayer coir fibres sound absorption panel. J Appl Sci 8(20):3709–3714. https://doi.org/10.3923/jas.2008.3709.3714
20. Hosseini Fouladi M, Md, Ayub, Nor JM, M. (2011) Analysis of coir fiber acoustical characteristics. Appl Acoust 72(1):35–42. https://doi.org/10.1016/j.apacoust.2010.09.007
21. Davern WA (1977) Perforated facings backed with porous materials as sound absorbers—an experimental study. Appl Acoust 10(2):85–112. https://doi.org/10.1016/0003-682X(77)90019-6
22. Lee F-C, Chen W-H (2001) Acoustic transmission analysis of multi-layer absorsbers. J Sound Vib 248(4):621–634. https://doi.org/10.1006/jsvi.2001.3825

23. Hosseini Fouladi M, Nor MJM, Md, Ayub, Leman ZA (2010) Utilization of coir fiber in multilayer acoustic absorption panel. Appl Acoust 71(3):241–249. https://doi.org/10.1016/j.apacoust.2009.09.003
24. Ali MM, Mohammed M (2010) Palm date fibres as sound absorber [Bachelor]. Faculty of Engineering, University of Khartoum, Department of Mechanical Engineering
25. Elwaleed AK, Mohamed NAN, Nor MJM, Tahir MFM, Zulkifli R (2012, November 26) Noise, vibration and comfort: selected, peer reviewed papers from the 4th international conference on Noise, Vibration and Comfort (NVC 2012). The 4th International Conference on Noise, Vibration and Comfort, Kuala Lumpur, Malaysia.
26. ISO 10534–2 (1998) Determination of sound absorption coefficient and impedance in impedance tubes – part 2: Transfer function method.

Chapter 13
The Date Palm as a Springboard for Circular Bioeconomy: A Biorefinery for Each Date Palm Byproduct

Abstract This chapter proceeds from the reality of each date palm plantation with its production: of dates and byproducts that can be sorted in site to different byproducts and from the necessity of rational utilization of these byproducts. Otherwise, these byproducts will turn into waste (as the situation is most predominantly now) being either open-field burnt causing considerable environmental pollution or sent to landfills!

Keyword Date palm · Circular bioeconomy · Biorefinery · Date palm byproducts · Products of annual pruning of date palms · Palm midribs · Palm leaflets · Empty fruit bunches · Petioles · Leaf sheaths · Spathes · Date kernels · Waste dates · Date palm trunks · Added value · Open-field burning · Landfills · New ethics · Direct use · Fermentation

13.1 The Date Palm Byproducts (DPBPs) Include

13.1.1 *Products of Annual Pruning of Date Palms*

- Palm Midribs
- Palm Leaflets
- Empty Fruit Bunches
- Petioles
- Leaf Sheaths fibers
- Spathes

13.1.2 *Date Kernels*

Date kernels are available at the industrial sites of date processing factories.

H. EL-Mously et al., *Date Palm Byproducts: A Springboard for Circular Bio Economy*, Materials Horizons: From Nature to Nanomaterials,
https://doi.org/10.1007/978-981-99-0475-4_13

13.1.3 Waste Dates

Waste dates are those dates unsuitable for human consumption. They are available during the processes of collection, sorting and packaging of dates.

13.1.4 Date Palm Trunks

The date palm trunks—as byproducts—are those trunks of unproductive date palms (or date palms heavily infested by the red worm weevil) at the sites of renewal of date palm plantations.

13.2 Estimation of the Annually Available Quantities of DPBPs on the World Level

13.2.1 Products of Annual Pruning

Assuming 140 million date palms [1] and 54.2 kg/palm [2] estimation of annual pruning (air-dry weight), the estimation of the annual products of pruning on the world level will reach 7.6 million tons.

13.2.2 Date Kernels

The date kernels on the world level have been estimated by 700,000 tons (i.e. 0.7 million tons) [3].

13.2.3 Waste Dates

The waste dates have been estimated by 5 to 30% of the dates [4]. Assuming a world production of dates 6.9 million tons [5]; the average quantity of date wastes will reach 1.2075 million tons.

The estimated quantity of date palm byproducts on the world level will be: $7.6 + 0.7 + 1.2075 \approx 9.51$ million tons.

13.3 The Present Status of DPBPs

The palm owners are generally unaware of the hidden economic and developmental potentialities of their DPBPs, which are viewed as a waste and a burden to be disposed.

The DPBPs are thus most dominantly treated as waste in most of the date producing countries. They are either open-field burnt or sent to the landfills. This leads in turn to considerable environmental pollution and a heavy burden on the landfills in date producing countries.

13.4 Significance of DPBPs

The expansion—and still expansion- of date palm plantations from their traditional locations in the main date producing countries (Iraq, Algeria, Iran, Saudi Arabia, Pakistan, Tunisia, Morocco, Egypt, UAE and Sudan) to 38 countries in Africa, Asia, Europe, North America and South America makes **DPBPs one of the most sustainable renewable materials bases for the realization of circular bioeconomy and sustainable development in the world.**

13.5 The Objectives of Developing a Separate Biorefinery for Each Date Palm Byproduct

13.5.1 Maximization of the added value from each date palm byproduct.

13.5.2 Generation of sustainable labor opportunities and hence realization of endogenous development of local communities where date palms are grown.

13.5.3 Avoidance of open-field burning of date palm byproducts, causing a considerable environmental pollution.

13.5.4 Avoidance of making pressure on landfills or creation of new landfills to get rid of date palm byproducts: a practice which is totally against the principles of sustainable development.

13.5.5 Creation of an economic stimulus for regular pruning of date palms. The regular pruning of date palms facilitates the inspection of the date palms and hence the avoidance of spreading of infestation by Red Weevil worm threatening date palm plantations due to neglect of pruning.

13.6 New Ethics or Steps Needed to Attain Successful Biorefineries for Date Palm Byproducts

13.6.1 Researchers should transcend their individualistic mindset and declare their ethical responsibility to scientifically describe what remains from the DPBP after extracting their products. This facilitates for them –or other researchers- the discovery of new products, that can be extracted from what remained from the DPBP along the pathway of refinery of each date palm byproduct.

13.6.2 Palm owners should by educated about the potentiality of extraction of new products from their DPBPs. They should be –via appropriate legislations- made responsible for delivering their date palm byproducts to the gate of their date palm plantations correctly packaged (e.g. midribs and spadix stems in bundles, petioles, leaf sheathes fibers, date kernels and waste dates in jumbo bags, etc.). This represents a very important condition for facilitating further procurement of these materials to different sites for further processing.

13.7 Examples of Biorefineries for Date Palm Byproducts

13.7.1 Midribs

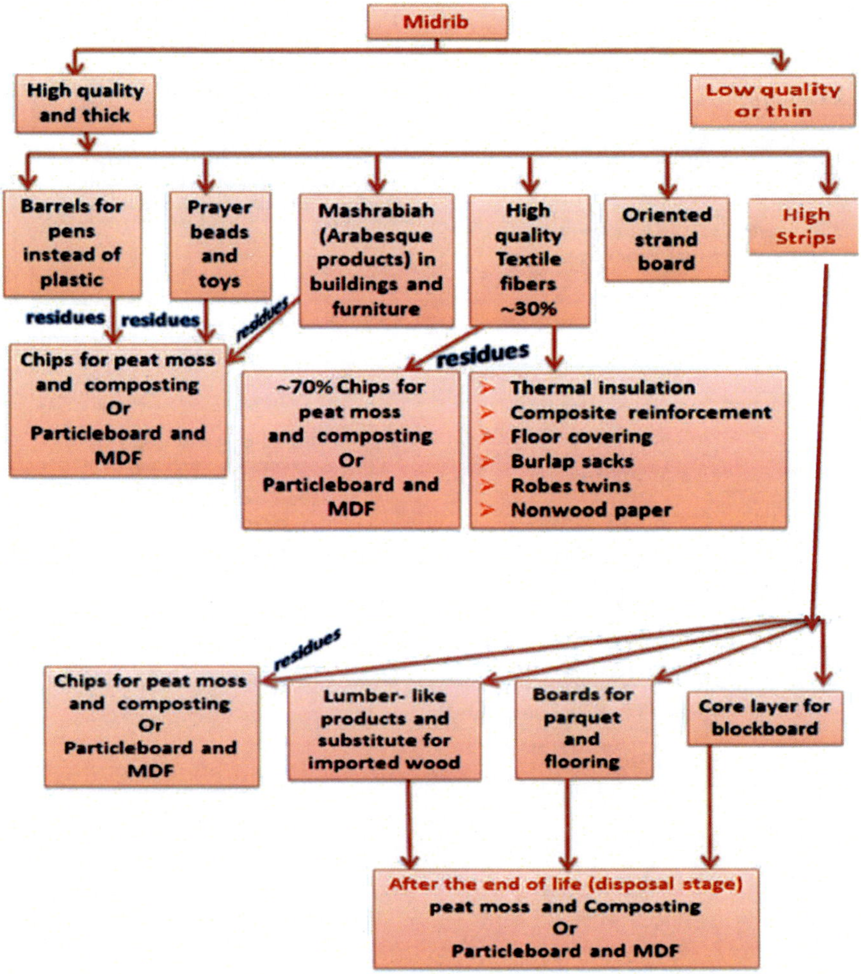

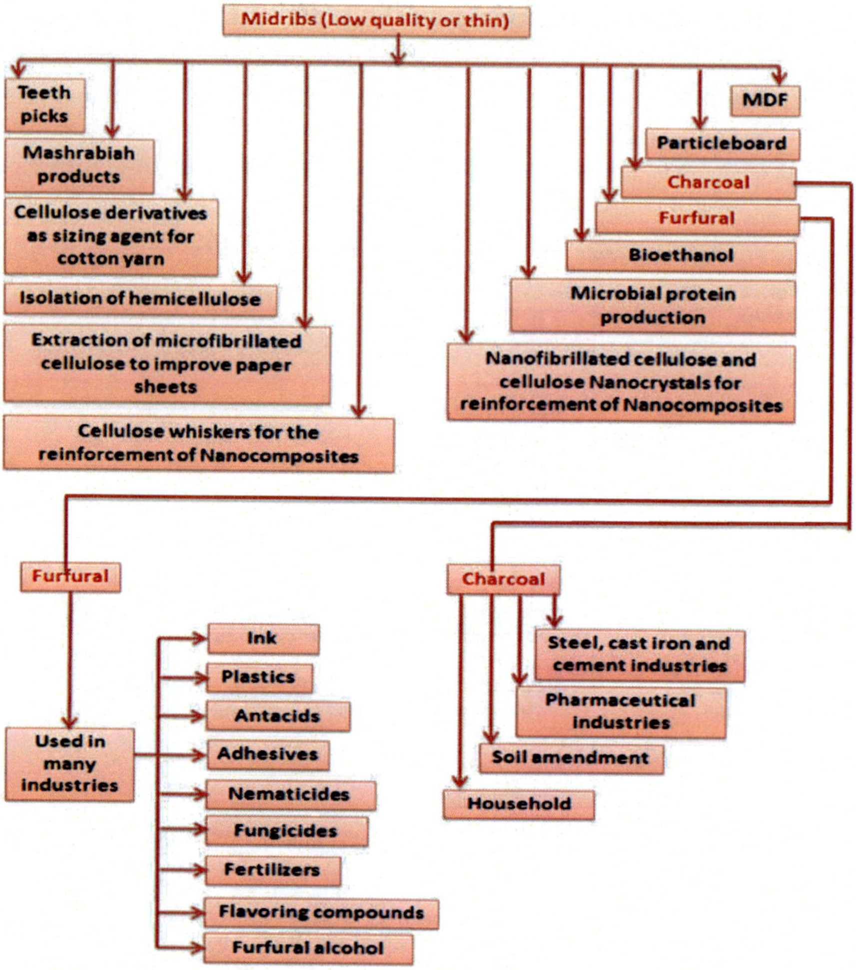
Midribs (Low quality or thin)
Teeth picks
Mashrabiah products
Cellulose derivatives as sizing agent for cotton yarn
Isolation of hemicellulose
Extraction of microfibrillated cellulose to improve paper sheets
Cellulose whiskers for the reinforcement of Nanocomposites
MDF
Particleboard
Charcoal
Furfural
Bioethanol
Microbial protein production
Nanofibrillated cellulose and cellulose Nanocrystals for reinforcement of Nanocomposites
Furfural
Used in many industries
Ink
Plastics
Antacids
Adhesives
Nematicides
Fungicides
Fertilizers
Flavoring compounds
Furfural alcohol
Charcoal
Steel, cast iron and cement industries
Pharmaceutical industries
Soil amendment
Household

13.7.2 Leaflets

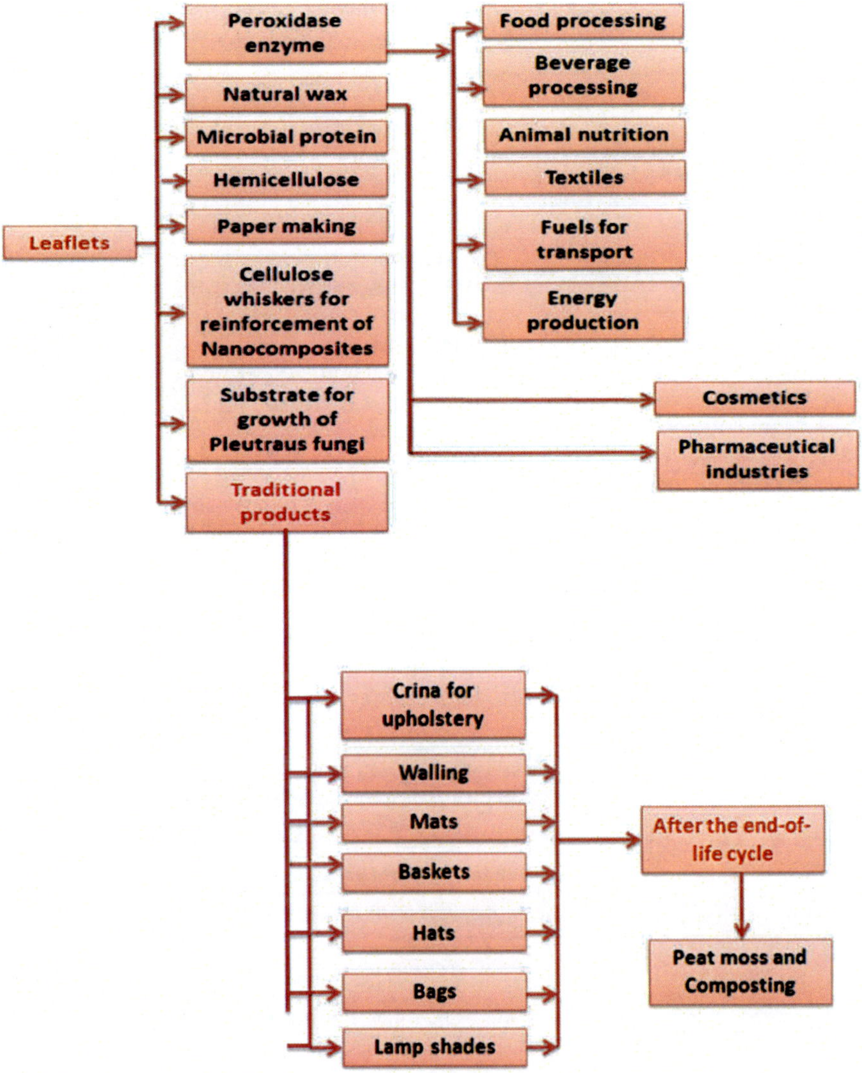

13.7.3 *Spadix Stem*

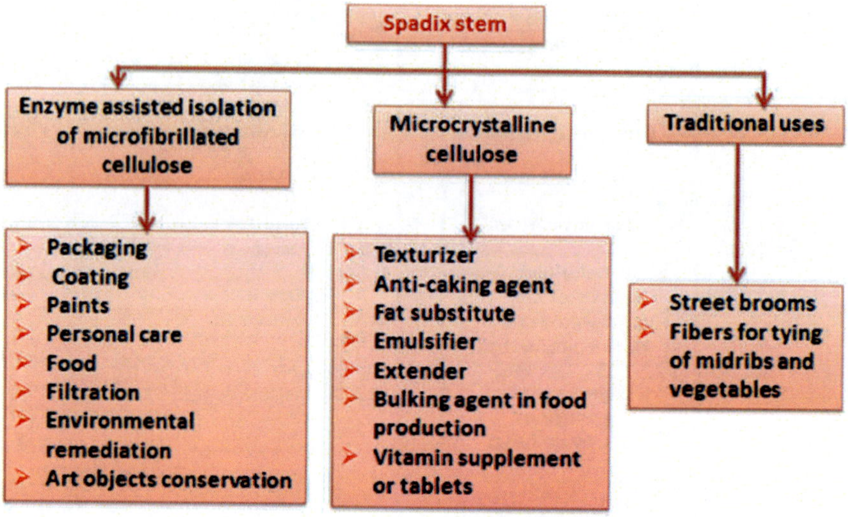

13.7.4 *Leaf Sheath Fibers*

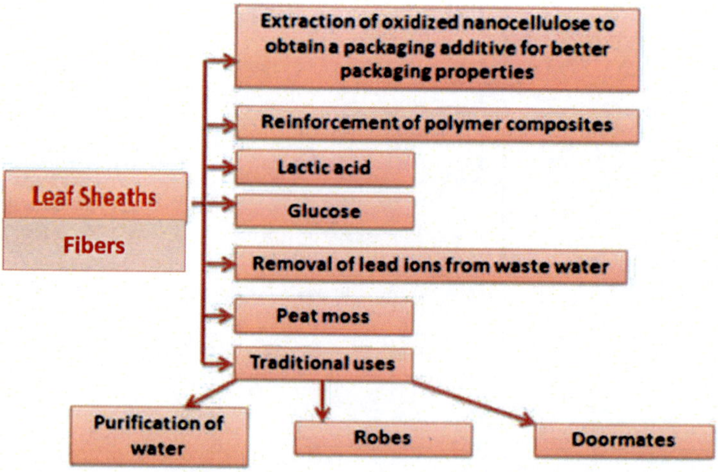

13.7.5 Date Palm Kernels

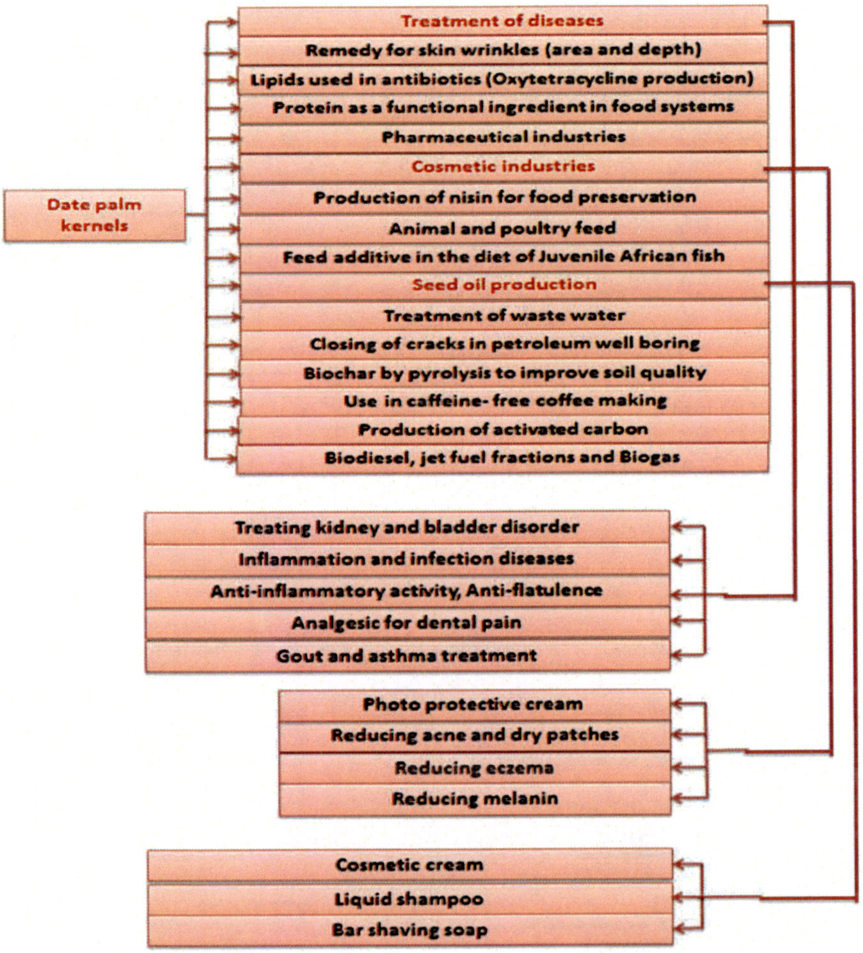

13.7.6 Waste Dates

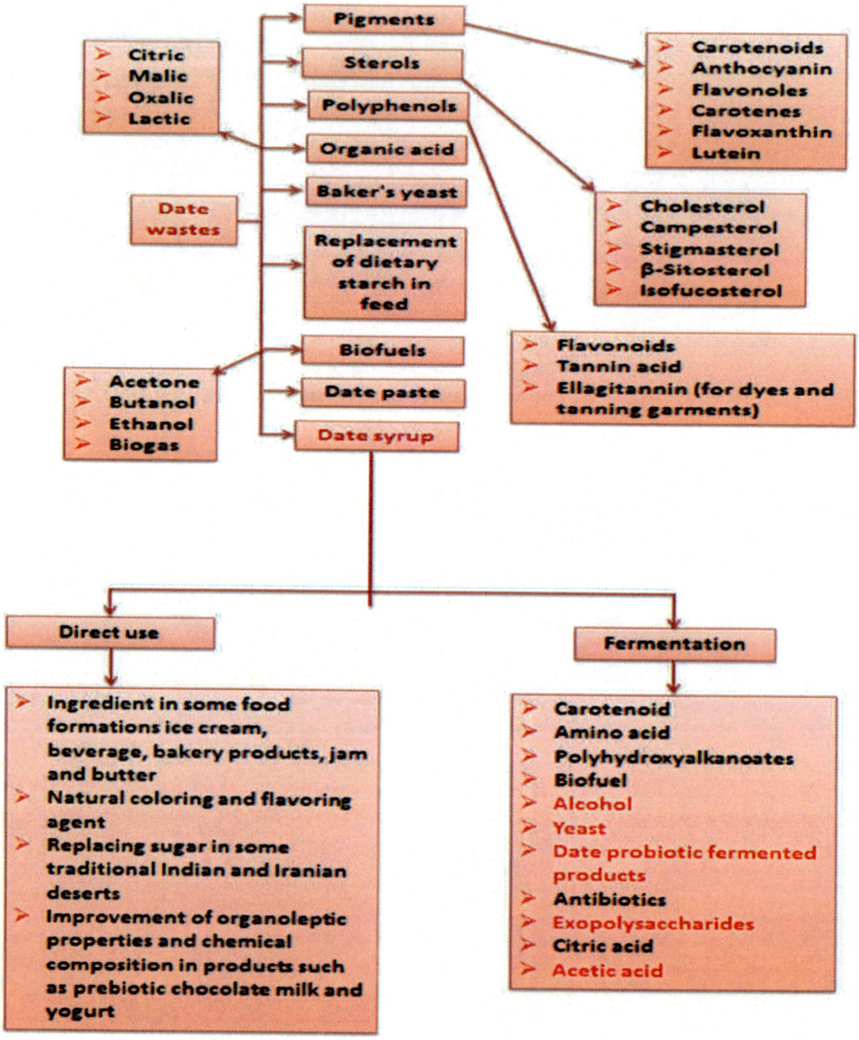

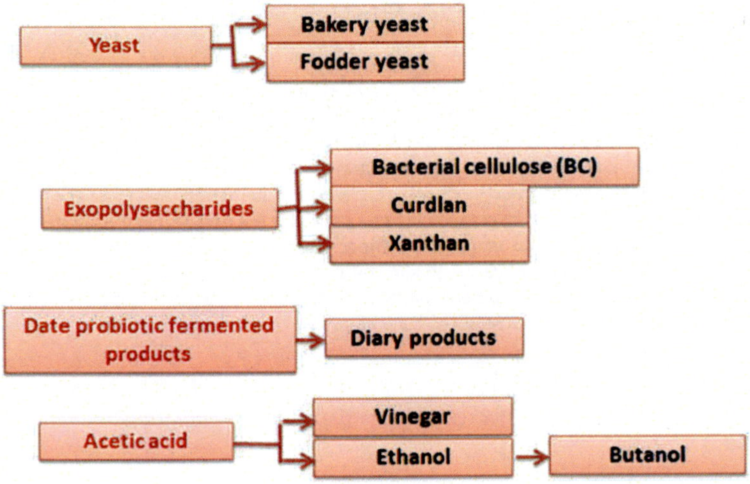
Yeast
Bakery yeast
Fodder yeast
Exopolysaccharides
Bacterial cellulose (BC)
Curdlan
Xanthan
Date probiotic fermented products
Diary products
Acetic acid
Vinegar
Ethanol
Butanol

13.7.7 *Trunks*

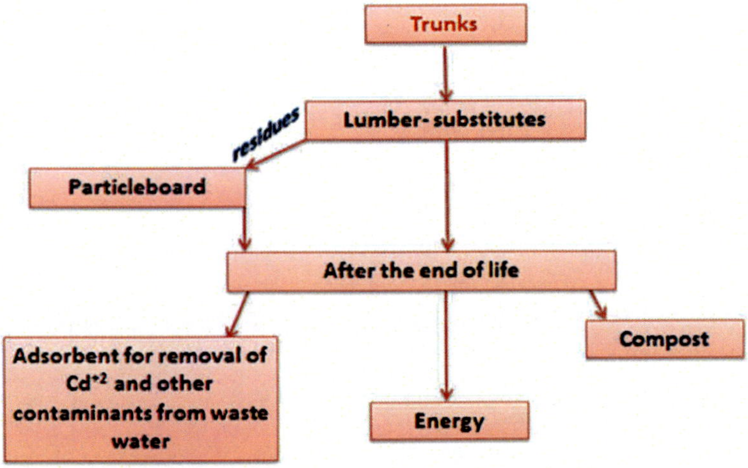
Trunks
Lumber- substitutes
residues
Particleboard
After the end of life
Adsorbent for removal of Cd^{+2} and other contaminants from waste water
Energy
Compost

References

1. Abdelouahhab zaid, Sandar Piesik (2020) The Khalifa awards report.
2. El Mously H, Elerian KM (2016) Project has been conducted by the Faculty of Engineering. Ain Shams Univ, In collaboration with the Ministry of Environment
3. Sherif Mehanny, Hamdy Ibrahim, Lamis Darwish, Mahmoud Farag, Abdel-Halim M. El-Habbak, Emad El-Kashif (2020) effect of environmental conditions on date palm fiber composites. In: Date palm fiber composites processing, properties and application.
4. Faris M AL-Oqla (2020) Evaluation and comparison of date palm fibers with other common natural fibers. In: Date palm fiber composites processing, properties and application.
5. Said Awad, Yonghui Zhou, Evina Katsou, Mizi Fan (2020) Polymer matrix systems used for date palm composite reinforcement. In: Date palm fiber composites processing, properties and application.

Appendix

Glossary of Date Palm Byproducts

1. **Leaf, Frond or Branch**
   The green elements growing from the trunk and forming the crown of the date palm, (Fig. A.1).
2. **Midrib, Rachis, Frond Stalk, Leaf Stalk or Twig**
   The central solid part or stalk of the date palm leaf, connected to the trunk and holding the leaflets on both sides, (Figs. A.1, A.2).
3. **Leaflets**
   The green flat parts of the leaf emerging from the midrib, responsible for the photosynthesis operation, by virtue of which the sun energy is transformed to chemical energy, (Figs. A.1, A.2).

Fig. A.1 Leaf, Midrib, Leaflet, Petioleb

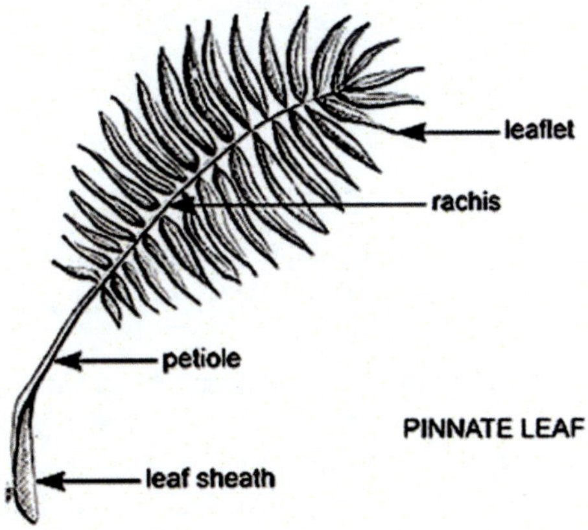

H. EL-Mously et al., *Date Palm Byproducts: A Springboard for Circular Bio Economy*, Materials Horizons: From Nature to Nanomaterials,
https://doi.org/10.1007/978-981-99-0475-4

Fig. A.2 Leaflet, Midrib

Fig. A.3 Petiole

4. **Petiole, Leaf base, Basal end or Frond base**
   Basal part of the leaf, Left on the trunk in pruning to be removed after drying, (Figs. A.2, A.3).
5. **Leaf Sheath, Fibrillum, Palm Coir**
   A structure at the base of a leaf's petiole that partly surrounds or protects the trunk and is composed of intersecting layers of fibers, Fig. (A.4).
6. **Spadix Stem, Peduncle, Fruit stalk or Fruit bunch stalk**
   The main unbranched stalk of date palm inflorescence, Fig. (A.5).
7. **Pedicels or Spikelets**
   The stalks of individual flowers in an inflorescence, Fig. (A.6).

Fig. A.4 Leaf Sheath

Fig. A.5 Spadix Stem

8. **Spines**
   Specialized leaflets converted into tough pointed pins, Fig. (A.7).
9. **Empty fruit bunch**
   The fruit stalk with the empty spikelets, to which the date were attached, Fig. (A.8).

Fig. A.6 Pedicels

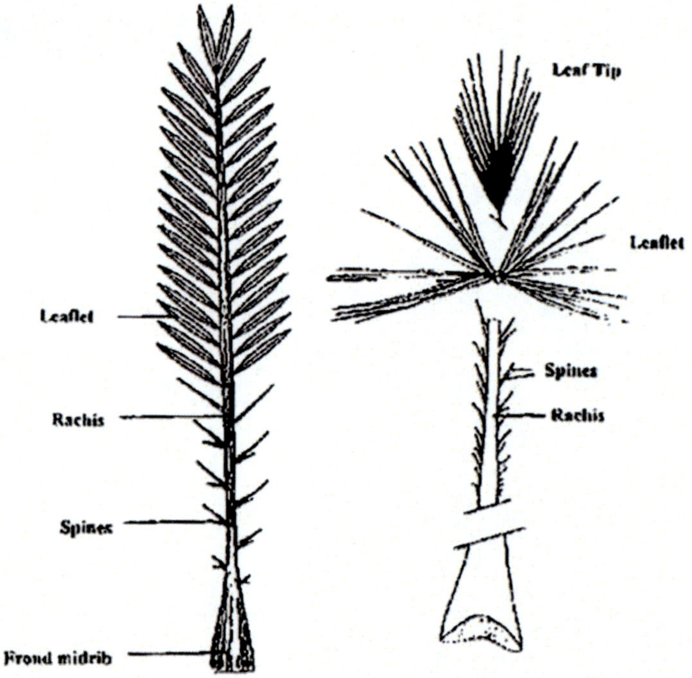

Fig. A.7 Spines

10. **Spathe**
    A large sheathing bract surrounding the inflorescence, Fig. (A.9).

Fig. A.8 Empty fruit bunch

11. **Trunk**
    The date palm trunk, also called stem, is vertical cylindrical and columnar of the same size all the way up without any ramification. The trunk becomes available upon natural or accidental death of the palm or by forced removal. Fig. (A.10).
12. **Date stone, Kernel, Pit or Seed**
    The seed is a hard body, usually ablong and vertically grooved. It is located inside the date fruit. Fig. (A.11).

Fig. A.9 Spathe

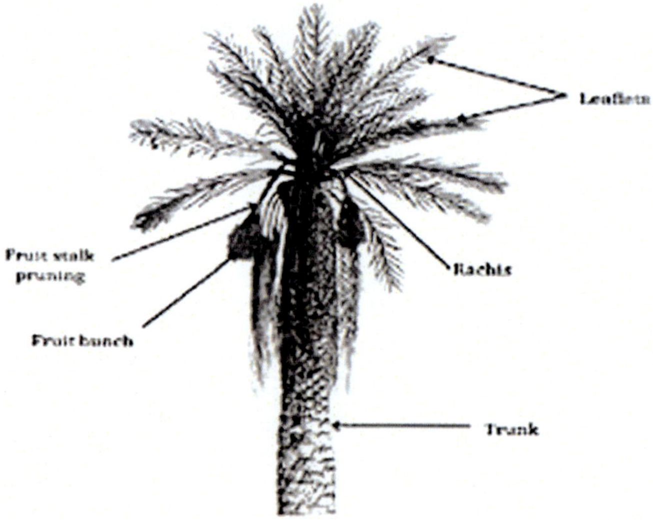

Fig. A.10 Trunk

Fig. A.11 Date stones